SUSPENDED

PARTICULATE

MATTER

IN

LAKES, RIVERS,

AND MARINE SYSTEMS

SUSPENDED PARTICULATE MATTER IN LAKES, RIVERS, AND MARINE SYSTEMS

Lars Håkanson

Department of Earth Sciences
Uppsala University
Sweden

Suspended Particulate Matter in Lakes, Rivers, and Marine Systems

ISBN-10: 1-932846-14-X
ISBN-13: 978-1-932846-14-0

Library of Congress Control Number: 2005922598

THE BLACKBURN PRESS
P. O. Box 287
Caldwell, New Jersey   07006
U.S.A.
973-228-7077
www.BlackburnPress.com

I would like to dedicate this book to the inner circle of people in the MOIRA team, Luigi Monte, Eduardo Gallego, John Brittain, and Dmitry Hofman. The work we carried out in that project on modelling, multi-attribute analysis, decision support systems, and GIS may be regarded as an opening of a door to a new garden of sustainable ecosystem management. I am sorry to say that many people have not discovered that door. But it is there, and this book draws much from the work carried out in the MOIRA project. The excellent cooperation and friendship in that group has meant a lot to me.

Contents

Prologue

The aim of this book is to address important questions related to the role of suspended particulate matter (SPM) in lakes, rivers, and marine systems based on a process-oriented ecosystem perspective. This is the key perspective in the new European Water Directive. The focus is on the factors regulating the SPM variability within and among aquatic systems, on empirical data and models, and on transport processes at the ecosystem scale.

This is not a "who did what" book regarding SPM in aquatic systems, but rather a book on "how it works." There is a long tradition among limnologists to carry out comparative studies and look for factors causing variations in key functional characteristics among lakes. This work follows that tradition expanding the comparative studies to include rivers and marine systems. Many limnologists, hydrologists, and marine ecologists act as "border patrol" in the sense that they guard their scientific territory. The author has a pious hope that this work may be seen as an argument against such attitudes.

There exist many physical/hydraulic models concerning SPM, especially in rivers, which often omit biota and quantify SPM fluxes at small time scales (minutes to days), small spatial scales (using partial differential equations), and utilize online climatological data. Such models will not be discussed in this work, which focuses on the many roles that SPM plays for the structure and functioning of aquatic systems at temporal scales of weeks to years. This focus will be further discussed in chapter 1.

Although SPM is an important variable, it is not regularly measured in many monitoring programs and studies, especially not in marine environments. One further motive for writing this book is to illustrate the importance of SPM in contexts related to the transport, biouptake and effects of water pollutants, to discuss how SPM regulates water clarity, and hence primary and secondary production, including production of zooplankton and fish. Sedimentation of SPM influences sediment redox conditions and factors related to the sediments as a historical archive; the redox conditions influence the survival of zoobenthos, a key food for fish, and an important factor regulating the age of sediments. A key argument in this book is that it is very difficult to understand the structure and function of aquatic systems without this knowledge on SPM.

There is also a history to this book. To make that history very short, it is based on results from several European Union (EU) and United Nations (UN) projects in radioecology (VAMP, MOIRA, ECOPRAQ, COMETES, and EVANET, several of these coordinated by Luigi Monte, ENEA, Rome) and from an EU project in limnology (Phytoplankton-on-line, coordinated by Werner Eckert, Kinneret Limnological Laboratory, Tiberias). But this work would probably not have been written at this stage if it were not for two new projects: (1) an INTAS project on saline lakes coordinated by Richard Kemp, University of Wales, and (2) Thresholds, an EU project on threats to European coastal ecosystems, coordinated by Carlos Duarte, CSIC University, Mallorca. The aim has been to try to write a state-of-the-art book on quantitative modelling of SPM in lakes, rivers, and coastal areas, which could also form a basis for the work in the new projects.

1. Background, Introduction, and Aim

This book is based on several papers published by the author and coworkers during the last five years (table 1.1). It is meant as a compilation or state-of-the-art work concerning suspended particulate matter (SPM) in aquatic systems. But why write an entire book on this subject? There are many good reasons. First, SPM regulates the two major transport routes: the dissolved transport in water (the pelagic route) and the particulate sedimentation (or benthic) route of all types of materials and contaminants. SPM in the water column is also a metabolically active component of aquatic ecosystems. The carbon content of SPM is crucial at lower trophic levels as a source of energy for bacteria, phytoplankton, and zooplankton (see Jørgensen and Johnsen, 1989; Wetzel, 2001; Kalff, 2002). SPM is also directly related to many variables of general use in water management as indicators of water clarity (e.g., Secchi depth, water color and the depth of the photic zone). Suspended particles will settle out on the bottom and the organic fraction will be subject to bacterial decomposition. This will influence the oxygen concentration and hence also the survival of zoobenthos, an important food for fish (Håkanson and Boulion, 2002). SPM influences primary production of phytoplankton, benthic algae, and macrophytes, the production and biomass of bacterioplankton, and hence also the secondary production (e.g., zooplankton, zoobenthos and fish). The effects of SPM on recycling processes of organic matter, major nutrients, and pollutants determine the ecological significance of SPM in any given aquatic environment. Understanding the mechanisms that control the distribution of SPM in rivers, lakes, and marine systems is an issue of both theoretical and applied concern, as physical, chemical and biological processes ultimately shape aquatic ecosystems.

Comparative studies in aquatic sciences often aim to find general factors regulating and explaining why systems differ in fundamental properties, such as trophic level and fish production. Fig. 1.1 gives a comparison between SPM values from marine systems, lakes, and rivers (these data emanate from different databases that will be used in this work, see appendix 9.1). Many factors (x variables) could potentially influence the variability in SPM (or any given y variable) among and within systems. The statistical analysis based on empirical data can be used to rank the importance of how the different x variables influence y. In these contexts, one must clearly differentiate between statistical and causal analyses. Statistical treatments can never

Table 1.1. Papers on SPM and related subjects (distribution coefficients and sedimentation) from our group at Uppsala University during the last five years that form the basis for this book. Note that there are more extensive literature lists in these papers.

Lakes	
Empirical SPM models	Lindström et al. (1999)
Dynamical SPM models	Håkanson (1999), Håkanson et al. (2000a), Malmaeus and Håkanson (2003)
Rivers	
Empirical SPM models	Håkanson et al. (2004b)
Dynamical mass-balance model	Håkanson (2004c, d)
Marine areas	
Empirical SPM model	Håkanson and Eckhéll (2004)
Dynamical SPM model	Håkanson et al. (2004a)
Related subjects	
Distribution coefficients	Johansson et al. (2001)
Internal loading in lakes	Håkanson (2004e)
Burial in lakes	Håkanson (2003b)
Water clarity, SPM, and Secchi depth in lakes	Håkanson and Boulion (2003)
Sedimentation and oxygen status in marine areas	Carlsson et al. (1999)

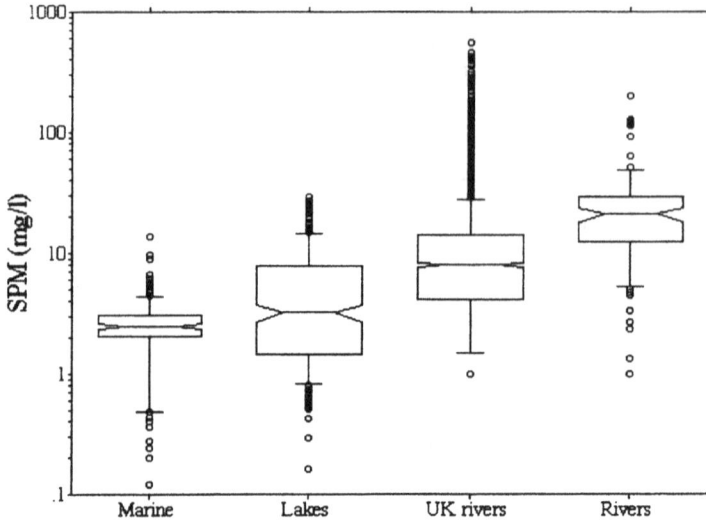

Figure 1.1. Comparison between SPM data from marine systems, lakes, and rivers (rivers in the United Kingdom and Europe). These data emanate from the database given in appendix 9.1. The box-and-whisker plots provide the medians, quartiles, 90th and 10th percentiles and outliers.

mechanistically "explain" why certain x variables end up with a high correlation toward y, but results from correlations and regressions can provide important information for further mechanistic interpretations and modelling. Regression analyses can be performed for many reasons, e.g., to compare modelled values with empirical data, to test hypotheses about relationships, and to develop statistical/empirical models. Many textbooks examine regression analyses (e.g., Draper and Smith, 1966; Mosteller and Tukey, 1977; Pfaffenberger and Patterson, 1987; Taylor, 1990; Newman, 1993).

From fig. 1.1, SPM seems to vary in a systematic way among and within marine systems, rivers, and lakes, and table 1.2 gives a statistical compilation of the data shown in the figure. An objective of this book is to try to explain and predict the variations shown in fig. 1.1 and table 1.2.

Table 1.2. Statistical compilation of data on SPM from marine systems, lakes, and rivers (from UK and Europe, respectively). References and presuppositions related to these data are given in appendix 9.1.

	Marine	Lakes	UK Rivers	European Rivers
Number of data (n)	196	386	3669	89
Minimum value (Min.)	0.12	0.16	1	1
Maximum value (Max.)	13.9	29.1	560	202
Mean (MV)	2.6	5.5	16.2	28.4
Median (M50)	2.5	3.2	8.0	21.0
Standard deviation (SD)	1.7	6.2	34.7	32.6
Coefficient of variation (CV)	0.65	1.07	2.14	1.06

To meet that objective, both statistical/empirical models and dynamical models will be discussed. Many of these models for SPM are, in fact, basically meant to be submodels in ecotoxicology and radioecology to quantify fluxes of toxic substances (Monte, 1997), in ecosystem modelling (e.g., to model bacterioplankton production; Håkanson and Boulion, 2002) or in the modelling of ecosystem effects of water pollutants (Håkanson, 1999). Many contaminants (such as radionuclides, heavy metals, and nutrients) may be removed from the water column due to their sedimentation with SPM and burial in the bottom sediments (IAEA, 2000). Fig. 1.2 illustrates that SPM may be regarded as a target variable in geosciences. However, in water management and aquatic ecology, the main focus may not be on SPM but on ecosystem effect variables related to the major threats to the aquatic systems. In such cases, SPM can be regarded as one of many important variables influencing the target ecosystem effect variables.

A dynamic approach to the problem of understanding the role of SPM in aquatic systems may start with a mass-balance approach calculating inflow to a system by using data on water discharge (Q in m^3/time) and SPM concentration (g dw/m^3; dw = dry weight). Then, to model the conditions within a defined system (a lake, a river stretch or a coastal area), one would quantify the key processes regulating the SPM concentration in water. Differential equations are often used to quantify fluxes (e.g., g X /time), amounts (g X) and concentrations (g X/m^3) of all types of materials (such as SPM, organic matter, toxins, and nutrients), but not generally ecosystem effect variables. Regressions based on empirical data are often necessary to relate concentrations of chemicals in water or sediments to target effect variables. In theory, both these model approaches (fig. 1.3A and B) may be used for the effect-load-sensitivity analyses (ELS; see Håkanson, 1999) provided that at least one operationally defined ecological effect variable relevant for the load variables(s) in question is included in the model. Ideally, the effect variable should express the production or biomass of defined functional organisms (preferably fish at the top trophic level, fig. 1.3C), which characterize a given system. Fig. 1.3D illustrates schematically that two systems with different SPM concentrations are likely to react differently to a change in the load of toxins and/or nutrients to the system. The classical approach to carry out environmental consequence analyses is to use (1) dynamic mass-balance models to predict concentrations of pollutants and (2) empirical models (like regressions) to link these concentrations to measured data on the effect variables. So, there are good reasons why these two modelling approaches are addressed in this book.

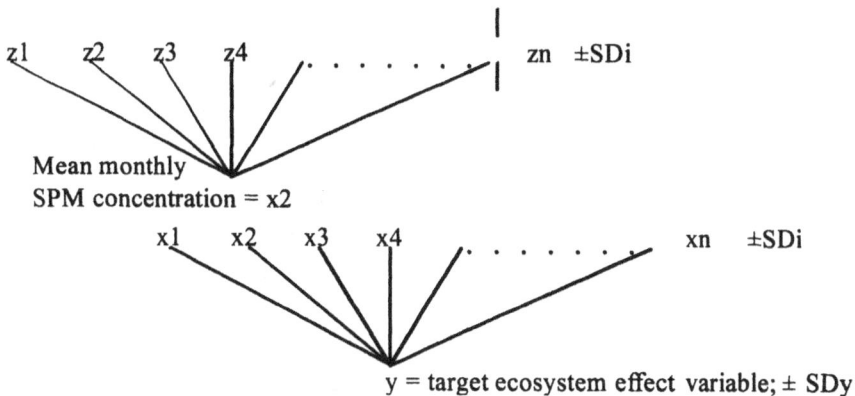

Fig. 1.2. Illustration of how a target variable in geosciences, such as the concentration of SPM, may be used as an x variable in predictive models in water management, where the y variables may express functional aspects of ecosystem status, such as water clarity, chlorophyll-a concentrations, and/or biomasses of functional groups of organisms.

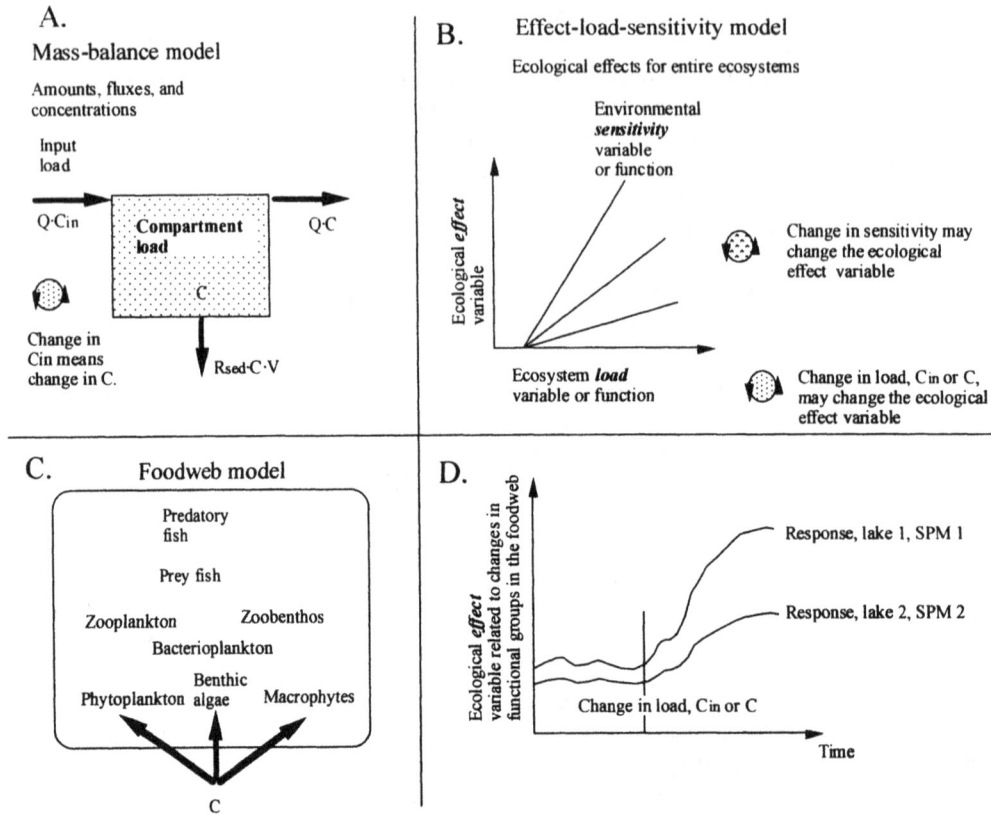

A.

Mass-balance model

Amounts, fluxes, and concentrations

Input load

Q·Cin   **Compartment load**   Q·C

C

Change in Cin means change in C.

Rsed·C·V

B.   Effect-load-sensitivity model

Ecological effects for entire ecosystems

Environmental *sensitivity* variable or function

Ecological *effect* variable

Ecosystem *load* variable or function

Change in sensitivity may change the ecological effect variable

Change in load, Cin or C, may change the ecological effect variable

C.   Foodweb model

Predatory fish

Prey fish

Zooplankton   Zoobenthos

Bacterioplankton

Benthic

Phytoplankton   algae   Macrophytes

C

D.

Ecological *effect* variable related to changes in functional groups in the foodweb

Response, lake 1, SPM 1

Response, lake 2, SPM 2

Change in load, Cin or C

Time

Fig. 1.3. Illustration of the fundamental difference between dynamic, mass-balance models (A) and effect-load-sensitivity models (ELS) based on regressions (B), and ELS models related to dynamic foodweb models (C) and (D). The figure also shows how changes in the load at a given time may cause different responses in the lake foodweb in systems with different SPM concentrations (lake 1 compared to lake 2). The wheels indicate that by means of remedial measures the load variable in dynamic models and the load and the sensitivity variables in ELS models may reduced.

The predictive accuracy of a model should always be stated. The predictive power is generally determined by means of statistical tests, where model predictions of the target y variable are compared to empirical data of the y variable. Models that provide r^2 values (r^2 = the coefficient of determination; r = the correlation coefficient) higher than about 0.75 (and statistical certainties, p values, lower than 0.05; Prairie, 1996) could generally be used in practice in water management for predictions in individual ecosystems. Models providing lower r^2 values (but still p values < 0.05) could be used for regional predictions, when predictive failure in individual ecosystems could be accepted (Håkanson and Peters, 1995). The operational range, the domain, of the model must always be explicitly given to avoid abuse of the model for ecosystems for which it was never intended. If dynamic (time-dependent) mass-balance models can meet these requirements, they would generally be preferable to statistical/empirical models because they can provide better understanding of mechanisms and processes.

A central problem in communications among scientists and a key reason for much misunderstanding has to do with scale. To model in great detail at the scale dealing with hourly or daily changes for target variables in water management is a very difficult task for one system, and to do this in a general, predictive manner for many systems is even more difficult. This is illustrated in fig. 1.4. This figure gives some important general concepts related to the "philosophy" behind the concept of scale in aquatic studies. The curve marked CV in fig. 1.4 illustrates how the uncertainty in empirical data increases if the data are collected during a day (a relatively small CV of 0.3; CV = the coefficient of variation; CV = SD/MV; SD = standard deviation; MV = mean value), during weekly sampling within a system (a higher CV, 0.35), monthly sampling (a still higher value, CV = 0.55) and for samples collected over a year (CV = 0.95). The curve called "accessibility" illustrates that the overall sampling effort would increase, and the money and work related to the accessibility of data needed to run and test a model would also increase, for models based on daily predictions (e.g., using series of measured climatological data) to models for long-term predictions based on fewer and more readily available data from standard maps and/or monitoring programs. The bolded curve is meant to illustrate one of many possible ways to define the optimal size of a predictive model accounting for these two factors, the uncertainty in the empirical data (as given by CV) and the accessibility of the data (as given by N; where N is simply the number of days used for sampling; note that it is not meaningful to determine a mean value or a standard deviation from a study with fewer than three samples). Any model developer or user would have a set of criteria to define the practical usefulness of the model; different users may have different criteria. Given the bolded curve for practical usefulness in this figure, one can note that models striving for short-term (hourly - daily) predictions may not be very useful due to too high demands related to data accessibility, and that models based on annual predictions would generally be suboptimal due to the high uncertainties in the empirical data. This motivates the scale in focus of the SPM models discussed and used in this book (fig. 1.5)

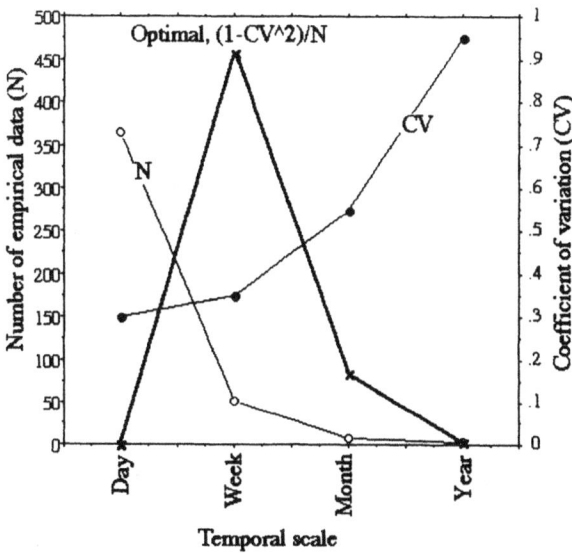

Fig. 1.4. Illustration of factors regulating the optimal size of practically useful predictive models in water management. The curve marked N illustrates the accessibility (number of data); the CV curve illustrates uncertainties in empirical data; and the bolded curve expresses the optimal temporal scale related to the CV curve and the N curve.

Fig. 1.6 is included here in the introduction to put SPM variations in aquatic systems into a more general context. It is well known for lakes (Håkanson and Jansson, 1983) that sedimentation increases from about zero at the water depth separating areas of fine sediment transport and accumulation, to maximum values at the deepest part of the basin, and the concentration of SPM increases due to this and by the fact that the volume (defined for a certain water depth) decreases with increasing water depths, a phenomenon often referred to as "sediment focusing" (fig. 1.6, left).

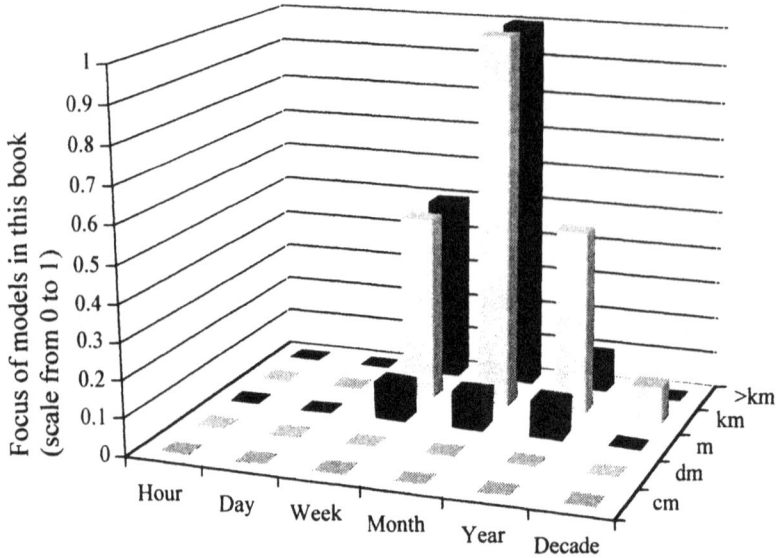

Fig. 1.5. Illustration of the focus of the models discussed in this book, i.e., on entire lakes, rivers or marine systems and weekly to monthly predictions.

SPM focusing in lakes

SPM focusing in coastal areas

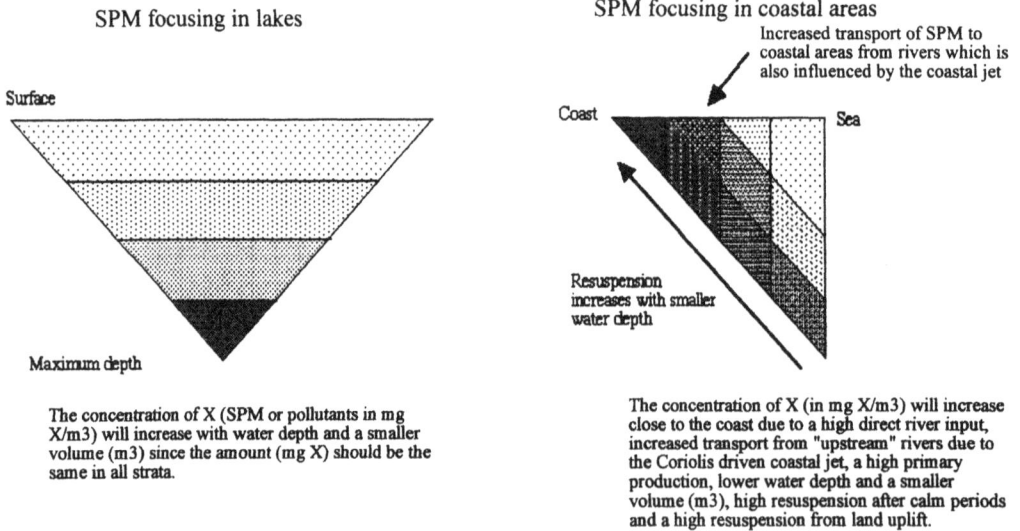

Increased transport of SPM to coastal areas from rivers which is also influenced by the coastal jet

Resuspension increases with smaller water depth

The concentration of X (SPM or pollutants in mg X/m3) will increase with water depth and a smaller volume (m3) since the amount (mg X) should be the same in all strata.

The concentration of X (in mg X/m3) will increase close to the coast due to a high direct river input, increased transport from "upstream" rivers due to the Coriolis driven coastal jet, a high primary production, lower water depth and a smaller volume (m3), high resuspension after calm periods and a high resuspension from land uplift.

Fig. 1.6. SPM focusing in lakes and coastal areas.

The concept "coastal focusing" may be used in analogy with sediment focusing in lakes (fig. 1.6). SPM (and related substances) generally appear with relatively high values in coastal areas where the river input of SPM may be significant and accentuated by the coastal current and small water depths. Resuspension will also add to the SPM fluxes. The pattern shown in fig. 1.6, is typical for SPM, but this pattern may be modified by many processes (see fig. 1.7) which is an attempt to provide a general mechanistic framework for SPM variations in coastal areas.

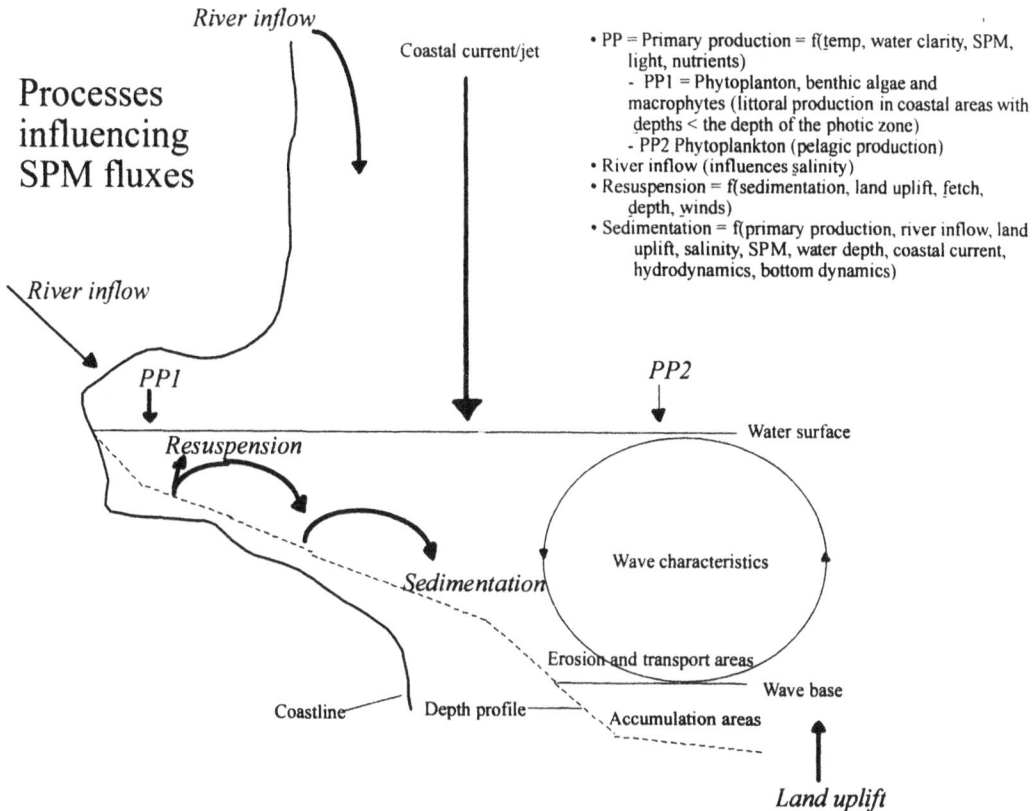

Processes influencing SPM fluxes

River inflow

Coastal current/jet

River inflow

- PP = Primary production = f(temp, water clarity, SPM, light, nutrients)
  - PP1 = Phytoplanton, benthic algae and macrophytes (littoral production in coastal areas with depths < the depth of the photic zone)
  - PP2 Phytoplankton (pelagic production)
- River inflow (influences salinity)
- Resuspension = f(sedimentation, land uplift, fetch, depth, winds)
- Sedimentation = f(primary production, river inflow, land uplift, salinity, SPM, water depth, coastal current, hydrodynamics, bottom dynamics)

PP1

PP2

Resuspension

Water surface

Wave characteristics

Sedimentation

Erosion and transport areas

Wave base

Coastline   Depth profile   Accumulation areas

Land uplift

Fig. 1.7. Processes (in italics) influencing SPM in coastal areas. The underlined factors are included in the model discussed in chapter 6 for open marine areas. Note that many of the processes and factors influencing SPM are interrelated in a complex manner. There is also a very dynamic exchange of water, SPM, and pollutants between the open sea and coastal areas.

- A high river inflow will increase SPM. Note that this may be from rivers close to the sampling site or from "upstream" rivers influencing the given site due to the coastal current/jet.

- Autochthonous production will increase SPM; the bioproduction is high in coastal areas where phytoplankton, benthic algae, and macrophytes contribute to the primary production; in open water areas, only phytoplankton dominate the primary production.

- Sedimentation is a key process to reduce suspended particles from the water mass.

- Resuspension of fine sediments from areas or erosion and transport, i.e., advective transport of matter deposited under calm conditions and then resuspended by wind-wave activity or slope processes.

- Land uplift related to the latest glaciation is also an important contributor to SPM in, e.g., the Baltic Sea.

This book will discuss process-based mass-balance models for SPM for lakes, rivers, and coastal areas. The models used and discussed here emanate from models developed in radioecology. During the last ten years, there has been a "revolution" in aquatic ecosystem modelling (see Håkanson, 2000, 2004a). The generality and predictive power of models for radionuclides, nutrients, metals, organics, and SPM have increased in a way that was inconceivable ten years ago. This new generation of dynamic mass-balance models (the LakeMab modelling approach discussed in this book) predict as well as one can measure - if one measures well. And yet, they are driven by readily available driving variables and have a general structure that applies to most types of substances in aquatic systems. The major reason for this development is, in fact, the Chernobyl accident. Large quantities of radiocesium (^{137}Cs) and radiostrontium (^{90}Sr) were released in April/May 1986 as a pulse. To follow the pulse of these radionuclides through ecosystem pathways has meant that important fluxes and mechanisms, i.e., ecosystem structures, have been revealed. It is important to stress that many of these new structures and equations are valid not just for radionuclides, but for most types of contaminants, e.g., for metals, nutrients, and organics. This means that the models, methods (of building and testing models) and equations used in this book for SPM in lakes, rivers, and coastal areas should be of great interest also to other ecosystem modellers. When validated, i.e., when the LakeMab model outputs on concentrations in water, sediments, on suspended particles, and in small planktivorous fish were blind tested against independent data from a wide domain of lakes, this modelling predicts very well. This will be discussed in greater detail later but it is exemplified in fig. 1.8 using data on radiocesium from the Chernobyl accident as a tracer. When 357 empirical data from 23 very different European lakes were compared to modelled values of ^{137}Cs concentrations, the coefficient of determination (r^2) was 0.96 and the slope 0.98. This is almost like an analytical solution. Fig. 1.8 also gives the 95% confidence intervals for the mean y value. The two confidence intervals are very close to the regression line.

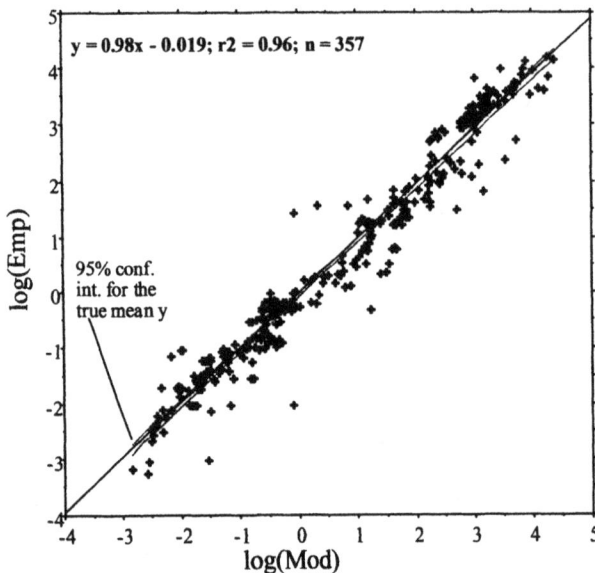

Fig. 1.8. Comparison between empirical data and modelled values for the mass-balance model for lakes (LakeMab) using radiocesium from the Chernobyl accident as a tracer (data from Håkanson et al., 2004b). This modelling approach is also used for SPM in lakes, rivers, and coastal areas. The figure gives the regression line and the 95% confidence intervals for the true mean y. Note that the slope and the r^2 value are almost perfect, 0.98 and 0.96, respectively. The regression is based on 357 empirical data from the 23 lakes covering a wide limnological range.

Fig. 1.9 shows one of the many data series included in fig. 1.8. The modelled values in this validation are close to the mean empirical values and within the uncertainty bands of the empirical data. It can also be noted from fig. 1.9 that there are major uncertainties in the empirical data. The modelling discussed in this book is not meant to predict the SPM concentration in every sampling bottle; it is meant to predict the mean concentrations in water on a monthly basis. The time and area compatibility of the empirical data and the modelled values should be as high as possible; they should represent the same thing. This modelling is based on a structure that is general and valid for all types of systems (rivers, coastal areas, and lakes). The processes, such as inflow, sedimentation, resuspension, diffusion, mixing, and outflow are general. There are also substance-specific components in some of the algorithms for the processes, which will be exemplified later in this book. This model has also been tested using sensitivity and uncertainty tests (Monte Carlo techniques; see Chapter 3) and validated (blind tested) with positive results for radiocesium also from the Khystym accident (see Håkanson and Sazykina, 2001), for radiostrontium in lakes (see Håkanson et al., 2002), for calcium from lake liming (see Håkanson, 2003a; liming is a major environmental industry in many countries suffering from acid rain) and for phosphorus in lakes and hence also for lake eutrophication (Håkanson and Boulion, 2002).

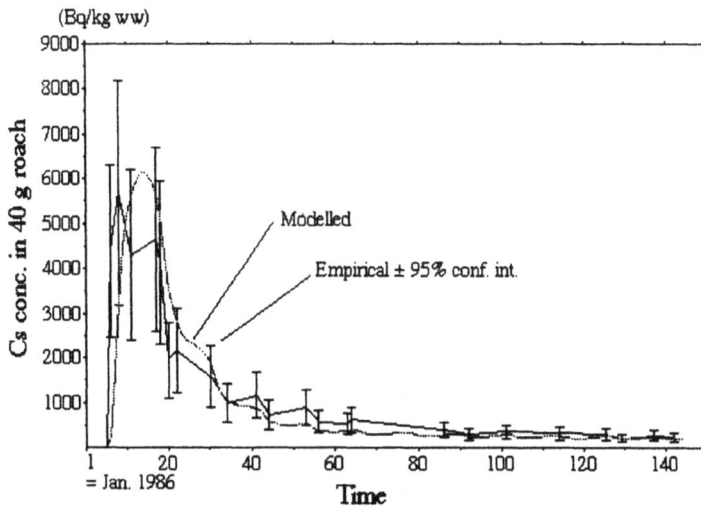

Fig. 1.9. The relationship between modelled values (using the LakeMab model) and empirical data for Cs concentrations in 40 g roach in Lake Siggefora, Sweden (from Håkanson, 2000). The figure also gives the 95% confidence intervals for the empirical data (i.e., a measure of the uncertainty in the empirical data).

Modelling of phosphorus in lakes is a central paradigm in limnology and lake management and fig. 1.10 gives an example of how the LakeMab model predicts TP concentrations in Lake Miastro, Belarus (see Håkanson and Boulion, 2002). It should be noted that to get the results in this figure, there was no "tuning" of the model, i.e., no changes in any of the model algorithms or model constants, and only changes in the obligatory lake-specific driving variables that are readily accessed from standard monitoring programs and/or maps.

The fact that the LakeMab model has been validated with good results for many substances and for many types of aquatic systems should lend some credibility to the following parts that aim to highlight dynamic mass-balance modelling for SPM using this modelling approach.

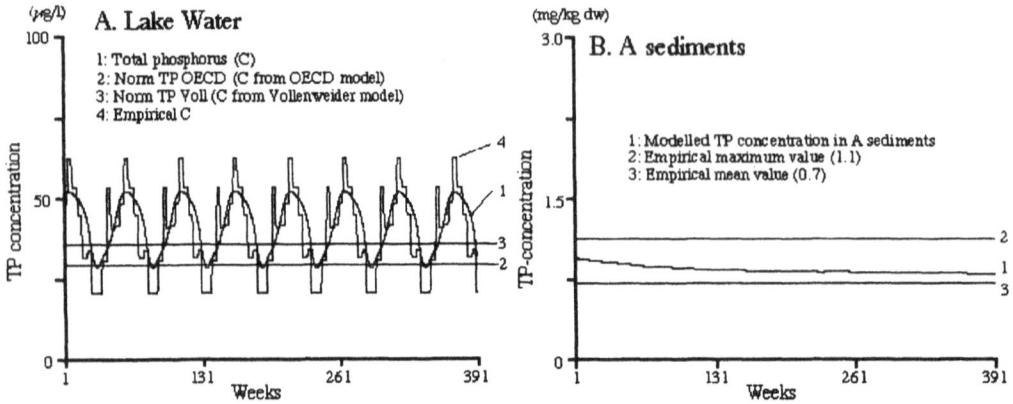

Fig. 1.10. Validations of the LakeMab model for (A) TP concentrations in water and (B) TP concentrations in A sediments in Lake Miastro, Belarus. In fig. A, there are three reference lines to the modelled values (given by curve 1), curve 2 gives TP concentrations from the empirical OECD model (OECD, 1982), curve 3, TP concentrations from the empirical Vollenweider model (Vollenweider, 1968) and curve 4, the empirical data for the lake. The empirical data are based on mean monthly values from the period 1980-90. Fig. B gives two empirical reference lines for the modelled TP concentrations.

1.1. Fresh water versus marine systems

In this introduction, it may be of interest to highlight some differences between freshwater and marine systems:

- The amount of SPM always depends on two main causes, allochthonous inflow and autochthonous production. In the Baltic Sea (which will be discussed in the following parts of this book), there is also another primary source, land uplift. Thousand-year-old sediments influence the ecosystem of the Baltic Sea today. When the old bottom areas rise after being depressed by the glacial ice, they will eventually reach the wave base, which is the water depth above which the waves can exert a direct influence on and resuspend the sediments. It is easy to imagine that storms leading to high waves have a major influence on coastal ecosystems. The wave base, or the critical depth separating bottom areas where transport processes dominate the bottom dynamic conditions from areas of continuous sedimentation of fine materials, depends on the effective fetch (see chapter 2), and the duration and velocity of the winds. During storms, the wave base may reach water depths of more than 50 meters in the Baltic Sea (see Håkanson, 1999). So, as a result of land elevation, the old sediments deposited hundreds and thousands of years ago will be resuspended. The land uplift in the Baltic Sea varies from about 9 mm/yr in the northern part of the Bothnian Bay to about zero for the southern part of the Baltic (see Voipio, 1981). This implies that large amounts of old glacial and postglacial sediments are eroded. In this way, the carbon, nitrogen, and phosphorus contained by these sediments, as well as metals and mineral particles, will again enter the ecosystem of the Baltic Sea, perhaps thousands of years after they were originally deposited onto the bottom in a considerably calmer environment. Measurements have demonstrated that as much as 80% of the material sedimenting onto the deep bottoms may be old, eroded material (see Jonsson, 1992). The new supply of material to the sea from rivers, direct discharges, and the bioproduction in the system only contribute about 20% of the amount annually deposited on the accumulation areas (A areas). The amount of matter deposited on the areas of erosion and transport (ET areas) may be resuspended by wind/wave action or slope processes, so resuspension is the fourth main process influencing the flux of matter in coastal areas. Resuspension is, however, a secondary process since it depends on the three primary processes. Land uplift is not a major primary source of matter in most lakes and coastal areas, but is, as has been stressed, a very important factor in the Baltic.

- The allochthonous inflow of matter to a lake is simply $Q \cdot C_{in}$ (Q = tributary water discharge in m^3/month; C_{in} = tributary concentration of SPM or a given pollutant in g/m^3). For marine coastal areas, the tributary inflow of matter is, by definition, important in estuaries, but outside the estuaries, the inflow is given by $Q_{sea} \cdot C_{sea}$, where Q_{sea} is the net inflow of water from the sea and/or adjacent coastal areas (m^3/month) and C_{sea} is the concentration of the pollutant in the water outside the coast (in g/m^3). This will be discussed in greater detail in chapter 6.

- Freshwater and saltwater systems are generally classified into different trophic categories reflecting differences in nutrient status, water clarity, primary and secondary production. Hypertrophic systems are extremely productive; eutrophic systems are productive, mesotrophic and oligotrophic systems have a lower production (table 1.3). In studies of anthropogenic eutrophication, target variables are mean, representative ecosystem values of chlorophyll-a (a practical, operational measure of algal biomass and an indicator of primary productivity), or Secchi depth (as a general index of water clarity) and oxygen concentration in the deep water zone. Richard Vollenweider (1968, 1976) presented the first useful load models for phosphorus for lakes in the late 1960s. Total phosphorus is since long recognized as the most crucial limiting nutrient for lake primary production in most but not all lakes (Ahlgren, 1970; Schindler, 1977, 1978; Bierman, 1980; Chapra, 1980; Boynton et al., 1982; Wetzel, 2001; Persson and Jansson, 1988; Boers et al., 1993). Nitrogen is often regarded as a key limiting nutrient in many marine areas. Both elements may be of vital importance in estuaries and brackish waters like the Baltic Sea (Redfield, 1958; Ryther and Dunstan, 1971; Nixon and Pilson, 1983; Howarth and Cole, 1985; Howarth, 1988; Hecky and Kilham, 1988; Ambio, 1990; Nixon, 1990). These questions will also be discussed in greater detail in chapter 6.

- The bottom dynamic conditions (E = Erosion, T = Transportation and A = Accumulation) in coastal areas also differ from those in lakes and rivers. The bottom dynamic conditions influence and are influenced by the hydrodynamic conditions, mixing, stratification, winds, etc. In sheltered coasts, A areas can be found at relatively small water depths, but coastal areas generally have smaller percentages of A areas and larger percentages of ET areas than lakes (Håkanson, 1999, 2000). This has profound consequences on the internal loading and many internal transport processes of SPM and water pollutants.

Table 1.3. Characteristic features in (A) lakes and (B) marine (here Baltic) coastal areas of different trophic categories (modified from OECD, 1982; Håkanson and Jansson, 1983 and Wallin et al., 1992).

A. Lake data; * = mean values for the growing season; ** = mean values for the spring circulation

Trophic level	Primary production (g C/m^2·yr)	Secchi (m)	Chl a (mg/m^3)	Algal volume* (g/m^3)	Total P **	Total N **	Dominant fish
					------(mg/m^3)------		
Oligot.	<30	>5	<2.5	<0.8	<10	<350	Trout, Whitefish
Mesot.	25-60	3-6	2-8	0.5-1.9	8-25	300-500	Whitefish, Perch
Eut.	40-200	1-4	6-35	1.2-2.5	20-100	350-600	Perch, Roach
Hypert.	130-600	0-2	30-400	2.1-20	>80	>600	Roach, Bream

B. Coastal area data; all data are based on mean summer values.

Trophic level	Secchi (m)	Chl a (mg/m^3)	Total N (mg/m^3)	Inorg N (mg/m^3)	Total P (mg/m^3)	Oxygen (mg/l)	Oxygen (%)
Oligotrophic	>6	<1	<260	<10	<15	>10	>90
Mesotrophic	3-6	1-3	260-350	10-30	15-25	6-10	60-90
Eutrophic	1.5-3	3-5	350-400	30-40	25-30	4-6	40-60
Hypertrophic	<1.5	>5	>400	>40	>30	<4	<40

- The salinity is of paramount importance to the number of species, as illustrated in fig. 1.11. It also influences the aggregation of suspended particles (see chapters 2 and 4), which is of particular interest in modelling SPM. The saltier the water, the greater the flocculation.

- A typical deposition in lakes is about 0.4 cm/yr (Håkanson and Peters, 1995). Of course, the value varies among and within lakes. Highly eutrophic lakes may have a net deposition of more than 1 cm/yr. Within lakes, the deposition generally increases from the wave base to the maximum depth (= sediment focusing; fig. 1.6). The deposition can be very large in limited deep holes, say several cm/yr. In marine areas, a higher deposition than in lakes is generally expected. There are several reasons for this; one being related to the coastal currents. The dominant water circulation along coasts in the northern hemisphere constitutes an anticlockwise cell, which distributes the settling particles, the suspended materials, and the pollutants in a typical pattern, reflecting the flow of the water. This anticlockwise cell is created by the rotation of the earth (the Coriolis force), which deflects any plume of flowing water to the right in relation to the direction of the flow in the northern hemisphere (and to the left in the southern hemisphere). The existence of these Coriolis-driven currents implies that relatively little material is transported to the deep open parts since these currents move the suspended particulate materials along and into the coasts. This also means that allochthonous matter from the rivers entering the sea, the autochthonous production in the coastal areas and the resuspension in the shallow coastal areas together with the dominating coastal currents, create an environment of high sedimentation within coastal A areas. There are reported cases of deposition of more than 10 cm/yr from the Stockholm archipelago in Sweden (Markus Meili and Per Jonsson, personal communication), and this agrees with the maximum value obtained using the model presented by Håkanson (1999) of 9.2 cm/yr. The mean deposition for Baltic coastal areas is 2-3 cm/yr, which is a factor of 5-10 greater than the characteristic value for lakes (0.4 cm/yr).

- The hydrodynamical conditions are generally much more dynamic in marine coastal areas than in lakes (a typical surface water retention time is 2-6 days for a coastal area and about 1 year for a lake; see Håkanson, 1999, 2000). This has implications for the conditions in the coastal areas, which are evidently greatly influenced by the conditions in the outside sea and/or adjacent coastal areas in direct contact with the sea. Fig. 1.12 gives a comparison (note the logarithmic scale of the y axes) between the predicted ^{137}Cs concentrations in a lake (from the Chernobyl fallout in 1986) and a coastal area for (A) water and (B) 10 g ww planktivores, if all the many characteristics for the lake and the coastal area are comparable, e.g., area (5 km^2), mean depth (9 m), size of catchment, catchment characteristics, fallout of radiocesium, and bottom dynamic conditions (A areas and ET areas). The main difference is the theoretical water retention time, which is 1 year for the lake and 0.16 months for the surface water for the given coastal area.

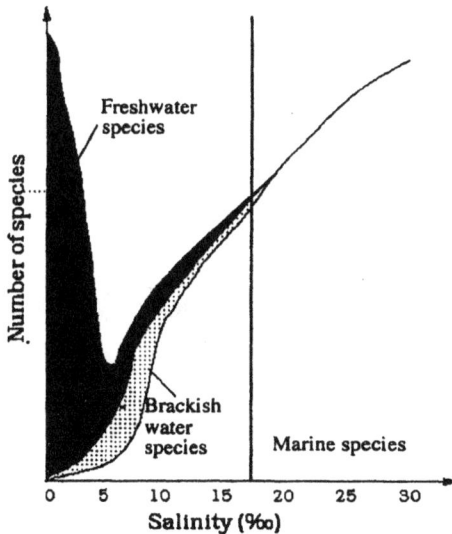

Fig. 1.11. The relationship between salinity and number of species. Redrawn from Remane (1934).

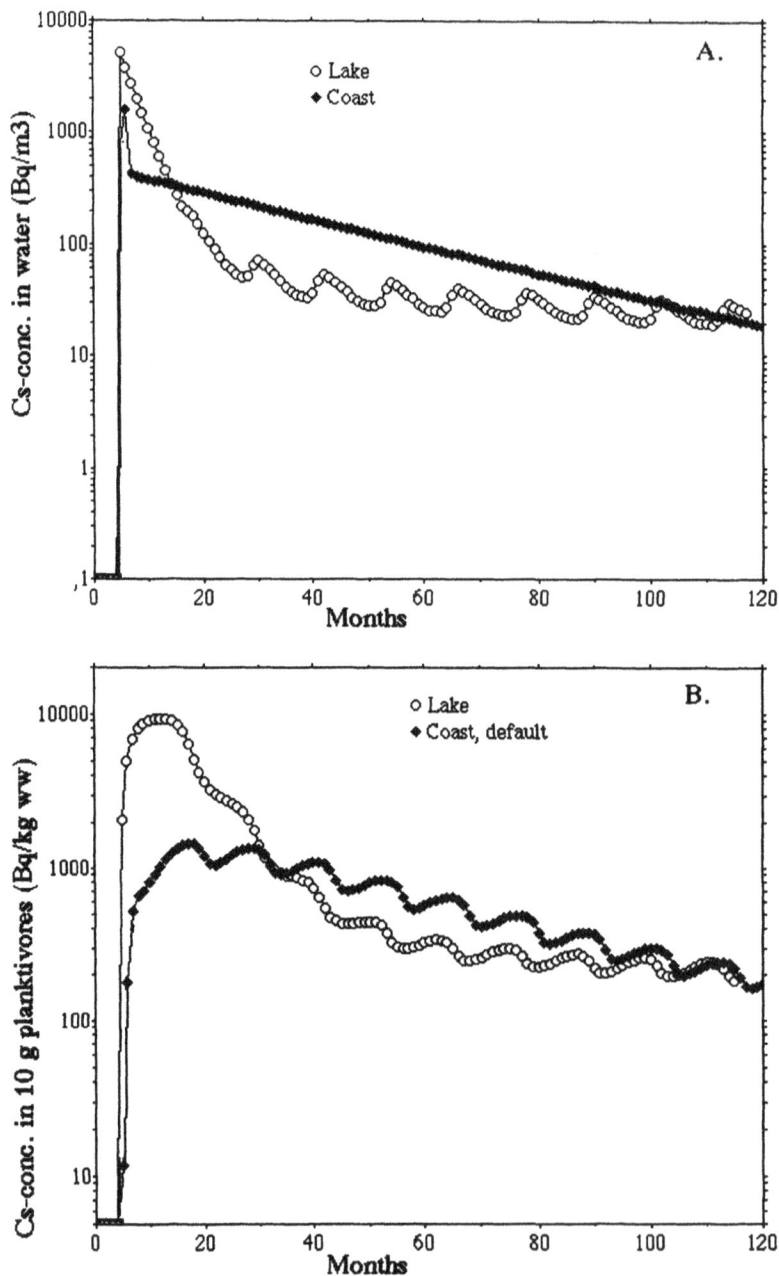

Fig. 1.12. Lake versus coast. A comparison (note the logarithmic scale of the y axis) between the predicted ^{137}Cs concentrations in a lake and a marine coastal area in (A) water and (B) 10 g wet weight planktivores. The theoretical water retention time is 1 year for the lake and 0.16 months for the coastal area, while all else is equal (modified from Håkanson, 2000).

- A very important question concerns the definition of coastal ecosystems in large lakes or marine areas. The question is where to place the boundaries toward the sea and/or adjacent coastal areas. It is crucial to use a technique that provides an ecologically meaningful and useful definition of the coastal ecosystem. How should one define this area so that parameters, like mean depth, can be relevant as model variables (x) to predict target y variables? The problem is illustrated in fig. 1.13B using data on Secchi depth (the variable to be predicted in this example; Secchi depth is a simple, useful operational measure of water quality) and mean depth (D_m, the x parameter which reflects resuspension processes). For lakes, there exists a significant ($r^2 = 0.38$, $p = 0.0001$ for 88 Swedish lakes) positive relationship between D_m and Secchi depth: The deeper the lake, the larger the bottom areas beneath the wave base, the less resuspension, the less suspended materials in the lake water, the clearer the water, and the larger the Secchi depth. This is logical. The mean depth has a significant meaning for an important ecosystem variable, Secchi depth. The entire lake is the defined ecosystem. But how would this apply for a marine coastal area? Is there a method to define the boundaries and establish coastal ecosystems where morphometric parameters like the mean depth have a meaning in predictive ecosystem models? This is illustrated in fig. 1.13A. In this example, there are three boundary lines, A, B and C. The mean depths of the enclosed areas are 4.5, 3.5, and 2.5 m, respectively. The exposure of the coastal area is the ratio between the section area and the enclosed coastal area (Ex = 100·At/Area, where At = the section area or the opening area toward the sea in km^2 and Area = the coastal area in km^2). The Ex values are quite different in the three cases, 0.05 for A, 0.1 for B and 0.2 for C, but the Secchi depth is the same in all three cases (2 m). Arbitrary borderlines can be drawn in many ways and the mean depths of such areas would be devoid of meaning in relation to target ecosystem variables. The approach in this work (from Håkanson et al., 1986 and Pilesjö et al., 1991) assumes that the borderlines are drawn at the topographical bottlenecks so that the exposure (Ex) of the coast from winds and waves from the open sea is minimized. It is easy to use the Ex value as a tool to test different alternative borderlines and to define the coastal ecosystem where the Ex value is minimal. If the coastal ecosystem is defined in this way, then there exists, as illustrated in fig. 1.13A, a weak but statistically significant negative relationship between Secchi depth and mean depth ($r^2 = 0.14$, $p = 0.08$ for 23 Baltic coastal areas where the tidal impact is negligible): The larger D_m, the more suspended materials will be retained in the coastal water, the more turbid the water, and the smaller the Secchi depth. This is also logical because coastal areas are, by definition, open to the outside sea (the section area, At > 0; if At = 0, then this is not a coastal area but a lake near the sea). For open coastal areas, a significant part of the fine materials suspended in the water can "escape" from the coastal area to the open water area or to surrounding coastal areas. This is not the case in the same way for lakes. So, coastal areas with small mean depths will generally have coarse bottom sediments (sand, gravel, etc.) with small amounts of fine materials causing high turbidity when resuspended.

It is always important to define the presuppositions of any model. When and where will it apply? The definition of the ecosystem boundaries is a crucial aspect for the models discussed in this book for coastal areas. Also note that concentrations of pollutants in marine coastal areas depend very much on the conditions outside the given coastal area, which are, evidently, very difficult to predict using a general, useful management model since there are great differences among coastal areas in this respect because of oceanographic conditions, topographical presuppositions, latitude, tidal characteristics, etc.

Defining ecosystem boundaries

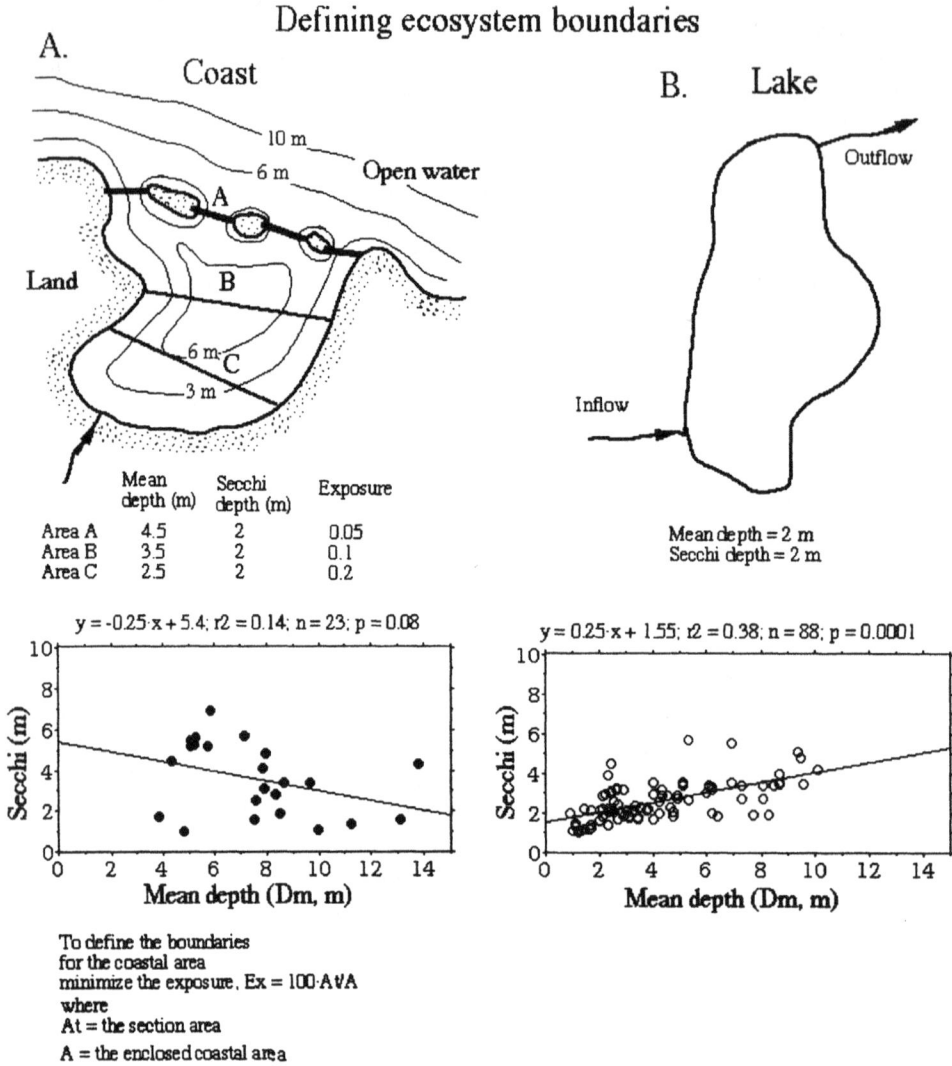

A. Coast

10 m
6 m   Open water
A
Land
B
6 m C
3 m

	Mean depth (m)	Secchi depth (m)	Exposure
Area A	4.5	2	0.05
Area B	3.5	2	0.1
Area C	2.5	2	0.2

$y = -0.25 \cdot x + 5.4; \; r2 = 0.14; \; n = 23; \; p = 0.08$

To define the boundaries
for the coastal area
minimize the exposure, Ex = 100·At/A
where
At = the section area
A = the enclosed coastal area

B. Lake

Outflow
Inflow

Mean depth = 2 m
Secchi depth = 2 m

$y = 0.25 \cdot x + 1.55; \; r2 = 0.38; \; n = 88; \; p = 0.0001$

Fig. 1.13. Illustration and rationale for the definition of ecosystem boundaries. The coastal ecosystem is defined by the borderline marked A, which gives a minimum exposure (Ex; the "topographical bottleneck method").

It is evident that coastal systems, just like lakes, are extremely complex. There are relatively few validated ecosystem-based models for chemical pollutants for marine coastal areas. One reason for this is, probably, that most scientists working in coastal and marine environments are not used to applying an ecosystem perspective, an approach that comes naturally to limnologists.

Chapter 2 will give several examples on why SPM is important in aquatic studies. Chapter 3 gives basic methodological aspects related to statistical analyses of SPM data, statistical modelling, dynamic modelling and model testing. Chapter 4 will discuss empirical and dynamical models for SPM in lakes; chapter 5 focuses on rivers; and chapter 6 on marine systems. The appendix (chapter 9) gives information about the utilized databases and other issues omitted from the main text.

2. Why SPM?

This chapter will first define SPM and then discuss fundamental concepts related to the role of SPM in aquatic systems.

2.1. Defining SPM

The total amount of a substance or a group of substances in the water is often separated into a particulate phase, the only phase subject to gravitational sedimentation, and a dissolved phase, generally the most important phase for direct biouptake. Operationally, the limit between the particulate phase and the dissolve phase is generally determined by means of filtration using a pore size of 0.45 µm. Evidently, this is an operational approach and many colloidal particles will pass such filters (see Boulion, 1994) and are, hence, operationally included in the dissolved fraction, although, they are not truly dissolved in a chemical or biological sense. SPM as determined in this way by filtration, drying and weighing, is sometimes (see Gray et al., 2000) also referred to as SSC, the suspended sediment concentration. Filtration is often a justifiable method from sedimentological, ecological, and mass-balance modelling perspectives.

Fig. 2.1 classifies SPM according to origin. Many factors are known to influence SPM in aquatic systems (Vollenweider, 1958, 1960; Carlson, 1977, 1980; Brezonik, 1978; OECD, 1982; Wetzel, 1983; Ostapenia et al., 1985; Preisendorfer, 1986; Boulion, 1994, 1997). The most important factors are:

1. Autochthonous production (i.e., the amount of plankton, feces, etc. in the water - more plankton, etc. means a higher SPM).

2. Allochthonous materials, such as the amount of colored matter (e.g., humic and fulvic substances).

3. The amount of resuspended material.

This is easy to state qualitatively, but more difficult to express quantitatively because these three factors are not independent: High sedimentation leads to high amounts of resuspendable materials; high resuspension leads to high internal loading of nutrients and increased production; a high amount of colored substances means a smaller photic zone and a lower production; a high input of colored substances and a high production would mean a high sedimentation, etc. The results presented by Wallin et al. (1992) show that the Secchi depth (a standard measure of water clarity and the amount of suspended particles causing light scattering) should be much greater than that observed if only plankton cells were responsible for the light extinction. This means that particles other than plankton cells are perhaps the most important factors for determining water clarity.

SPM is generally a complex mix of substances of different origins with different properties (particulate size, form, density, specific surface area, capacity to bind pollutants, etc.). SPM may be divided into an organic fraction (POM) and an inorganic one (PIM, particulate inorganic materials). Total organic matter (TOM) is generally divided into particulate organic (POM) and dissolved organic (DOM) fractions. Normally, POM is about 20% of TOM, but this certainly varies among and within systems (Ostapenia, 1987, 1989; Velimorov, 1991; Boulion, 1994). Normally, about 4% of POM is living matter and the rest is dead organic matter (detritus). About 80% of TOM is generally in the dissolved phase, and of this about 70% is conservative in the sense that it does not change due to chemical and biological reactions in the water mass.

Particles in lake water

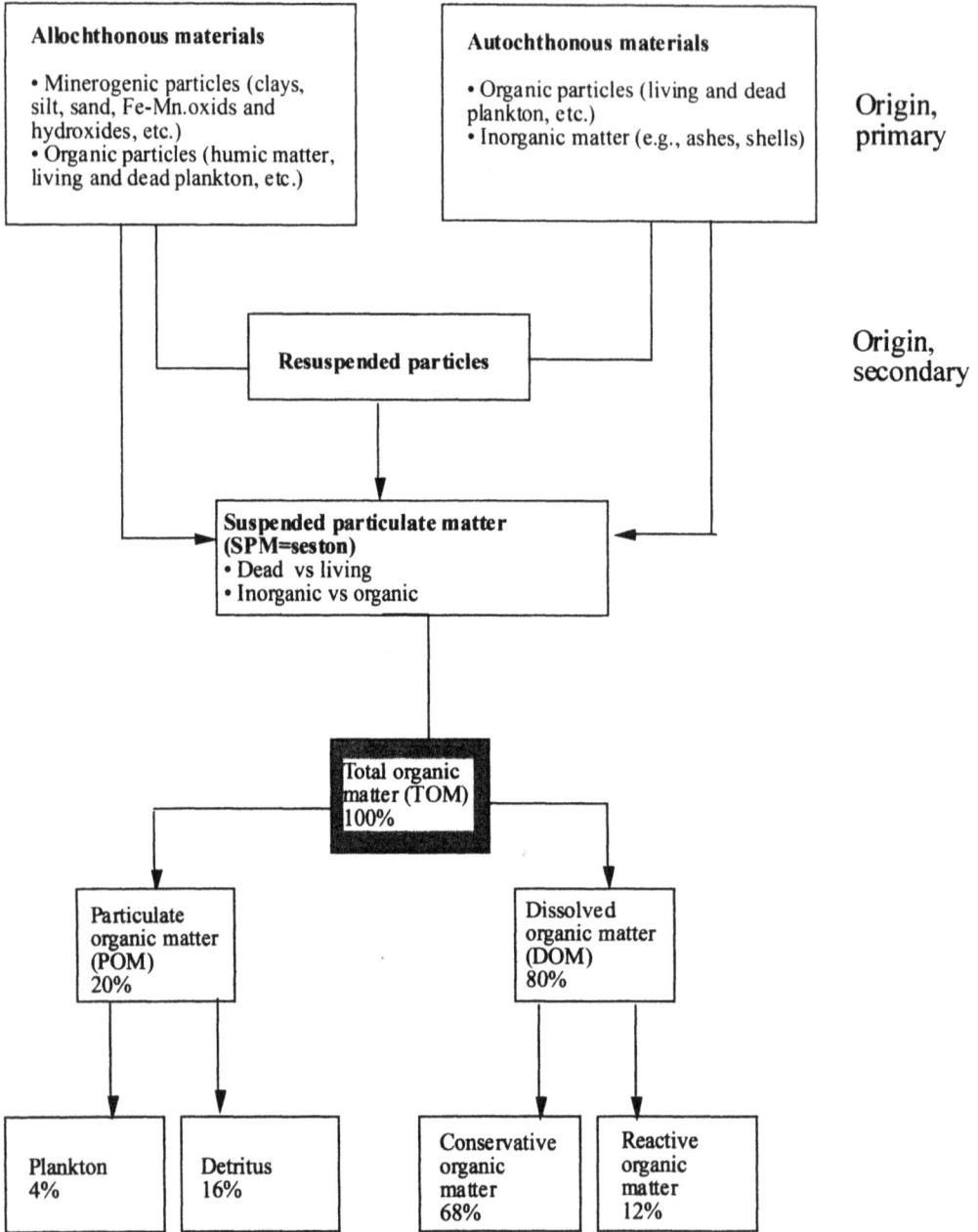

Fig. 2.1. Classification of particles in water (see Dubko, 1985; Ostapenia, 1985).

The ratio between DOM and POM is known to depend on the trophic status of the system and Ostapenia (1987, 1989) has presented the following empirical regression on that matter for lakes:

$$DOM/POM = 21.4 \cdot Chl^{-0.47}$$
$$(r^2 = 0.93; n = 23)$$
(2.1)

where the chlorophyll-a concentration (Chl) is given in µg/l. Eq. 2.1 indicates that DOM exceeds POM by a factor of 20-60 in oligotrophic lakes (if Chl is 0.1 to 1.0 µg/l), and in eutrophic lakes (if Chl is between 10 and 100 µg/l), the ratio is 2.5-7.

Total organic matter (TOM) seems to vary less than POM, as indicated by the following empirical regression, which is based on data from 17 lakes covering a wide trophic range (from Boulion, 1994):

$$POM = 0.023 \cdot TOM^{1.82}$$
$$(r^2 = 0.90; n = 17)$$
(2.2)

where POM and TOM are given in mg/l. Note that the exponent is higher than unity. This means that if TOM increases from 5 to 40 mg/l (a factor of 8), then POM would increase from 0.4 to 19 mg/l (a factor of 50).

Seston is another term for suspended particulate matter (SPM), and fig. 2.2 gives an overview of how seston may be differentiated into various categories, such as particulate organic matter (POM; about 60% of seston is POM in normal lakes), and particulate inorganic matter (PIM). About 50% of POM is detritus, i.e., dead plankton, remains of macrophytes, etc. Phytoplankton dominates among the plankton groups in many systems.

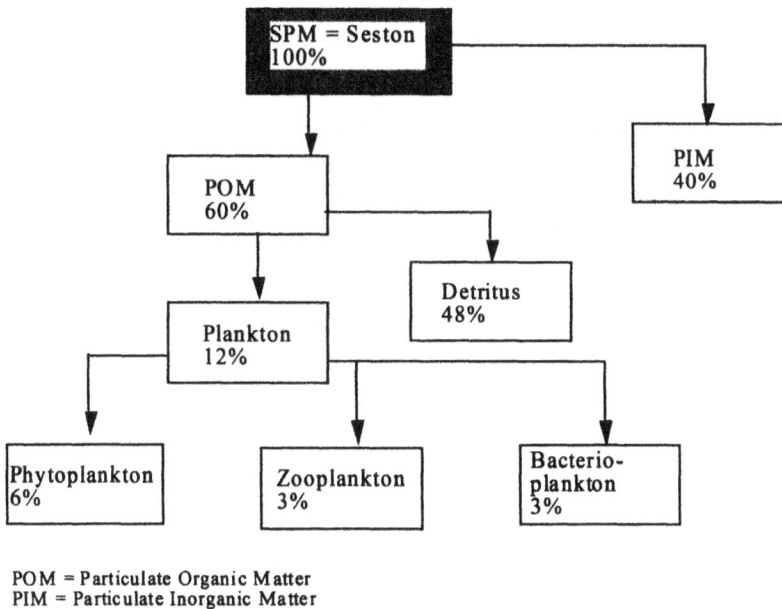

POM = Particulate Organic Matter
PIM = Particulate Inorganic Matter

Fig. 2.2. Composition of seston = SPM (see also Ostapenia, 1989; Velimorov, 1991).

Many factors influence the SPM variability among and within rivers, lakes, and marine areas (fig. 2.3):

1. Sampling phase in relation to high or low tributary water discharge in rivers and estuaries
2. Stormy or calm period
3. Excessive rain or no rain
4. High/low temperature; high/low bioproduction
5. High/low point source emissions close to sampling sites
6. Dredging
7. Boat resuspension and trawling.

SPM influences many fundamental properties in aquatic systems, but it does not, however, influence "everything." Fig. 2.3 is included here to stress one further point: To analyze the importance of SPM in relation to other factors, such as properties of the catchment area, climatological factors, etc., that may influence fundamental water variables, it is important to realize that any variable (y) will depend on factors causing seasonal variations within systems and on factors causing variations in mean, median, or characteristic values among systems. SPM values vary within as well as among systems.

2.2. SPM and ecosystem function

The following sections will exemplify why it is important to study SPM in aquatic systems.

2.2.1. SPM and water clarity

Values of Secchi depth are easy to understand by the general public - clear waters with large Secchi depths seem more attractive than turbid waters. Many factors are known to influence the Secchi depth (Vollenweider, 1958, 1960; Ahlgren, 1973; Carlson, 1977, 1980; Brezonik, 1978; OECD, 1982; Wetzel, 1983a; Ostapenia et al., 1985; Preisendorfer, 1986; Boulion, 1994, 1997). Water clarity is a fundamental variable in aquatic studies since it regulates primary production of phytoplankton, benthic algae, and macrophytes (see section 2.2.2).

Fig. 2.4A and B shows the very strong and logical relationship between lake Secchi depth and the concentration of suspended particulate matter (SPM in mg/l) based on 573 individual samples from 23 lakes (covering a wide limnological domain from oligotrophic to highly eutrophic conditions; see appendix 9.1 for data). Fig. 2.4C illustrates a similar regression using data for marine (Baltic) coastal areas. One can note some interesting differences:

- As already shown (fig. 1.1), marine and brackish waters generally have a higher clarity than lakes and rivers and the variability in SPM values is generally significantly smaller in marine systems (table 1.1). Fig. 2.4 also indicates that the slope of the regression line is smaller (-0.59 compared to -1.12). Evidently, the range in the coastal data used here is relatively small since the data in fig. 2.4C emanate from the Baltic Sea, but the correlation is highly significant ($r^2 = 0.80$) even in this narrow range; it is likely that the r^2 value would increase had the range been wider.

• The mechanistic reason for this difference between lakes and marine systems is related to the fact that the salinity of the system will influence the settling velocity of suspended particles (SPM): The higher the salinity, the greater the aggregation of suspended particles, the bigger the flocs and the faster the settling velocity (Kranck, 1973, 1979; Lick et al., 1992). This will be discussed further in chapters 4 and 6 in connection with the dynamic model for SPM for lake and marine coastal areas.

2.2.2. SPM, water clarity, and primary production

This section will illustrate the role of SPM and Secchi depth for primary production of phytoplankton, benthic algae, and macrophytes.

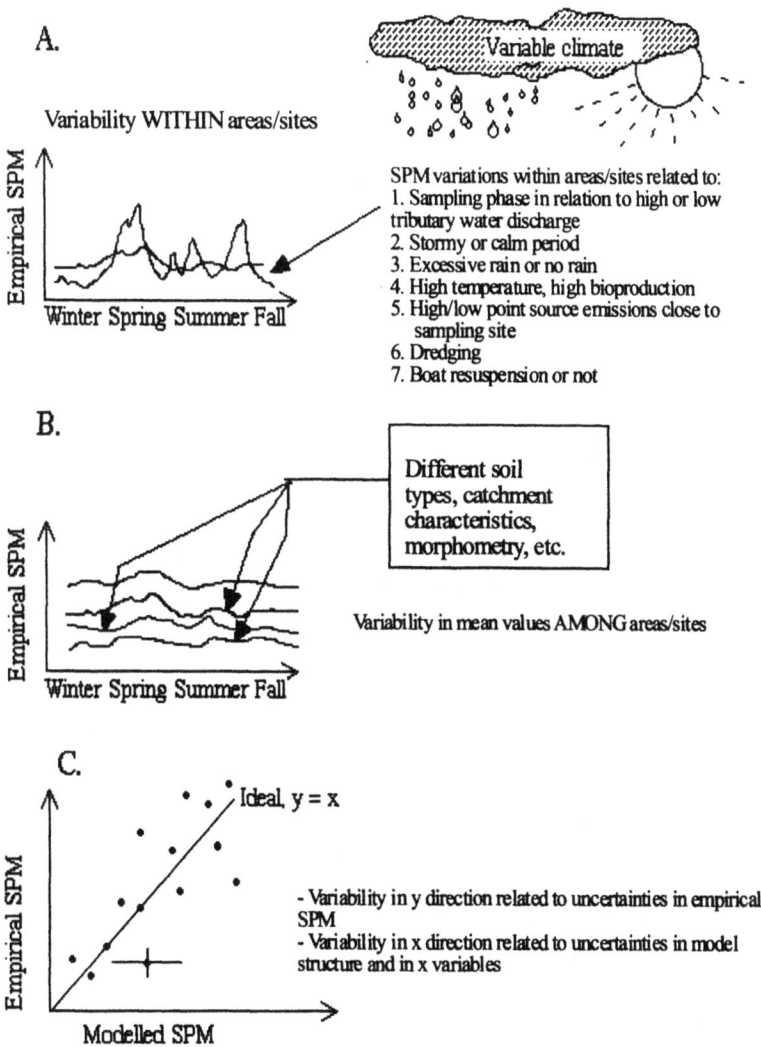

Fig. 2.3. Illustration of factors influencing within and among site variability in SPM.

$y = -1.123x + 0.993; r^2 = 0.784; n = 573$ **data from 23 lakes**

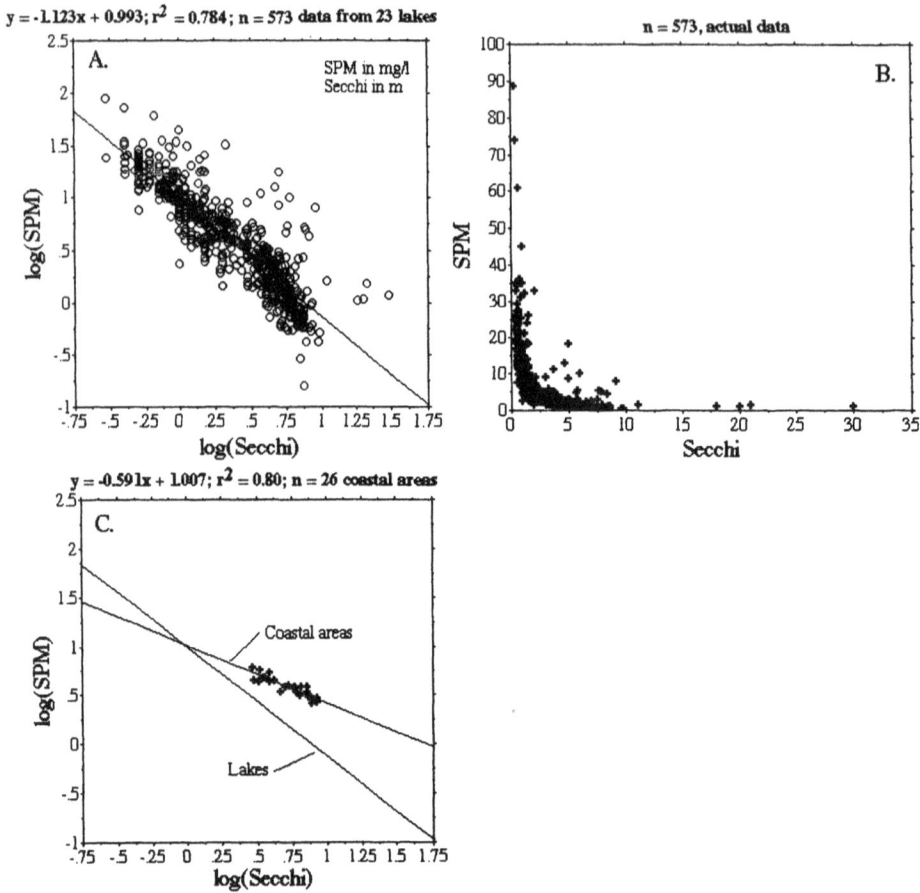

Fig. 2.4. The relationship between Secchi depth (in m) as a standard measure of water clarity and (figures A and B) the SPM concentration (mg/l) based on 573 data points from 23 lakes (data from appendix 1) covering a wide limnological domain (from highly eutrophic to oligotrophic conditions). Fig. C gives the corresponding regression for 26 Baltic coastal areas (data from appendix 9.1) and a comparison between the two regression lines for lakes and coastal areas. The figure gives the regression line, r^2 = coefficient of determination and n = number of data.

2.2.2.1. Phytoplankton

Calculations of primary phytoplankton production in the water, PrimP, require information about the actual, effective depth of the photic zone (D_z in m). But how should D_z be determined? From the comprehensive literature on this important topic (Vollenweider, 1958, 1969; Ahlgren, 1973; Brezonik, 1978; Bannister, 1979; Smith, 1979; Vollenweider and Kerekes, 1980; Peters, 1986; Håkanson and Boulion, 2002), one can argue that the simplification to approximate the effective depth of the photic zone with the Secchi depth is the best and simplest alternative; and Secchi depth is, as stressed, directly and mechanistically related to SPM. A key question is: Which value for the depth of the photic zone should be used in the following calculation of annual primary phytoplankton production (PrimP in g ww/m^2·yr) based in daily primary production (PrimP' in μg C/l·day; table 2.1)?

$$PrimP = PrimP' \cdot 2 \cdot 5 \cdot T_{DUR} \cdot D_z \cdot 0.001 = PrimP' \cdot 1.5 \cdot Sec \qquad (2.3)$$

where T_{DUR} is the duration of the growing season in days (fig. 2.5).

T_{DUR} is often set to 150 days for Nordic systems. Vollenweider (1958, 1969) argued that to address this issue one must account for k, the extinction coefficient and F, Atkins coefficient ($\approx$ 2.3) where $F \approx k \cdot Sec$. He (and also Boulion, 1985a, 1994) also argued that:

$$PrimP_{MV} \cdot D_z = a \cdot PrimP_{opt} \cdot Sec^b \qquad (2.4)$$

where a and b are close to 1, $PrimP_{MV}$ is the primary production as mean value for photic zone, and $PrimP_{opt}$ is the primary production under conditions of light saturation. Generally, light extinction with water depth is given the following equation, which is graphically illustrated in fig. 2.6:

$$I(x) = I_0 \cdot e^{-kx} \qquad (2.5)$$

where

$I(x)$	= a measure of the amount of light at water depth x;
I_0	= the amount of light at the initial reference water depth (x = 0);
k	= the light extinction coefficient.

Table 2.1. Compilation of equations and presuppositions for calculating primary phytoplankton production in lakes from chlorophyll and Secchi depth. To transform PrimP′ in µg C/l·day to PrimP in g ww/m²·yr, a factor of 2 (often 1/0.45) may be used to calculate g C to g dw; the factor 5 (if the water content is set to 80% ww) is used to calculate g dw to g ww; the effective depth of the photic zone is given by the Secchi depth; and the duration of the growing season is set to 150 days. Also note that primary production in highly eutrophic lakes will be reduced by self-shading of algal cells. Since phytoplankton primary production (PrimP) also depends on temperature, light conditions, water clarity, bioavailable fraction of phosphorus, etc., the classes given for PrimP are only meant to illustrate typical ranges, and the characteristic values given for PrimP are meant to be simple focal values (modified from Håkanson and Boulion, 2002).

Chl limits ($\mu g/l$)	PrimP′ (μg C/l·d)	Effective depth of photic zone $D_z \approx Sec$	PrimP (g ww/m²·yr)	Characteristic PrimP (g ww/m²·yr)
0.1	<4.4	>4.6	<125	50
1	32	2.6-7.9	125-380	200
10	270	1.3-3.6	380-1400	800
100	2400	0.4-1.8	1400-6400	3200
1000	>2400	<0.8	>6400	8000*

Equations and presuppositions:
$PrimP' = (2.13 \cdot Chl^{0.25} + 0.25)^4$
$PrimP = PrimP' \cdot 2 \cdot 5 \cdot 150 \cdot D_z \cdot 0.001 = PrimP' \cdot 1.5 \cdot Sec$
* = shading effects of algal cells; so, $3200 \cdot 4 = 12800$ is not used here but rather 8000; the shading effects at low Secchi depths imply that PrimP may be estimated from the following modified expression: If Sec < 1 m, then $PrimP \approx PrimP' \cdot 1.5 \cdot Sec^2$.

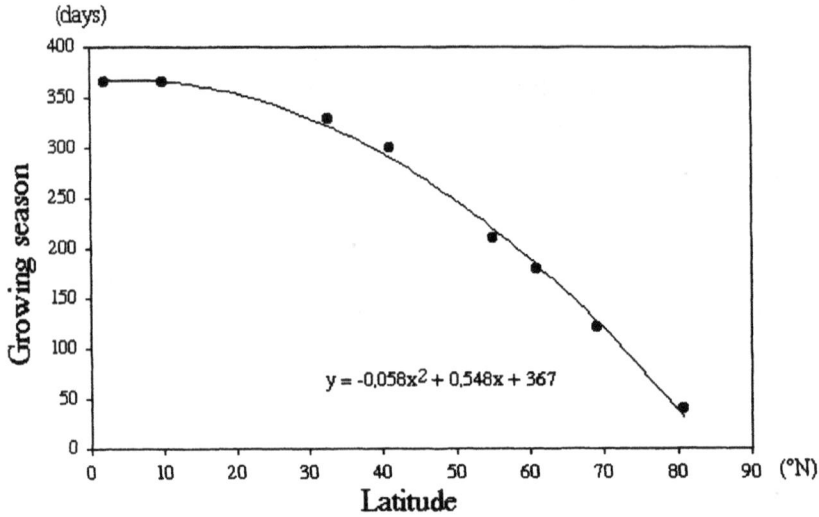

Fig. 2.5. The relationship between empirical data on the duration of the growing season and latitude (from Håkanson and Boulion, 2002).

From fig. 2.6, it is evident that if the maximum depth of the photic zone is being used to calculate primary production, one would obtain much too high PrimP values. The figure shows that 50% of the initial light amount remains at much smaller water depths.

Note that most of the many studies on the relationships between light conditions, Secchi depth, the depth of the photic zone, and primary production concern conditions at given sampling sites, and not mean characteristic conditions in a given system, which is the scale of most models discussed in this book.

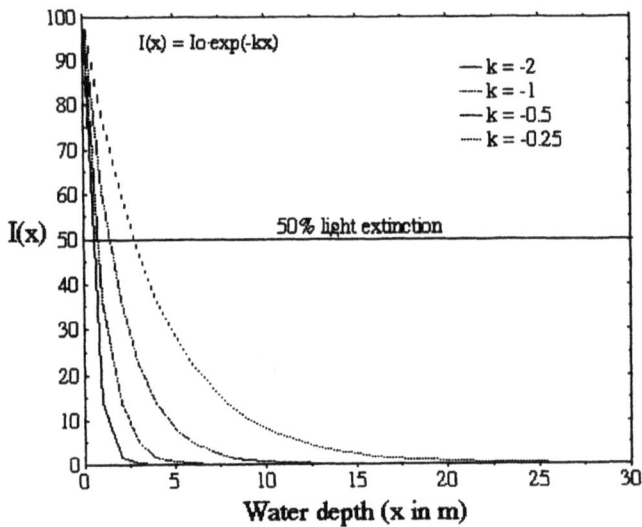

Fig. 2.6. Illustration of changes in light extinction with water depth.

2.2.2.2. Benthic algae

Obviously, the horizontal and vertical distributions of benthic algae strongly depend on the distribution of the illuminated substrates. In comparison with macrophytes and phytoplankton, benthic algae are confined to a relatively thin surficial sediment layer, within which the concentration of cells can be very high (Wetzel, 1983). Organisms on underwater substrates form heterogeneous and complex associations and colonize almost all types of substratum in the littoral zone (Vollenweider, 1969; Nikulina, 1979; Makarevich, 1985; Lalonde and Downing, 1991; Anokhina, 1999). The terminology used for the various algae living on different substrates varies (Vollenweider, 1969; Wetzel, 1964, 1983). Generally, the term "periphyton" is applied for all forms of vegetation (with the exception of macrophytes) growing on submerged materials. The underwater substratum can be sediments, stones, boats, constructions, and living organisms.

The littoral primary production often dominates the total primary production in shallow systems. Deep systems usually have only a relatively small littoral zone. A comparison of the littoral primary production in five oligotrophic lakes indicated a large contribution of benthic algae (Loeb et al., 1983), but also a logical decreasing contribution of benthic algae to the total primary production with increasing water depth. Analyzing data for lakes from different regions in the former Soviet Union, Boulion (2001) has also shown that the contribution of benthic algae to the total littoral primary production depends on two important limiting factors: The morphometry of the littoral zone and the optical properties of the water.

Håkanson and Boulion (2002) presented an empirical model for the ratio between the production of benthic algae (PR_{BA}) to the total production of benthic algae plus phytoplankton (PR_{tot}), using data from nine lakes and 42 sampling sites. It is interesting to note (fig. 2.7) that there is a very significant relationship between the ratio, PR_{BA}/PR_{tot} for the individual sampling sites and the ratio between the sampling depth (D in m) and the Secchi depth (Sec in m):

$$(PR_{BA}/PR_{tot}) = -85.7 \cdot (D/Sec)0.5 + 115.7 \qquad\qquad (2.6)$$
$$(n = 42; r^2 = 0.75)$$

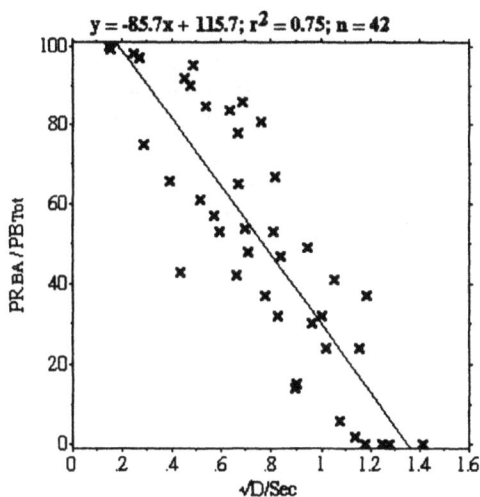

Fig. 2.7. The relationship between empirical data from individual sampling sites in lakes, on the y axis, the ratio between production of benthic algae to total production (of benthic algae plus phytoplankton), and on the x axis, the ratio between sampling depth (D in m) and Secchi depth (Sec in m). From Håkanson and Boulion (2002).

As much as 75% ($r^2 = 0.75$) of the variability among these 42 sites can be statistically explained by variations in the D/Sec ratio and there seems to be no significant differences among the lakes in this respect. The equation is evidently only applicable for D/Sec ratios yielding predictions of PR_{BA}/PR_{tot} smaller than 100%.

Eq. 2.6 is valid for individual sampling sites and demonstrates the logical relationship between the production of benthic algae in relation to primary production of benthic algae plus phytoplankton. The larger the sampling depth relative to the Secchi depth, the smaller the production of benthic algae compared to the phytoplankton production.

2.2.2.3. Macrophytes

For the determination of one of the most fundamental properties of aquatic systems, the trophic status, the basic attention is generally given to phytoplankton production. However, the macrophytes can contribute significantly to the total primary production, especially in shallow lakes. Sometimes macrophyte production exceeds phytoplankton production (Wetzel, 1983). The macrophytes intercept nutrients and keep the nutrients bound for long periods. Consequently, they can influence water quality (Pokrovskaja et al., 1983). It is also very important to emphasize that the evolution of any lake (even oligotrophic) is closely connected with the overgrowing by rooted plants (Beeton and Edmondson, 1972). Macrophytes also provide an important protective environment for small fish.

To determine the relative role of macrophytes and phytoplankton in primary productivity, it is necessary to study the development of these plants relative to morphometric and hydro-optical properties of water bodies. Vorobev (1977) analyzed data from 229 lakes of the Vologda district (Russia) and shown that the areal cover by macrophytes (MA_{cov}, % of lake area) is related to the ratio between the Secchi depth (Sec in m) and the mean depth (D_m in m):

$$MA_{cov} = 50 \cdot (Sec/D_m) \tag{2.7}$$

So, if $Sec/D_m = 0.25$, the covering should be close to 12.5%; if $Sec/D_m = 1$, it averages 50%, etc.

Note that it may be difficult to clearly define the ratio between macrophytes to epiphytes and benthic algae. The dominance of macrophytes or benthic algae depends on the characteristics of the bottom substratum. If it is stony, then the benthic algae dominate. If it is soft, the rooted macrophytes (plus epiphytes) may shade the benthic algae. At higher latitudes, the littoral bottom areas are often covered by a stony substratum with benthic algae. At lower latitudes, the littoral areas often consist of relatively soft sediments and they tend to have a higher macrophyte cover.

It has been shown (Scheffer, 1990; Scheffer et al., 2000) that in a given lake with macrophytes, there can be two alternative stable states, one with turbid waters, high SPM concentrations, and a minimum of macrophytes, the other with clear water conditions, which maximizes macrophyte production and cover.

Håkanson and Boulion (2002) used empirical data from many lakes (table 2.2), and statistical modelling to rank the factors influencing the variability among the lakes in macrophyte cover (MA_{cov}). Those results are shown in table 2.3.

Table 2.2. Macrophyte production and cover; data from 35 lakes. Note that 1 kcal ≈ 0.52 g dw ≈ 1.56 g ww for macrophytes (from Håkanson and Boulion, 2002).

No	Lake	Region	Lat. °N	Area km²	D_m m	D_{max} m	Secchi m	MA_{cov} %	PRMA g ww/m²·yr	References
1	Chedenjarvi	Karelia	62.0	0.65	3.4	6.6	0.4	10	37	Gorbunova et al., 1973
2	Big Kharbey	Vorkuta, R	67.5	21.3	4.6	18.5	2.5	5	132	Vlasova et al., 1973; Kochanova, 1976
3	Krasnoe	Karelian I	60.8	9.13	6.6	14.6	2.1	7	211	Andronikova et al., 1973
4	Glubokoe	Moscow district	55.8	0.59	9.3	32	1.8	8	178	Scherbakov, 1967
5	Drivjaty	Belarus	55.7	32.6	5.2	11.8	2.0	20	132	Zakharenkova, 1970
6	Arakhley	Chita district, S	52.0	58.2	10.4	16.7	6.7	43.7	264	Zolotareva, 1981; Nazarova & Shishkin, 1981
7	Dusja	Lithunia	54.5	23.2	14.6	32.4	3.2	13.6	1188	Manjukas, 1973
8	Big Eravnoe	Burjatia	52.7	99.5	2.0	?	bottom	85.6	924	Neronova & Karasev, 1977
9	Sosnovskoe	"	52.7	23.7	2.5	?	bottom	100	1220	"
10	Little Eravnoe	"	52.7	56.2	1.5	?	bottom	100	1848	"
11	Isinga	"	52.7	30.0	1.5	?	bottom	91.4	1294	"
12	Big Kharga	"	52.7	29.5	1.0	?	bottom	100	1848	"
13	Little Kharga	"	52.7	6.5	0.8	?	bottom	100	1848	"
14	Karakhul	Kazakhstan	43.5	1.4	1.5	3.8	1.5	50	10560	Khusainova et al., 1973
15	Onega	Karelia	61.5	10340	29.5	120	4.0	0.29	1	Raspopov, 1973; Dotsenko & Raspopov, 1982
16	Marion	Canada	49.3	0.13	2.4	6.0	?	20	238	Efford, 1972
17	Mikolayskoe	Poland	53.7	4.6	11	27	3.5	19	436	Kajak et al., 1972
18	Tchad	Africa	13.0	24000	3.5	5.5	?	100	12672	Leveque et al., 1972
19	Naroch	Belarus	54.8	79.6	9.0	24.8	5.3	30	784	Winberg et al., 1972
20	Miastro	"	54.8	13.1	5.4	11.3	1.6	17	178	"
21	Batorino	"	54.8	6.3	3.0	5.5	0.8	23	87	"
22	Kiev r	Ukraina	51.0	925	3.5	?	2.2	32	232	Gak et al., 1972; Priymachenko, 1983
23	Lacha	Vologda district, R	61.0	345	1.6	5.0	1.1	18.3	462	Raspopov, 1978
24	Vozhe	"	61.0	418	1.4	5.0	1.1	48	792	"
25	Rybinskoe r	Volga River	58.5	4550	5.6	28.0	1.5	16.7	401	Ekzertsev, 1958; Ekzertsev & Dovbnja, 1973
26	Gorkovskoe r	Volga River	57.0	1610	5.5	?	1.2	1.4	33	"
27	Nogon-Nur	Mongolia	47.9	20	1.0	1.5	1.0	50	2640	Boulion, 1985b
28	Vechten	Netherlands	52.0	0.047	6.0	11.9	4.0	9	98	Best, 1982
29	Sjargozero	Karelia	63.6	10.54	7.8	22	3.0	4.5	20	Kljukina & Freindling, 1983
30	Sonozero	"	63.6	9.53	3.7	?	1.8	4.9	11	"
31	Sukkozero	"	63.3	8.9	1.9	?	bottom	3.8	8	"
32	Gormozero	"	63.3	2.36	3.0	?	1.8	8.1	18	"
33	Tuhkozero	"	63.6	1.58	6.2	?	5.0	11.3	53	"
34	Vjagozero	"	63.6	1.29	1.4	?	bottom	21.7	59	"
35	Torosjarvi	"	63.6	0.61	3.3	?	3.0	17.6	71	"

R =Russia; I = Isthmus; S = Siberia, r = reservoir; D_m = mean depth; D_{max} = maximum depth

Table 2.3. Stepwise multiple regression analyses using the data for the parameters given in table 2.2 to calculate the macrophyte cover (MA_{cov} in % of lake area) using statistical methods (transformations, etc.) as given by Håkanson and Peters (1995). A1 is the area shallower than 1 m water depth; n = 19 lakes. From Håkanson and Boulion (2002).

Step	F	r^2	x variable	model
1	18	0.52	Sec/D_m	$y=1.944+4.825 \cdot x_1$
2	7	0.67	$90/(90\text{-Lat})$	$y=6.757+3.83 \cdot x_1-1.57 \cdot x_2$
3	4	0.74	$\sqrt{D_{max}}$	$y=8.31+2.57 \cdot x_1-1.50 \cdot x_2-0.286 \cdot x_3$
4	4	0.84	$\log(A1)$	$y=10.49+1.502 \cdot x_1-1.993 \cdot x_2-0.422 \cdot x_3+0.490 \cdot x_4$

The macrophyte database includes data on MA_{cov} from 35 lakes covering a wide limnological domain (table 2.2), large and small lakes (from 0.047 to 24000 km^2), from latitudes 13°N (Lake Tchad in Africa) to 67.5°N (Lake Big Kharbey, Vorkuta, Russia), and deep and shallow lakes (the maximum depth varies from 1.5 to 120 m). The macrophyte cover varies from 0.29% to 100%. The light conditions are important for the macrophytes, and the database includes information on Secchi depth, which varies from 0.4 m to 6.7 m. In many lakes, the Secchi depth reaches its maximum value, the maximum depth.

Results of stepwise multiple regressions are given in table 2.3. This table gives a ranking of the factors influencing the MA_{cov} variability among the lakes studied.

- Sec/D_m is the most important factor to statistically explain the variation among these lakes in MA_{cov}; $r^2 = 0.52$ for n = 19 (there are only complete data for the stepwise multiple regression for 19 lakes).

- The next important factor for MA_{cov} is latitude, which is evidently related to lake temperature. If latitude is added, 67% of the variation in MA_{cov} can be statistically accounted for among the lakes.

- The third factor is maximum depth; the deeper the lake, the smaller MA_{cov}.

- The fourth factor is the area of the lake above 1 m, A1; $r^2 = 0.84$.

Fig. 2.8 gives a 3D diagram relating the two most important model variables, Sec/D_m and latitude, to MA_{cov}. This diagram shows how these two model variables influence MA_{cov} when D_{max} and A1 are held constant.

It is interesting to conclude the very strong influence of SPM/Secchi depth, mean depth, and latitude on how the macrophyte cover varies among lakes.

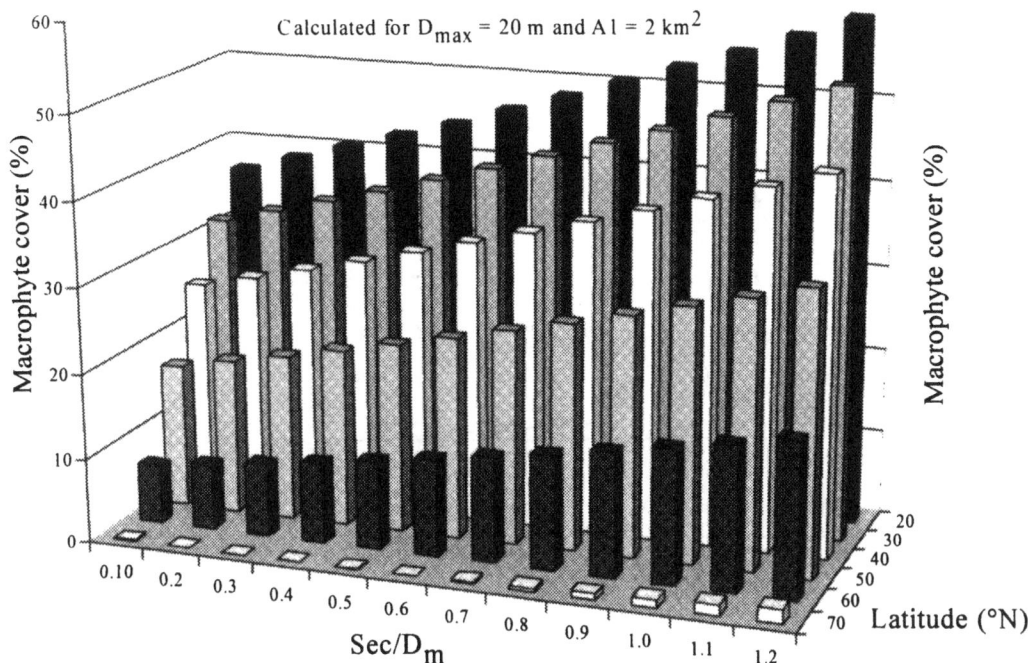

Fig. 2.8. 3D diagram illustrating how the ratio Sec/D_m and latitude influence macrophyte cover for a lake with a maximum depth of 20 m with the 2 km^2 lake area above 1 m water depth (from Håkanson and Boulion, 2002).

2.2.3. SPM and bacterioplankton

The main role that bacteria play in aquatic ecosystems is to decompose dead organic matter, which, as discussed, is a significant part of SPM. In fig. 2.9, bacteria is found in the entire water mass, although the highest bacterial biomasses often appear close to the bottom and near the water surface. This is a "normal" situation. For example, Kuznetsov (1970) showed that there may be significant differences among lakes, and seasonally within lakes depending on stratification. There also exists a direct dependence between the biomass and/or growth of bacteria and the biomass and/or production of phytoplankton (Overbeck, 1972; Aizaki et al., 1981; Bird and Kalff, 1984; Currie, 1990; Conan et al., 1999).

To illustrate the role of SPM for bacteria, a simulation experiment will be conducted using the LakeWeb model. This is a general foodweb model, which also incorporates the LakeMab model (i.e., the mass-balance model for SPM, nutrients, metals, and radionuclides). The biotic part of the LakeWeb model is based on functional groups, which appear in all aquatic systems. The model is described in appendix 9.2. The LakeWeb model has been critically tested with positive results against very comprehensive empirical data and this should lend credibility to the following simulations, which first aim to illustrate how changes in SPM influence water clarity and bacterioplankton biomass, then how these changes affect the functional groups included in the model (phytoplankton, benthic algae, macrophytes, herbivorous zooplankton, predatory zooplankton, zoobenthos, prey fish, and predatory fish). The LakeWeb model has also been tested by means of extensive sensitivity and uncertainty analyses (using Monte Carlo techniques) for both uniform and characteristic coefficients of variation for all model variables to quantify and rank the uncertainties influencing the model predictions (Håkanson and Boulion, 2002).

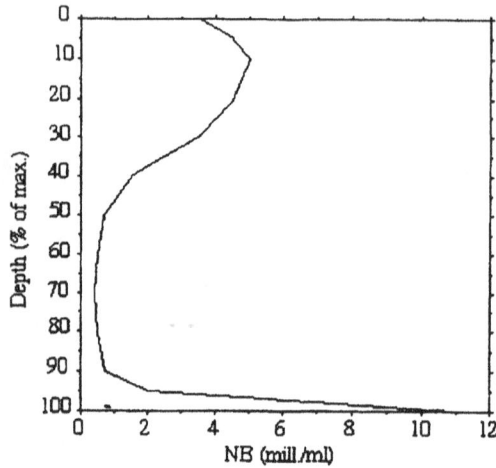

Fig. 2.9. Characteristic vertical distribution of the number of bacterioplankton (NB) in a mesotrophic lake (modified from Wetzel, 2001).

This simulation experiment is based on the following presuppositions:

- All simulations concern a lake with an area of 1 km^2, a mean depth of 10 m, a maximum depth of 20 m with no changes in the limnological state variables. The total phosphorus concentration is kept at 10 μg/l (typical for oligotrophic lakes), lake pH is 7 (a normal value for many lakes), and lake color (an expression of humic level) is constant at 50 mg Pt/l.

- Lake salinities have been changed in 2-year steps from 0, 5, 10 15 to 20% (fig. 2.10A) and the corresponding changes have been modelled for many variables. The salinity will influence the aggregation of suspended particles (as already discussed), which will reduce SPM in the water (fig. 2.10A), and hence increase Secchi depth and water clarity (fig. 2.10B). This experiment is meant to illustrate how changes in SPM influence the system, as if there were significant inflows of saltwater on a 2-year basis to the given lake while all else would be held constant. In a way, this may be regarded as a sensitivity analysis highlighting how changes in salinity and the related changes in SPM influence the structure and function of the system.

- Modelled values will also be compared to values calculated using empirical models. These reference values are called "norms" since they give empirically based normal reference values to the modelled values. Initially, a good correspondence between the norm values and the modelled values is expected, but when the salinity increases, there should be differences since the empirical models (table 2.4) mainly apply for softwater lakes.

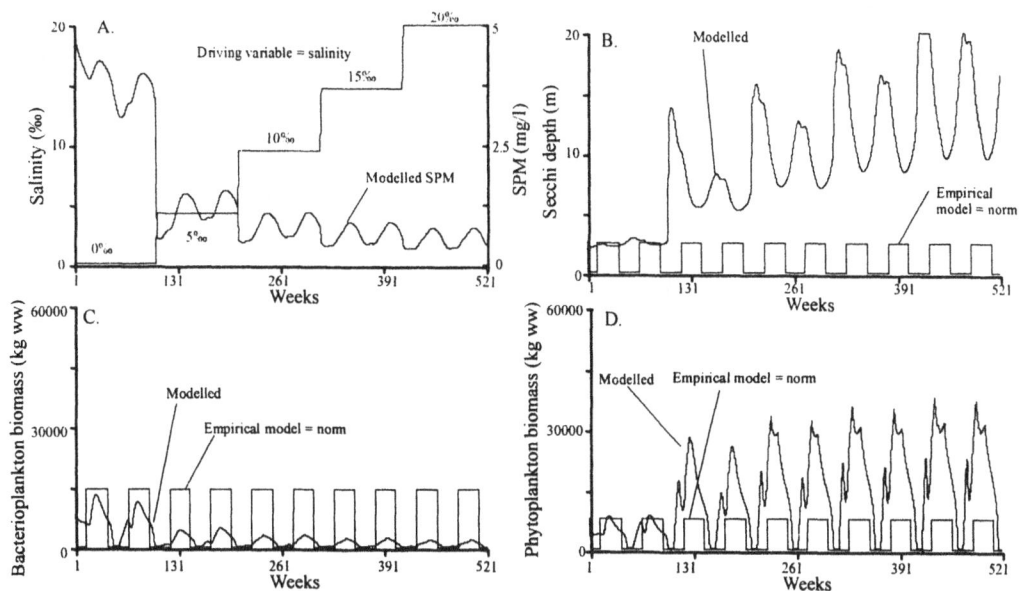

Fig. 2.10. Simulations using the LakeWeb model to illustrate how five stepwise 2-year changes in salinity (0, 5, 10, 15, and 20%) would influence modelled SPM values (fig. A); fig. B gives the corresponding modelled Secchi depths using the LakeWeb model and Secchi depths calculated using an empirical model (the "norm"; all the following "norms" are from table 2.4); fig. C gives corresponding data for bacterioplankton biomass; and fig. D. similar data for phytoplankton biomass. These calculations use data for a lake with an area of 1 km^2, a mean depth of 10 m, a maximum depth of 20 m, a pH of 7, a mean concentration of total phosphorus of 10 µg/l, and a color value of 50 mg Pt/l, which corresponds to a typical glacial, boreal lake from northern landscapes).

Fig. 2.10 shows:

- SPM decreases significantly, from about 4 mg/l to about 1 mg/l when the salinity increases from 0 to 20%.

- This means that the Secchi depth increases from about 2.5 m to about 10-15 m. Note that the values during the first year in each 2-year step calculations are not well adjusted to the new situation. The modelled curves illustrate the dynamic response of the system to these sudden changes in salinity and SPM. For the first 2-year period, there is also an excellent correspondence between the norm values and the modelled values for the Secchi depth.

- Fig. 2.10C gives the corresponding values for bacterioplankton biomass. There is a good correspondence between modelled values and norm values for the initial period and the significant decrease in bacterioplankton biomass as a result of the reduced SPM concentrations, since the organic fraction of SPM is the "fuel" for bacterioplankton: The lower the SPM concentration, the lower the bacterioplankton biomass, if all else is constant. This is easy to conclude on a qualitative basis, but here it is given quantitatively.

- The increase in water clarity (fig. 2.10B) will increase the depth of the photic layer and also increase phytoplankton production and biomass (fig. 2.10D).

- The increase in water clarity will also markedly increase the biomass of benthic algae (fig. 2.11A) and the macrophyte production (fig. 2.11B), as already discussed using empirical data.

- It is also interesting to note the good correspondence between modelled values and norm values for the initial period for these functional groups of organisms in these simulations.

Table 2.4. Regressions based on extensive empirical data used as normative values (= norms) in the testing of the LakeWeb model. Note that some variables may be predicted with great precision (high r^2), others with much less. From Peters (1986), OECD (1982), Vorobev (1977) and Håkanson and Boulion (2002). Lake characteristic TP in µg/l. PrimP is primary production (in g ww/m^2·yr); Mac$_{cov}$ = Macrophyte cover of lake bed (in % lake area); Sec$_{MV}$ = mean annual Secchi depth (in m); D$_m$ = mean lake depth (in m); Col = lake color (mg Pt/l); n = number of lakes used in the regression, ww = wet weight and dw = dry weight.

y value	Equation	Range for x	r^2	n	Units
Equations based on phosphorus used as norms in LakeWeb:					
Chlorophyll (summer mean)	$=0.28 \cdot TP^{0.96}$	2.5-100	0.77	77	mg ww/m^3
Chlorophyll (summer max.)	$=0.64 \cdot TP^{1.05}$	2.5-100	0.81	50	mg ww/m^3
Chlorophyll (weekly mean)	$=0.5 \cdot TP$ [(0.64+0.28)/2 ≈ 0.5; (0.96+1.05)/2 ≈ 1)]				mg ww/m^3
Max. primary prod. (TP>10)	$=20 \cdot TP-71$	7-200	0.95	38	mg /m^3·d
Max. primary prod. (TP<10)	$=0.85 \cdot TP^{1.4}$				mg C/m^3·d
Mean primary prod. (TP>10)	$=10 \cdot TP-79$	7-200	0.94	38	m C/m^3·d
Mean primary prod. (TP<10)	$=0.85 \cdot TP^{1.4}$				mg C/m^3·d
Phytoplankton	$=30 \cdot TP^{1.4}$	3-80	0.88	27	mg ww/m^3
Phytoplankton	$=30 \cdot TP^{(1.4-0.1 \cdot (TP/80-1))}$				mg ww/m^3
Bacterioplankton (Col < 50)	$=0.90 \cdot TP^{0.66}$	3-100	0.83	12	mill./ml
Bacterioplankton	$=(1+0.25 \cdot (Col/50-1))) \cdot 0.90 \cdot TP^{0.66}$ (if color > 50)				mg Pt/l
Zooplankton, herbivores	$=0.77 \cdot 38 \cdot TP^{0.64}$	3-80	0.86	12	mg ww/m^3
Zooplankton, predators	$=0.23 \cdot 38 \cdot TP^{0.64}$ (the distr. coeff. is 0.77)				mg ww/m^3
Zoobenthos	$=810 \cdot TP^{0.71}$	3-100	0.48	38	mg ww/m^2
Zoobenthos	$=(1-0.5 \cdot (TP/100-1)) \cdot 810 \cdot TP^{0.71}$ (if TP >100)				mg ww/m^2
Fish	$=590 \cdot TP^{0.71}$	10-550	0.75	18	mg ww/m^2
Fish yield	$=7.1 \cdot TP$	8-550	0.87	21	mg ww/m^2·yr
Other equations used as norms in LakeWeb:					
Fish yield	$=0.0023 \cdot PrimP^{0.9}$	170-14000	0.64	66	mg ww/m^2·yr
Prey fish	$=0.73 \cdot$ fish biomass				mg ww/m^2
Predatory fish	$=(1-0.73) \cdot$ fish biomass				mg ww/m^2
Macrophyte cover	$= 0.50 \cdot (Sec_{MV}/D_m)$			229	%
Macrophytes	$=1.37 \cdot \log(Mac_{cov})+3.58$				g ww/m^2·yr
Zooplankton, herbivores	$=0.15 \cdot (PrimP \cdot 1000)^{0.86}$	13-15000	0.61	42	g ww/m^2·yr
Zooplankton, predators	$=0.076 \cdot (PrimP \cdot 1000)^{0.84}$	2-3000	0.43	42	g ww/m^2·yr

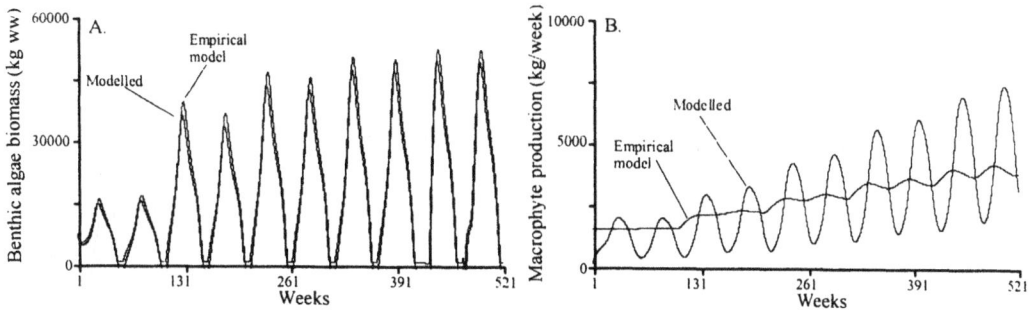

Fig. 2.11. Simulations using the LakeWeb model to illustrate how five stepwise 2-year changes in salinity (0, 5, 10, 15, and 20%; fig. 2.10A) would influence SPM and modelled values of (A) benthic algae and (B) macrophyte production. Modelled values are compared to "norm" values calculated using empirical models (from table 2.4). These calculations use data for a lake with an area of 1 km², a mean depth of 10 m, a maximum depth of 20 m, a pH of 7, a mean concentration of total phosphorus of 10 μg/l and a color value of 50 mg Pt/l, which corresponds to a typical glacial, boreal lake from northern landscapes).

In the following, similar results will be given for zooplankton, zoobenthos, and fish.

2.2.4. SPM and secondary production

2.2.4.1. Zooplankton

The LakeWeb model uses two functional categories of zooplankton, herbivorous zooplankton, which do the work of eating bacterioplankton, and phytoplankton. Herbivorous zooplankton are also consumed by predatory zooplankton and prey fish. So, since bacterioplankton biomass decreases (fig. 2.10C) and phytoplankton biomass increases (fig. 2.10D) in this simulation experiment, it is interesting to look for compensatory effects for herbivorous zooplankton. The results are given in fig. 2.12A. There is a small increase in the biomass of herbivorous zooplankton, but no major changes. There is good correspondence between norm values and modelled values for the whole simulation period for herbivorous zooplankton.

Fig. 2.12B gives the results for predatory zooplankton and for this functional group a more marked increase in biomass is seen. The explanation for this is shown in fig. 2.12D. The predation pressure is lower because the fish biomass becomes lower in this gradient.

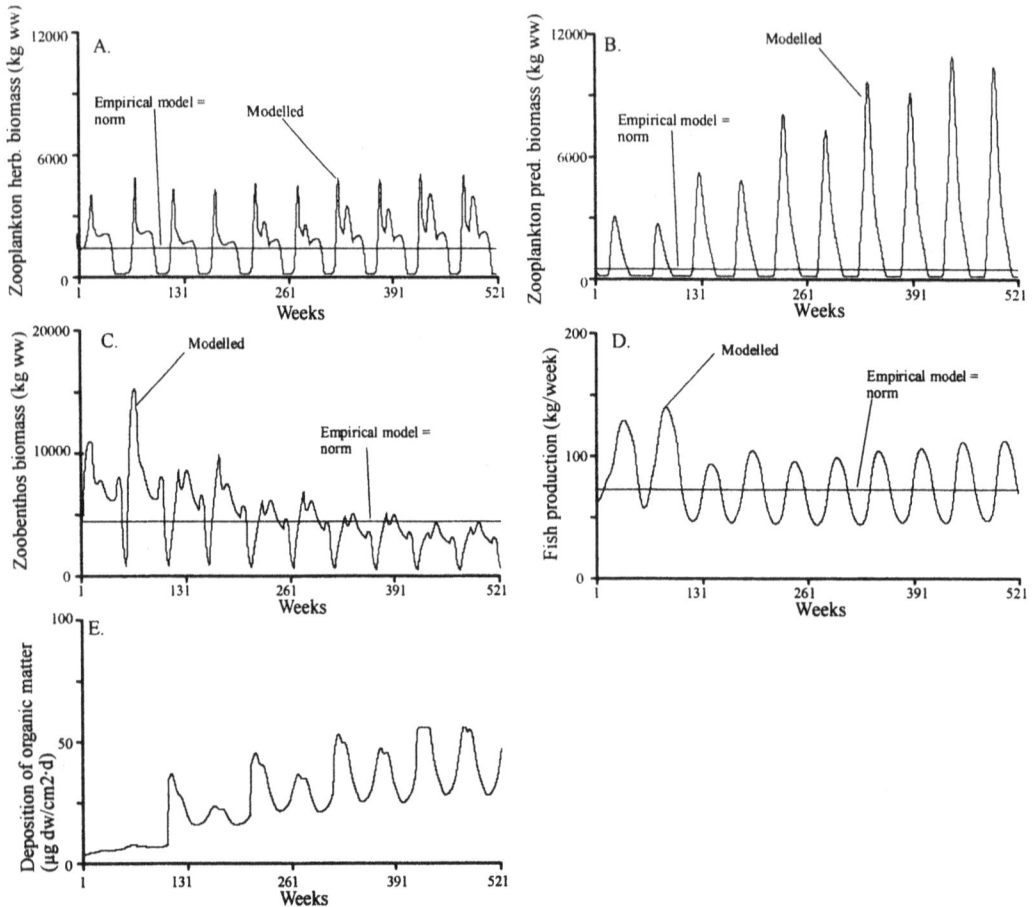

Fig. 2.12. Simulations using the LakeWeb model to illustrate how five stepwise 2-year changes in salinity (0, 5, 10, 15, and 20%; fig. 2.10A) would influence modelled SPM values and modelled values of (A) biomass of herbivorous zooplankton, (B) biomass of predatory zooplankton, (C) zoobenthos biomass, (D) total fish biomass and deposition on accumulation areas of organic matter, and (E) deposition of organic matter. Modelled values are compared to "norm" values calculated using empirical models (from table 2.4). These calculations use data for a lake with an area of 1 km², a mean depth of 10 m, a maximum depth of 20 m, a pH of 7, a mean concentration of total phosphorus of 10 µg/l, and a color value of 50 mg Pt/l.

2.2.4.2. Zoobenthos

The particles settling on accumulation areas (A areas) are, by definition, not further resuspended by physical forces. However, such particles can be eaten by zoobenthos, and zoobenthos by fish, so these particles can be transported up to the lake water again in this manner. They can also be transported by migrating zooplankton between the sediments and the water column. These particles will also be mineralized and decomposed by sediment-living bacteria and then the newly formed dissolved phase can migrate up to the lake water phase by diffusive processes. Bioturbation, i.e., the mechanical mixing of sediments by sediment-living organisms, can also contribute to the transport of settled particles both back to the water and downward to the biologically passive sediments.

Generally, various types of zoobenthos live in sediments down to about 5-15 cm sediment depth (fig. 2.13). This upper part of the sediment column is biologically active in the sense that the bottom fauna can influence the physical, chemical, and biological conditions of the sediments and cause bioturbation. However, the larger animals (macro- and meiofauna) generally die if the oxygen concentration becomes lower than about 2 mg/l. This will stop the bioturbation and laminated (layered and unmixed) sediments may appear. The continuous sedimentation will cause the sediment layer to grow upward so that the bioturbation limit, i.e., the limit between the upper biological layer and the lower biopassive (or geological) layer moves upward (see Håkanson and Jansson, 1983, for further information on recent sedimentary processes).

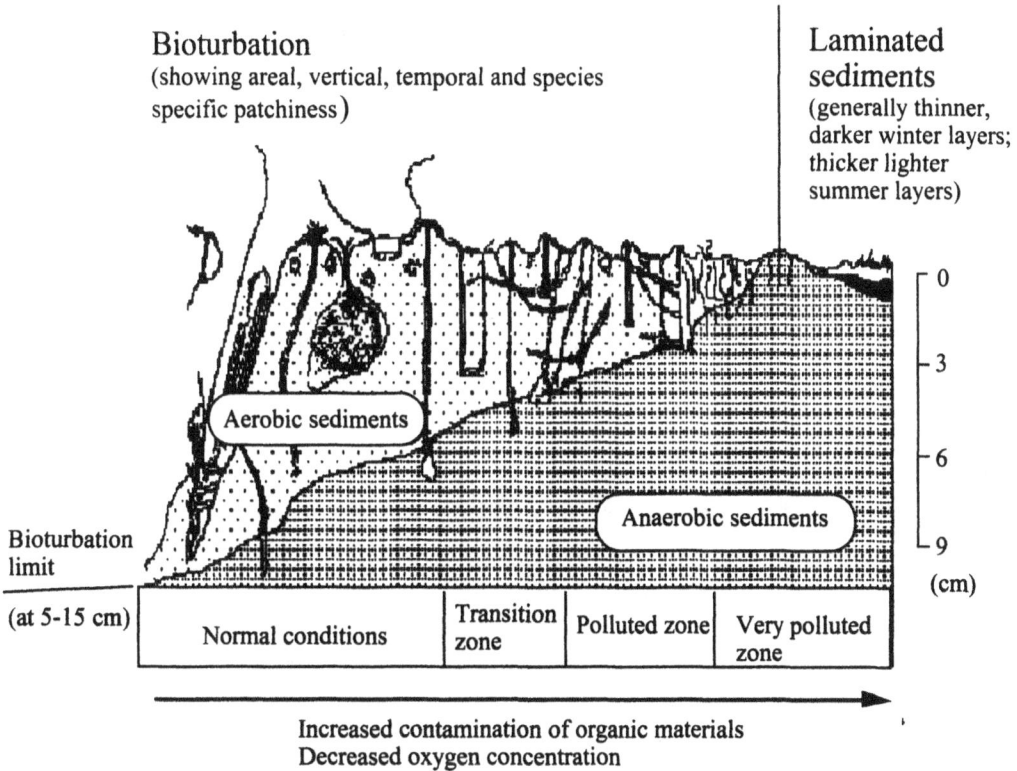

Fig. 2.13. Bioturbation and laminated sediments. Under aerobic (= oxic) conditions zoobenthos may create a biological mixing of sediments down to about 15 cm sediment depth (the bioturbation limit). If the deposition of organic materials increase and hence also the oxygen consumption from bacterial degradation of organic materials, the oxygen concentration may reach the critical limit of 2 mg/l, when zoobenthos die, bioturbation ceases, and laminated sediments appear (figure modified from Pearson and Rosenberg, 1976).

Fig. 2.12E illustrates how the deposition of organic matter on accumulation areas would increase with salinity in the simulation experiment. The aggregation and flocculation of the suspended particles in the water will imply a higher bacterial decomposition, a higher oxygen consumption at the sediment/water interface, and lower oxygen concentrations in this zone which means that more zoobenthos are likely to die. The reduction in zoobenthos biomass (fig. 2.12C) will influence the fish community. Again there is a situation with compensatory effects; predatory zooplankton eaten by prey fish increase and zoobenthos also eaten by prey fish decrease.

2.2.4.3. Fish

The results for fish production (of both prey and predatory fish) are shown in fig. 2.12D. This is an interesting slightly U-shaped curve with initial relatively high production values (which correspond to the norm values) and again higher values at the end of the simulation period reflecting the total response of the changes for herbivorous and predatory zooplankton and zoobenthos.

So, section 2.2.4 has exemplified and illustrated several important roles of SPM for the structure and function of aquatic systems, from primary producers to fish.

2.2.5. SPM and transport of pollutants

Many polluting substances have a considerable particle affinity (Broberg and Andersson, 1989; Riise et al., 1990; Madruga and Cremers, 1997; Konitzer and Meili, 1997; Konoplev et al., 1997). This is demonstrated very clearly by the information compiled in table 2.5, which shows correlation coefficients between SPM and:

• the metals aluminium, iron, manganese, total nitrogen, total phosphorus, and PO_4 P,
• standard water variables such as alkalinity, chloride, conductivity, DOC, pH, POC, salinity, and sulphate,
• variables expressing water clarity and primary production such as chlorophyll, NVSS (non-volatile suspended solids), Secchi depth and turbidity, and
• variables expressing key lake morphometric and catchment area properties related to inflow and resuspension of SPM (and water pollutants), such as mean depth, dynamic ratio, and tributary water discharge.

Table 2.5. Relationships between SPM (m/l) and environmental variables (†), correlation coefficients, r (-), number of data, n, and statistical level of significance, p. The statistics are given for individual data (ind), summer mean values (summer), annual mean values (annual) or monthly mean values (month). "Log" indicates that the statistical analysis of the relationship is based on logarithmic transformations of both SPM and the variable. From Johansson et al. (2001).

Variable	Variable range	SPM range	r (‡)	n	System(s)	p	Reference
$LogAl_{ind}$	0.109-8.875[#]	0.41-162.1[#]	+0.97	94	24 Streams	p<0.001	Cuthbert and Kalff, 1993
Fe_{ind}	0.004-1.3	0.07-1.87	+0.82	19	19 Lakes	*	Watras et al., 1995b
$LogFe_{annual}$	0.122-0.346	1.27-4.85	+0.65	15	15 Lakes	p<0.05	Johansson (unpublished)
$LogFe_{ind}$	0.018-1.200	0.73-5.90	+0.73	15	15 Lakes	p<0.05	Data from Watras et al., 1998
$LogFe_{ind}$	0.012-5.017[#]	0.41-162.1[#]	+0.97	140	24 Streams	p<0.05	Cuthbert and Kalff, 1993
$LogMn_{annual}$	0.019-0.124	1.27-4.85	+0.79	15	15 Lakes	p<0.05	Johansson (unpublished)
Mn_{ind}	0.008-0.211	0.41-162.1	+0.66	154	24 Streams	p<0.05	Cuthbert and Kalff, 1993
$LogTN_{annual}$	356-10206	0.97-64.7	+0.90	9	9 Lakes	p<0.05	Johansson (unpublished)
$LogTP_{annual}$	5-60	0.3-40	+0.86	26	26 Lakes	p<0.05	Lindström et al., 1999
TP_{annual}	6-4700	3-80000	+0.67	301	301 Rivers	p<0.01	Canfield and Bachmann, 1981
$LogTP_{summer}$	13-381	2.2-62.8	+0.73	28	28 Lakes & res	p<0.05	Data from Knowlton and Jones, 1993
$Log(PO_4\ P)_{annual}$	0.86-158	0.97-64.7	+0.89	10	10 Lakes	p<0.05	Johansson (unpublished)
$LogAlk_{annual}$	0.074-0.213	1.27-4.85	+0.81	15	15 Lakes	p<0.05	Johansson (unpublished)
$LogCl_{ind}$	0.1-3.7	0.73-5.90	+0.64	15	15 Lakes	p<0.05	Data from Watras et al., 1998
$LogCond_{annual}$	1.98-39.9	0.97-64.7	+0.75	9	9 Lakes	p<0.05	Johansson (unpublished)
$LogDOC_{ind}$	1.98-20.13	0.73-5.90	+0.77	15	15 Lakes	p<0.05	Data from Watras et al., 1998
DOC_{annual}	1.5-30	3.4-8.6	+	11	11 Tributaries	*	Hurley et al., 1998b
DOC_{ind}	1.9-6.0	0.6-4.5	+0.84	15	1 Stream	p<0.05	Benoit and Rozan, 1999
pH_{annual}	5.10-8.50	0.3-40	+0.43	26	26 Lakes	p<0.05	Lindström et al., 1999
pH_{annual}	5.67-8.53	0.97-64.7	+0.64	10	10 Lakes	p<0.05	Johansson (unpublished)
pH_{ind}	4.51-8.82	0.02-35.3	+0.80	86	2 Rivers	p<0.05	Data from Benoit, 1995
$LogPOC_{annual}$	2.44-4.24	0.97-64.7	+0.99	7	7 Lakes	p<0.05	Johansson (unpublished)
POC_{ind}	0.4-6.4	2-59	+0.73	32	1 River	*	Hurley et al., 1998a
$Salinity_{ind}$	1-3	1-55	-	28	1 Estuary	*	Shiller and Boyle, 1991
$Salinity_{ind}$	2.4-29.5	1.9-11	-	42	1 Estuary	*	Muller et al., 1994
$LogSO_{4ind}$	0.71-6.06	0.73-5.90	-0.74	15	15 Lakes	p<0.05	Data from Watras et al., 1998
$Chl\ a_{ind}$	0.1-13.5	0.55-6.03	+0.70	46	23 Lakes	*	Watras et al., 1995a
$LogChl\ a_{ind}$	0.8-259	0.6-59.8	+0.95	87	8 Lakes	p<0.05	Johansson (unpublished)
$LogChl\ a_{ind}$	1.08-9.62	0.73-5.90	+0.64	15	15 Lakes	p<0.05	Data from Watras et al., 1998
$LogChl_{summer}$	3-99	2.2-62.8	+0.48	28	28 Lakes & res	p<0.05	Data from Knowlton and Jones, 1993
$LogNVSS_{summer}$	0.8-51.4	2.2-62.8	+0.93	28	28 Lakes & res	p<0.05	Data from Knowlton and Jones, 1993
$LogSecchi_{ind}$	0.3-6.7	0.6-59.8	-0.95	102	8 Lakes	p<0.05	Johansson (unpublished)
$LogSecchi_{month}$	2.8-8.5	2.5-6.3	-0.90	26	21 Baltic coasts	p<0.05	Data from Håkanson, 1999
$LogSecchi_{summer}$	0.3-3.6	2.2-62.8	-0.94	28	28 Lakes & res	p<0.05	Data from Knowlton and Jones, 1993
$LogSecchi_{summer}$	0.2-4.7	1.2-64.3	-0.93	94	94 Reservoirs	*	Jones and Knowlton, 1993
$LogTurb_{ind}$	7-171	0.1-9.4	+0.71	142	15 Lakes	p<0.05	Johansson (unpublished)
$LogTurb_{annual}$	23-94	1.27-4.85	+0.95	15	15 Lakes	p<0.05	Johansson (unpublished)
$LogTurb_{ind}$	0.09-104.8[#]	0.41-162.1[#]	+0.97	142	24 Streams	p<0.05	Cuthbert and Kalff, 1993
$LogTurb_{ind}$	2.4-47.5	2.3-35.3	+0.85	39	1 River	p<0.05	Data from Benoit, 1995
$Turb_{ind}$	100-2000	20-650	+0.84	101	1 Stream	p<0.0001	Kronvang et al., 1997
$LogDm$	1.7-15.4	2.2-62.8	-0.70	28	28 Lakes & res	p<0.05	Data from Knowlton and Jones, 1993
$LogDR$	0.07-7.88	0.3-40	+0.79	26	26 Lakes	p<0.05	Lindström et al., 1999
$LogQ_{ind}$	0.02-251.0[#]	0.41-162.1[#]	+0.71	71	24 Streams	p<0.05	Cuthbert and Kalff, 1993
Q_{ind}	24-94	2-13	+0.91	11	1 River	*	Shafer et al., 1997
Q_{ind}	51-410	3-130	+0.98	10	1 River	*	Bortoli et al., 1998

† Al = total aluminium concentration (mg/l); Alk = alkalinity (meq/l); Cl = chloride concentration (mg/l); Chl a = chlorophyll a concentration (μg/l); Cond = conductivity (mS/m); D_m = mean depth (m); DOC = dissolved organic concentration (mg/l); DR = lake dynamic ratio (-) (= $\sqrt{(Area)}/D_m$, Area in km^2 and D_m the mean depth in m); Fe = total iron concentration (mg/l); Mn = total manganese concentration (mg/l); NVSS = nonvolatile suspended solids (mg/l); pH = H_3O^+ concentration; PO_4 P = molybdate reactive phosphorus concentration (μg/l); POC = particulate organic carbon concentration (mg/l); Q = stream water discharge (m^3/s); Salinity (%); Secchi depth (m); TN = total nitrogen concentration (μg/l); TP = total phosphorous concentration (μg/l); Turb = turbidity (e.g., NTU, TU).

‡ For relationships where the correlation coefficient was not given, or possible to calculate, only the slopes (+/-) are given.

Total range, the given relationship may use a smaller number of data.

* No data.

All these relationships appear with significant correlations toward SPM in streams and rivers, lakes and reservoirs, estuaries, and marine coastal areas.

This means that the behavior of all substances with a particular phase in aquatic systems is influenced by SPM (fig. 2.14). In addition, the retention of a given substance in a system will be governed by the turnover time of the water; i.e., the longer the water turnover time, the larger the proportion of the initial load is retained within the system.

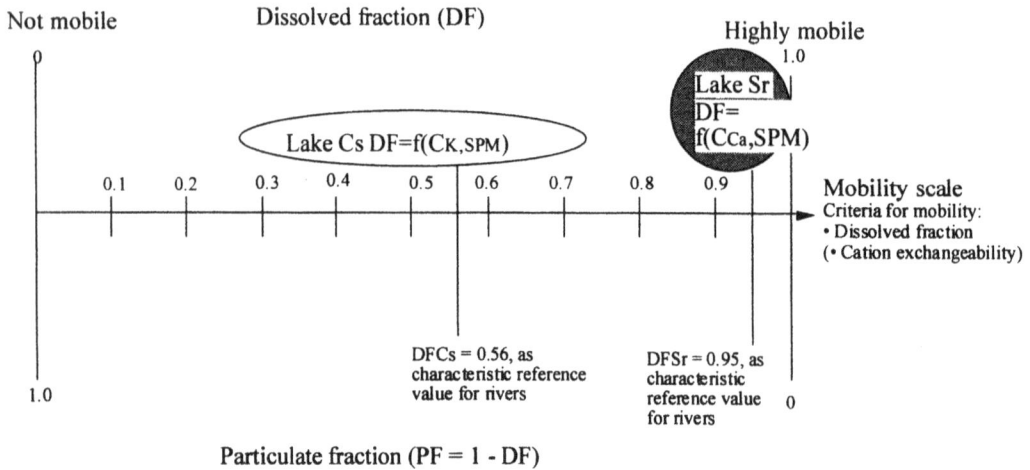

Fig. 2.14. Schematic illustration of the mobility scale related to the dissolved fraction - the more of the substance in dissolved phase, the more mobile it is. The dissolved fraction, in this example, for radiocesium is a function of the concentration of suspended particulate matter (SPM) and the potassium concentration of the water (C_K). A general characteristic DF value for ^{137}Cs in river water is 0.56, which is about 10% higher than the characteristic DF value for lake water. The DF value for radiostrontium is a function of SPM and the calcium concentration (C_{Ca}). A default DF value of 0.95 may be used for ^{90}Sr in river water. DF = 0 for SPM. Modified Håkanson et al. (2002).

A very important part of most mass-balance models for chemical substances is the distribution coefficient. The distribution coefficient is also often called the partition or partitioning coefficient and gives the particulate fraction (PF) or the dissolved fraction (DF = 1 - PF). The organic particulate fraction may be subject to mineralization by bacterioplankton. Traditionally, the K_d concept is used in these contexts; Kd is the ratio between the particulate (C'_{par} in g/kg dw) and the dissolved (C_{diss} in g/l) phases, i.e., $K_d = C'_{par}/C_{diss}$. K_d is often given in l/kg (Santschi and Honeyman 1991; Erel and Stolper, 1993; Benoit et. al., 1994; Warren and Zimmerman, 1994; Wehyenmeyer, 1996; Gustafsson and Gschwend, 1997). This means that the dissolved fraction (D_{diss}) can be written as:

$$D_{diss} = 1/(1 + K_d \cdot SPM \cdot 10^{-6}) \qquad (2.8)$$

where SPM is the amount of suspended particulate matter in the water in mg/l. It is essential to distinguish between the dissolved and the particulate fractions for all substances. It is especially important to do so for the key nutrients in water management, phosphorus and nitrogen, since phytoplankton take up the dissolved fractions and only the particulate fractions can settle out by gravity. This means that there are different transport routes for the two fractions. The SPM concentration influences the distribution of the nutrients into these two fractions. The settling velocity in m/yr for the particulate fraction may be turned into a sedimentation rate (dimension 1/time) by division with the mean depth of the system or a defined part of the system. The sedimentation rate regulates sedimentation, and hence also internal loading in the given system.

So, K_d describes particle affinity and represents the chemical equilibrium of numerous processes such as sorption onto particulate matter (see Weber et al., 1991), precipitation, and dissolution (Salomons and Förstner, 1984). Depending on the reversibility of these processes, it should be noted that K_d should not be regarded as a constant, but rather as a variable. Factors influencing the K_d equilibrium are, e.g., pH (Balistrieri and Murray, 1983; Tessier et al., 1989; You et al., 1989; Young and Harvey, 1992), salinity (Koelmans and Lijklema, 1992; Turner et al., 1993; Turner, 1996, 1999), SPM (Li et al., 1984; Hawley et al., 1986; Balls, 1989; You et al., 1989; Yan et al., 1991; Muller et al., 1994; Quémerais et al., 1998), redox conditions (Balistrieri et al., 1992; Pohl and Hennings, 1999), biogenic Si (Boyle and Birks, 1999) and the concentration of dissolved organic matter (DOC) (Watras et al., 1995a, b; Watras et al., 1998, Shafer et al., 1999). Examples of substances for which K_d have been either determined or modelled are trace metals (Balls, 1988; Honeyman and Santschi, 1988; Balls, 1989; Benoit et al., 1994; Benoit, 1995; Watras et al., 1995a; Turner, 1996), organic micropollutants (Turner et al., 1999; Zhou et al., 1999), phosphorus (Håkanson, 1999; Håkanson and Johansson, 2004) and radionuclides (Santschi and Honeyman, 1991; Carroll and Harms, 1999). Examples of published empirical K_d models are: (1) cadmium and 1,2,3,4-tetrachlorobenzene (Koelmans and Lijklema, 1992), (2) cadmium, copper and zinc (Sung, 1995), (3) mercury (Lindström, 2000), and (4) phosphorus (Håkanson and Johansson, 2004). In spite of (or maybe because of) their larger simplicity, statistical models might yield as good predictions, or even better, than models based on thermodynamics (e.g., Koelmans and Lijklema, 1992; Koelmans and Radovanovic, 1998).

The K_d concept has, as will be shown in chapter 3, some functional and "statistical" shortcomings that make the K_d approach problematic in contexts of mass-balance modelling.

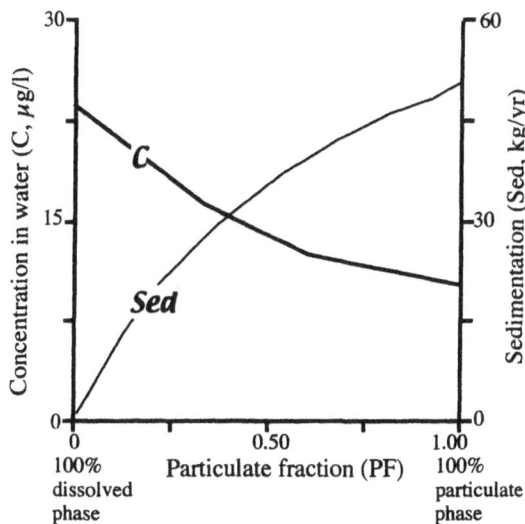

Fig. 2.15. Illustration of the important role of the particulate fraction of phosphorus in mass-balance calculations (annual simulations for a lake with an area = 1 km²; mean depth = 10 m; catchment area = 10 km²; mean annual precipitation = 650 mm/yr; mean tributary concentration = 26 µg/l of total phosphorus; and a settling velocity of 5 m/yr for the particulate fraction).

Fig. 2.15 illustrates the very important role that the PF value plays for the concentration of phosphorus in water and for sedimentation of phosphorus. It is evident that the PF value influences the concentration in lake water: The higher particulate fraction, the less in dissolved form, and the higher the sedimentation and the lower the concentration in water, and vice versa. This is a general principle valid for all substances. PF = 1 for SPM.

2.2.6. SPM, sedimentation, and resuspension

Geochemically, fine cohesive materials behave differently compared to coarse (friction) materials. From the basic Stokes' equation for settling particles (eq. 2.9), as well as for convenience, the limit between coarse and fine materials can be set at a particle size of medium silt.

Stokes' law expresses the settling velocity (v) as:

$$v = ((d_w - d_p \cdot g \cdot d^2)/(18 \cdot AV \cdot FR)) \qquad (2.9)$$

where

v = the settling velocity (usually in cm/s);
d_p = the particle density (usually in g dw/cm^3; g dw = gram dry weight);
d_w = the density of the lake water (often set to 1 g ww/cm^3; g ww = gram wet weight);
g = the acceleration due to gravity (980.6 cm/s^2);
d = the particle diameter (in cm);
AV = the coefficient of absolute viscosity (obtained from standard tables; 0.01 poise at 20°C);
FR = the coefficient of form resistance (set to 1 for spheres; Hutchinson, 1967).

Stokes' law (Stokes, 1851) is depicted in fig. 2.16. The behavior of materials that follow Stokes' law (i.e., particles with a diameter between about 0.01 and 0.0001 cm) differs from the coarser friction material and the still finer material. The sedimentological behavior of the friction material is closely linked to the grain size of the individual particles (Hjulström, 1935; Einstein, 1950; Allen, 1970); the sedimentological behavior of the very fine materials is governed by Brownian motion. These latter particles are so small that they will not settle individually, but will do so if they form larger flocs or aggregates that are dense enough to settle according to Stokes' law (Kranck, 1973, 1979; Lick et al., 1992). The cohesive materials that follow Stokes' law are very important, since they have a great affinity for pollutants (fig. 2.17). This group includes many types of detritus, humic substances, and plankton. All play significant roles in aquatic systems (Salomons and Förstner, 1984; Wetzel, 2001).

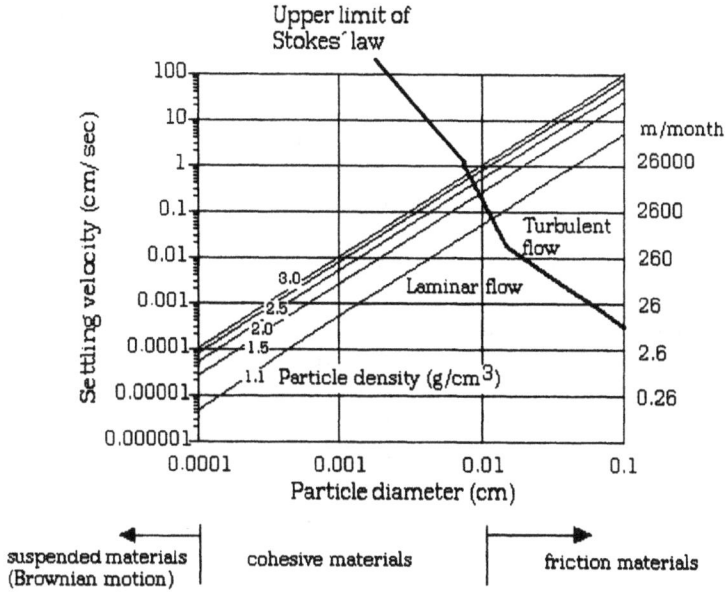

Fig. 2.16. The relationship between the settling velocity (v; of spherical particles) in water, particle diameter, and particle density (at 20°C) as given by Stokes' law.

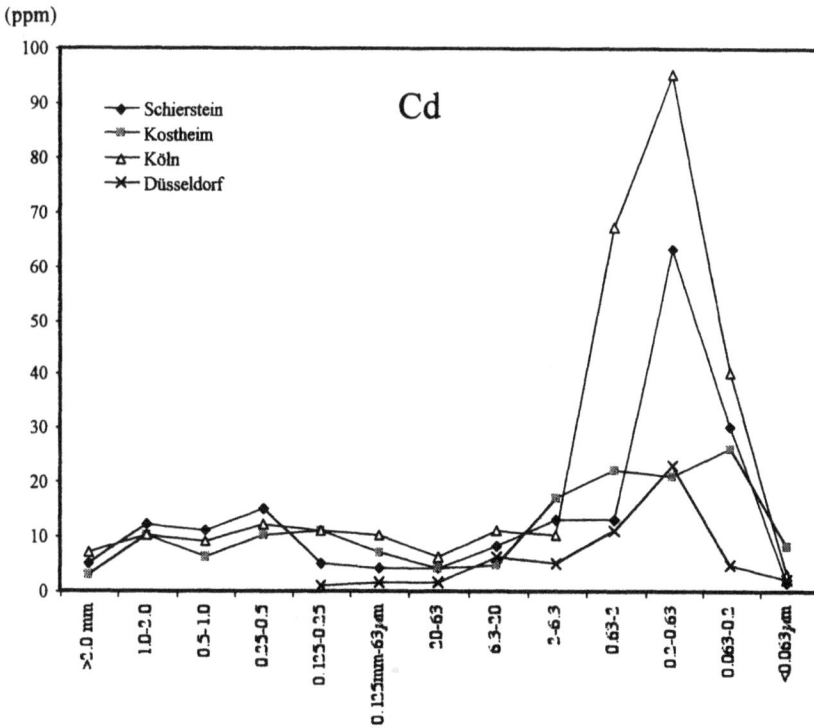

Fig. 2.17. Grain-size dependencies of cadmium concentrations in sediment samples from the German rivers (modified from Förstner and Salomons, 1981).

The settling velocity (v) of a given particle, aggregate, or particulate pollutant, and its distribution in an aquatic system, depends on the density, size, and form of the particle (and on the hydrodynamics of the flow of water in the system). If the particle density, d_p, is close to 1, if the form factor, FR, is large, and if the diameter, d, is small, the settling velocity, v, may be very small. If v is close to 0, the particle or aggregate is conservative in the sense that it may not be deposited in the system. In many lakes (see Håkanson and Jansson, 1983), Cl and alkalinity are often conservative substances, and color, organic matter, total P, Si, particulate P and SPM are more reactive.

The grain size and/or the composition of the material are often used as criteria to distinguish different sediment types (Sly, 1978). One can also differentiate between different sediment types by means of functional criteria (like erosion, transportation, and accumulation) of coarse sediments (friction material) or fine sediments (cohesive material). For further reading on physical sediment water interactions, see Golterman et al. (1983). In geoecological contexts, it is common to focus on the finer materials most easily set in motion/resuspension and having the highest capacity to bind pollutants (Thomas, 1972; Thomas et al., 1972). In defining the bottom dynamic conditions (erosion, transportation and accumulation), this work uses the following definitions (from Håkanson, 1977):

- Areas of erosion (E) prevail where there is no apparent deposition of fine materials but rather a removal of such materials, e.g., in shallow areas or on slopes; E areas are generally hard and consist of sand, gravel, consolidated clays, and/or rocks.

- Areas of transportation (T) prevail where fine materials are deposited periodically (areas of mixed sediments). This bottom type generally dominates where wind/wave action regulates the bottom dynamic conditions (fig. 2.18). It is sometimes difficult in practice to separate areas of erosion from areas of transportation.

- Areas of accumulation (A) prevail where the fine materials are deposited continuously (soft bottom areas). These are the areas (the "end stations") where high concentrations of pollutants may appear (table 2.4).

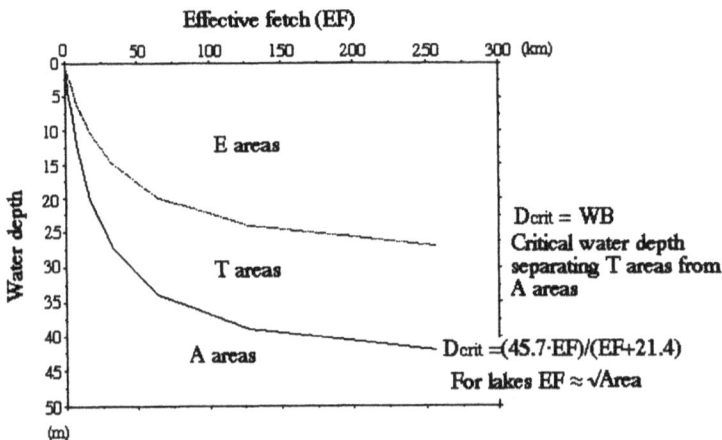

Fig. 2.18. The ETA diagram (erosion transportation-accumulation; from Håkanson and Jansson, 1983) illustrating the relationship between effective fetch, water depth, and bottom dynamic conditions. The dynamic modelling discussed in this book uses the critical water depth or the wave base (D_{crit} = WB) as a general criteria in mass-balance calculations to differentiate between surface water and deep water.

The generally hard or sandy sediments within the areas of erosion (E) often have a low water content, low organic content and low concentrations of nutrients and pollutants (table 2.6). The conditions within the T areas are, for natural reasons, variable, especially for the most mobile substances, like phosphorus, manganese and iron, which react rapidly to alterations in the chemical "climate" (given by the redox potential) of the sediments. Fine materials may be deposited for long periods during stagnant weather conditions. In connection with a storm or a mass movement on a slope, this material may be resuspended and transported up and away, generally in the direction toward the A areas in the deeper parts, where continuous deposition occurs. Thus, resuspension is a most natural phenomenon on T areas.

Table 2.6. The relationship between bottom dynamic conditions (erosion, transportation, and accumulation) and the physical, chemical and biological character of the surficial sediments. A. Example from Lilla Ullevi Bay (Lake Mälaren), Sweden, giving mean values, coefficients of variation and number of analyses from this freshwater system; B. Example giving characteristic values from marine coastal areas based on data from 11 Baltic areas (from Håkanson et al., 1984). ww = wet weight; dw = dry weight.

A. Lake	Erosion n = 15	Transportation n = 10	Accumulation n = 14
PHYSICAL PARAMETERS			
Water depth, m	13.0 (0.41)	17.5 (0.31)	31.6 (0.25)
Water content (% ww)	32.6 (0.28)	67.4 (0.14)	94.1 (0.024)
Bulk density (g/cm^3)	1.71 (0.087)	1.26 (0.079)	1.03 (0.019)
Organic content (loss on ignition, % dw)	4.6 (0.48)	10.7 (0.43)	24.3 (0.10)
NUTRIENTS (mg/g dw)			
Nitrogen	0.6 (0.67)	3.4 (0.35)	10.7 (0.14)
Phosphorus	0.8 (0.50)	2.8 (0.75)	1.6 (0.31)
Carbon	0.5 (1.0)	22.7 (0.74)	10.4 (0.16)
BENTHIC BIOMASS (mg ww/m^2)	1000-2000	3000-4000	6000-7000
CHEMICALLY MOBILE ELEMENTS (see also P)			
Iron (mg/g dw)	24.6 (0.42)	53.5 (0.27)	41.3 (0.077)
Manganese (mg/ g dw)	0.8 (1.0)	3.5 (0.74)	2.5 (0.60)
METALS			
Zinc (µg/g dw)	41 (0.46)	111 (0.24)	189 (0.090)
Copper (µg/g dw)	18 (0.50)	31 (0.42)	59 (0.10)
Nickel (µg/g dw)	23 (0.35)	40 (0.20)	57 (0.18)
B. Marine coastal area			
PHYSICAL PARAMETERS			
Water content (% ww)	< 50	50 - 75	> 75
Organic content (% dw)	< 4	4 - 10	> 10
NUTRIENTS (mg/g dw)			
Nitrogen	< 2	10 - 30	> 5
Phosphorus	0.3 - 1	0.3 - 1-5	> 1
Carbon	< 20	20 - 50	> 50
METALS			
Iron (mg/g dw)	< 10	10 - 30	> 20
Manganese (mg/g dw)	< 0.2	0.2 - 0.7	0.1 - 0.7
Zinc (µg/g dw)	< 50	50 - 200	> 200
Chromium (µg/g dw)	< 25	25 - 50	> 50
Lead (µg/g dw)	< 20	20 - 30	> 30
Copper (µg/g dw)	< 15	15 - 30	> 30
Cadmium (µg/g dw)	< 0.5	0.5 - 11.5	> 1.5
Mercury (ng/g dw)	< 50	50 - 250	> 250

Table 2.7 gives a compilation of sedimentation rates and fall velocities for many standard variables in water management (data from Håkanson and Jansson, 1983). Substances that only appear in dissolved phase, such as alkalinity, chloride, and calcium in the example from Lake Ekoln, Sweden, are generally referred to as conservative or nonreactive, i.e., substances which do not change (settle, evaporate or are being transformed in the lake). Most allochthonous particles and pollutants, which are transported to the lake from the catchment, have a particulate fraction (PF) higher than zero and are distributed in typical patterns with lobes of decreasing concentrations with distance from the source of pollution, or from the river mouth (fig. 2.19).

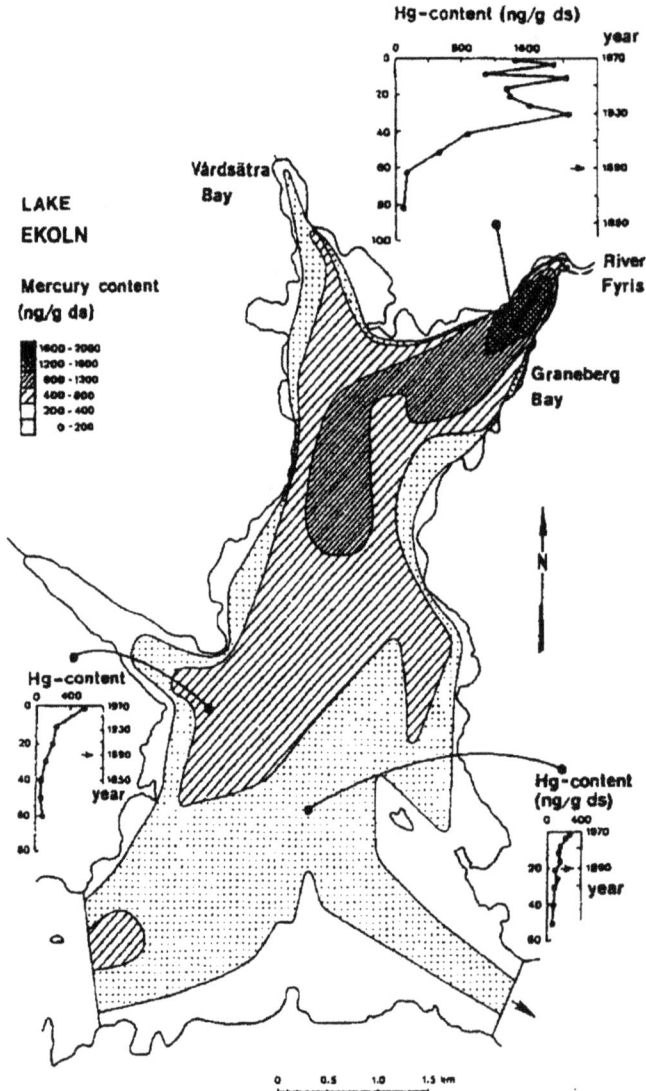

Fig. 2.19. The mercury contamination in sediments in Lake Ekoln, Sweden. The sediment age has been determined from data on sedimentation in sediment traps. The main tributary to this lake is River Fyris and mercury is transported to the lake from different sources (hospitals, dentists, laboratories, etc.) mainly from the city of Uppsala. Figure modified from Håkanson and Jansson (1983).

Table 2.7. Sedimentation rates (1/yr) and fall velocities (m/yr) for standard lake variables based on measured data on inflow and outflow for Lake Ekoln, Sweden (area = 18.6 km^2; mean depth = 19 m), for the period 1967-1977 (from Håkanson and Jansson, 1983).

	Variable	Sedimentation rate	Fall velocity
conservative	Alkalinity	0	0
	Chloride	0	0
	Calcium	0	0
	Sodium	0.005	0.1
	Magnesium	0.048	0.9
	Sulfate	0.048	0.9
	Potassium	0.048	0.9
	Color	0.082	1.6
	Organic nitrogen	0.15	2.9
	Phosphate P	0.16	3.0
	Total N	0.22	4.2
	Nitrate N	0.22	4.2
	Total P	0.24	4.6
	Nitrite N	0.25	4.8
	Silicon	0.26	4.9
	Particulate P	0.32	6.1
	SPM	0.39	7.4
	Secchi depth	0.40	7.6
Reactive	Ammonia N	0.41	7.8

Lakes of different size and form will not have the same sedimentation rates for one and the same substance since the sedimentation rate is basically defined by the ratio between the fall velocity (v in m/yr) and the mean depth (D_m in m).

Fig. 2.20 gives the maximum r^2 in regressions based only on lake morphometric parameters on the y axis (lake morphometric parameters are, e.g., mean depth, lake area, volume, etc.) and the sedimentation rates for the variables given in table 2.7 on the x axis. The sedimentation rates in fig. 2.20 emanate from empirical measurements from a 10-year period of both input and output to Lake Ekoln. There is a significant, logical and positive relationship between the sedimentation rate, which expresses how the substance is retained in the system – the higher the rate the more of the substance is retained in the lake, and the r^2 value from the regression models based on lake morphometric parameters. Conservative substances, like alkalinity, are not really influenced by lake morphometry. Reactive substances with sedimentation rates higher than zero, on the other hand, may be strongly influenced by morphometry. The information in fig. 2.20 provides an important interpretational key to how variables vary within and among aquatic systems. It is especially interesting to focus the attention on SPM or Secchi depth. This will also be explained in more detail in the following chapters.

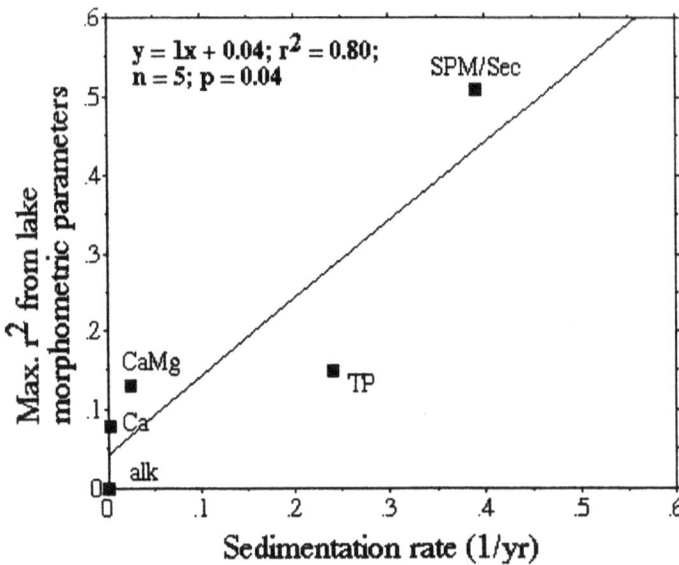

Fig. 2.20. The relationship between the maximum r^2 values obtained for empirical regression models based on lake morphometric parameters versus the sedimentation rate (1/yr) for five lake variables (SPM/Secchi depth, total phosphorus concentrations, hardness = CaMg, calcium concentrations and alkalinity) using data for Lake Ekoln, Sweden. For this lake there has been a full mass balance so that both the inflow and the outflow have been measured for the period 1967-1977. The theoretical lake water retention time for Lake Ekoln is 2.07 years. The lake area is 18.6 km^2 and the mean depth 19 m (data from Håkanson and Jansson, 1983).

Internal lake processes (e.g., resuspension, diffusion, mixing, and mineralization) affect SPM very much: The larger the mean depth, the lower the value of SPM (Håkanson and Boulion, 2002), which by definition settles out in lakes. There are, however, many compensatory and seemingly conflicting arguments to explain the role of morphometry in predicting SPM – SPM depends on the presence of humic substances, which have a low sedimentation rate (table 2.7); Secchi depth depends on total phosphorus concentrations (TP), because TP increases primary production; SPM settles out in aquatic systems and this will reduce the Secchi depth; Secchi depth depends on resuspension, which increases Secchi depth and depends on sedimentation.

The regression in fig. 2.20 illustrates the role of lake morphometric parameters in predicting SPM. Fig. 2.21 gives three categories of lake variables in a lake-type triangle: (1) variables that depend very much on catchment characteristics at the top, (2) variables that depend much on lake morphometric parameters at the right side of the base; and (3) variables mainly depending on "other causes" (such as climate) at the left side of the base. All water variables can be distributed in this triangle and this tells much about the factors causing variations in the given variable. Conductivity, hardness (CaMg), calcium, iron, color, pH, alkalinity, SPM/Secchi depth, and TP concentrations may be identified in the diagram.

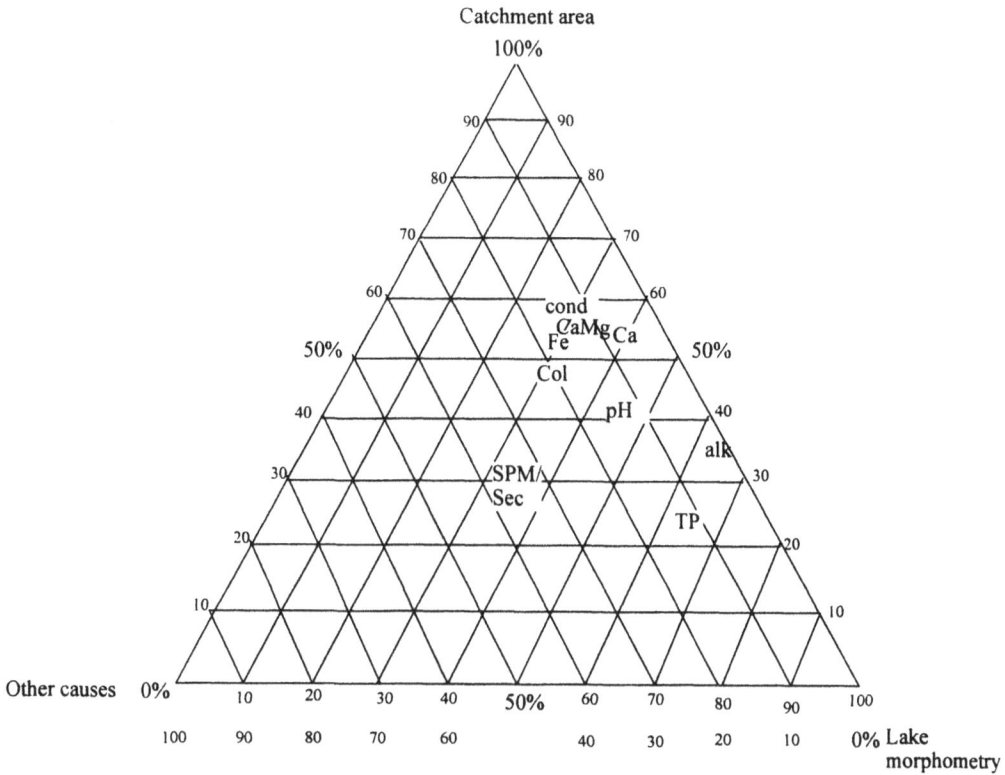

Fig. 2.21. Diagram showing how catchment area parameters (top), lake morphometric parameters (down, right) and other factors (down left; these factors include water chemical variables, climatological variables causing seasonal changes, etc.) influence predictions of the given water variables (conductivity, hardness = CaMg, calcium, iron, alkalinity, pH, total phosphorus and SPM/Secchi depth). Figure modified from Håkanson (2004b).

To place a given variable in the triangle, one needs data on the maximum r^2 from regressions based only on lake morphometric parameters, maximum r^2 from catchment area parameters, maximum r^2 from both lake morphometric and catchment parameters and characteristic CV for the given variable so that the highest reference r^2 (r_r^2) can be obtained. This gives that about 35% of the variability in mean annual values of lake SPM may be linked to variations among lakes in morphometric parameters, 35% may be related to variations in catchment characteristics and 30% to other causes, mainly climatological factors.

From this diagram, one can note that SPM/Secchi depth is more related to lake morphometry than the other given variables. The interpretational key explains why, and it has to do with the fact that SPM settles out in lakes.

It should be stressed that fine suspended materials are rarely deposited as a result of simple vertical settling in natural aquatic environments. The horizontal velocity component is generally at least ten times larger, sometimes up to 10,000 times larger, than the vertical component for fine materials or flocs that settle according to Stokes' law (see Bloesch and Burns, 1980; Bloesch and Uehlinger, 1986). The literature on general sedimentological processes in aquatic systems (Muir Wood, 1969; Thomas et al., 1976; Dyer, 1979; Postma, 1982; Golterman et al., 1983; Håkanson and Jansson, 1983) show that:

1. River action dominates the sedimentological properties in estuaries, where deltas may be formed if the amount of sandy materials carried by the tributaries is large

enough.   Generally, the rate of sedimentation and the grain size decrease logarithmically with the trajectory distance from the mouth of the river.

2. Wind/wave action generally dominates the bottom dynamic conditions in shallow areas. The rate of sedimentation decreases from the wave base and with increasing water depth below the wave base. The coarsest materials (sand, gravel, etc.) are often found in shallow waters.

3. Current action (unidirectional flows) dominates in rivers and in certain lake areas. Then the "Hjulström curve" (Hjulström, 1935; fig. 2.22) gives the relationship between critical erosion and critical deposition of materials.

4. Slope-induced (gravity) turbidity currents appear on bottoms inclining more than about 4-5% (Håkanson, 1977), and bioturbation generally prevails in oxic sediments, where the macro- and meiofauna cause a mixing of the sediments.

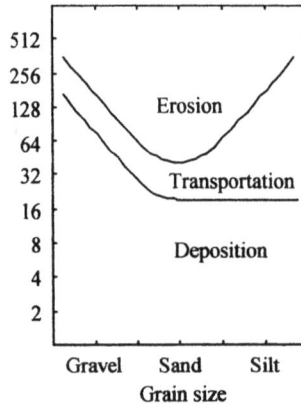

Fig. 2.22. The Hjulström (1935) diagram illustrating the relationship between critical erosion (upper curve) and critical deposition (lower curve) velocities for various grain sizes for unidirectional flow conditions.

This section has demonstrated the important and different roles that SPM plays in aquatic systems. This knowledge is, the author argues, fundamental for a proper understanding of how aquatic systems work.

2.2.7. SPM and the state variables TP, pH, and color in lakes

The most important nutrients in aquatic ecosystems are phosphorus and nitrogen. Total phosphorus (TP) has long been recognized as the nutrient most likely to limit lake primary productivity (Schindler, 1977, 1978; Chapra, 1980; Wetzel, 1983). The concentration of TP in the lake water is a powerful predictor in limnology (table 2.4) and, as stressed, one of the traditional key variables used to classify lakes (table 2.1). It should also be noted that the production in very humic lakes is not only or mainly a matter of the supply phosphorus. Humic lakes can have a substantial production of bacteria and zooplankton driven by allochthonous carbon and nitrogen (see Jonsson, 1997).

Lake pH is also a very important variable that influences the entire ecosystem. Many animals accustomed to a circum-neutral pH (pH ≈ 7) cannot reproduce and survive in acidified lakes.

Some, like crayfish, molluscs, crustaceans and snails, are very sensitive to changes in pH, whereas other animals, like perch (*Perca fluviatilis*) and pike (*Esox lucius*), are less sensitive. The literature on anthropogenic acidification of land and water, its ecological damage and its economical consequences is extensive (Likens et al., 1979; Ambio, 1976; Overrein et al., 1980; Monitor, 1981, 1991; and Merilehto et al., 1988). There are many models and modelling approaches to address the acidification of aquatic and terrestrial environments and to propose remedial measures for acidification (Eliassen and Saltbones, 1983; Sverdrup, 1985; Warfvinge, 1988). Ivanova (1997) has reported that zooplankton exists at pH from 3.5 to 10.5. Reductions in species number can be observed in lakes with pH of about 5.5. It seems that pH of about 5.5 may be boundary condition for the structure of the zooplankton community.

There are also many publications on colored substances in aquatic systems (Shapiro, 1957; Gorham et al., 1983, 1986; Christman and Gjessing, 1983; Aiken et al., 1985; Hayes et al., 1989; Allard et al., 1991). There are several reasons for this interest. Colored substances reduce light transmission through water (Kirk, 1976; Spence, 1982), affect fluxes, bioavailability and ecological effects of nutrients (Peterson, 1991), natural elements, like Fe, Al and Pb (Stumm and Morgan, 1981; Abrosov, 1982; Baccini et al., 1982) and contaminants (Wershaw et al., 1969; Peterson, 1991; Nilsson and Håkanson, 1992), affect plankton metabolism (Jackson and Hecky, 1980; Janus and Vollenweider, 1981; Benner et al., 1986), and may inhibit secondary production (Rasmussen and Kalff, 1987).

So, lake TP, color, and pH may be regarded as limnological state variables in the sense that these three variables may be used to predict several more complicated biological and ecosystem variables (Håkanson and Peters, 1995). Lake TP, pH, and color are also determined by independent methods and they play different roles in aquatic systems, but they are anyhow related. There are many reasons for this: All of them are generally, at least partly, transported to lakes by the same tributaries, many internal processes, like resuspension and mixing, affect them in similar ways, and they are all transported out of the lake together with all other dissolved and suspended substances.

The studies by Nürnberg and Shaw (1998) and Håkanson and Lindström (1997) quantified and ranked many variables of significance to predict how lake water clarity (Secchi depth and/or SPM) varies among lakes. The most important variables were: Lake color (expressing allogenic input of different types of humic materials), TP and lake temperature (measures of production of autogenic materials). An empirical model based on data from 935 lakes covering a very wide range gives Secchi depth (Sec in m) as a function of TP (in µg/l) and lake color (Col in mg Pt/l) as:

$$\log(Sec) = 1.41 - 0.49 \cdot \log(TP) - 0.30 \cdot \log(Col) \tag{2.10}$$

This model is illustrated by the graph in fig. 2.23, which gives TP on the x axis and lake color on the y axis.

The trophic status of a lake is usually estimated by values of primary production (chlorophyll-a concentrations) (Thienemann, 1928; Ohle, 1956, 1958; Hutchinson, 1957; Elster, 1958; Rodhe, 1958; Winberg, 1960; Walker, 1979; Aizaki et al., 1981; Milius, 1982, for further information). Håkanson and Boulion (2002) defined a Trophic State Index (TSI), accordingly:

$$TSI = 25 \cdot (\log(Chl) + 1) \tag{2.11}$$

TSI = 100 for Chl = 1000 µg/l (the limit for hypertrophic lakes; Chl = the mean, lake characteristic value for the growing season), TSI = 75 for Chl = 100 µg/l, the limit between eutrophic and hypertrophic lakes, etc. (fig. 2.24). Note that if Chl < 0.1, then TSI should be set to 0. So, TSI is an index that varies from 0 to 100.

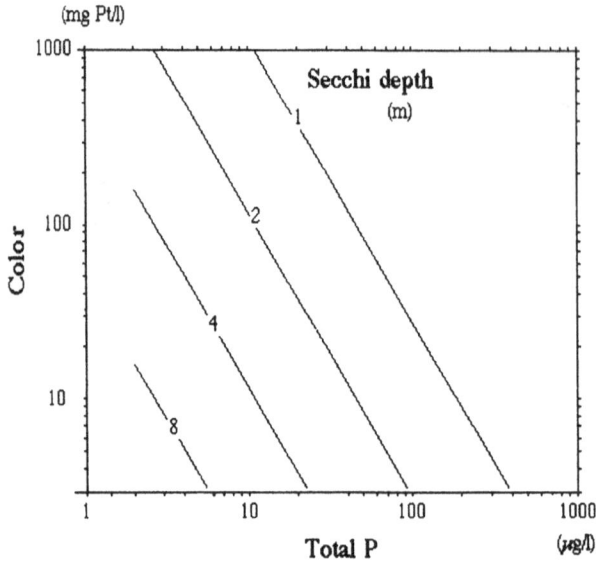

Fig. 2.23. The relationship between lake color, TP concentrations and Secchi depth (from Håkanson and Boulion, 2002).

An analogous humic state index is based on color values. The set-up using categories of autotrophy on the x axis and categories of allotrophy on the y axis to classify lakes is classical in limnology (see Wetzel, 2001). Using lake color values as a criteria for lake humic state, a Humic State Index (HSI) is defined on a scale from HSI = 100, hyperhumic, to HSI = 0, ultraoligohumic (color values lower than 3 would be very rare, and if color < 3 mg Pt/l, then HSI = 0) related to the lake characteristic mean values of lake color, accordingly:

$$HSI = (100/3) \cdot \log(Col - 3) \tag{2.12}$$

Col = 1000 (higher values than 1000 mg Pt/l would be most exceptional) gives HSI = 100; Col = 300 gives HSI = 82, etc. (fig. 2.24). Note again that if Col < 3 mg Pt/l, then HSI should be set to 0. The following color values represent the different humic classes: 3, 10, 30, 100, and 300 mg Pt/l for oligo-, meso-, poly- and hyperhumic lakes.

Fig. 2.24 demonstrates that there exist lakes in most of the "boxes" defined by the humic and trophic categories, expect that there are no hypertrophic and ultraoligohumic lakes and no polyhumic/hyperhumic and oligotrophic lakes. This does not exclude that such lakes exist, it only indicates that such lakes are rare.

Since so many biological/ecological characteristics of lakes can be related to the three state variables, it is also interesting to see how they depend on SPM. Fig. 2.25A gives regressions (based on data from Johansson and Håkanson, 2004) on SPM versus lake TP concentrations (using data from 146 samples), fig. 2.25B gives the same thing using not TP but particulate phosphorus (PP in µg/l), fig. C gives a similar regression for dissolved phosphous (DP in µg/l), fig. 2.25D gives this for particulate phosphorus expressed in mg/g dw and fig. 2.25E gives SPM versus the particulate fraction (PF).

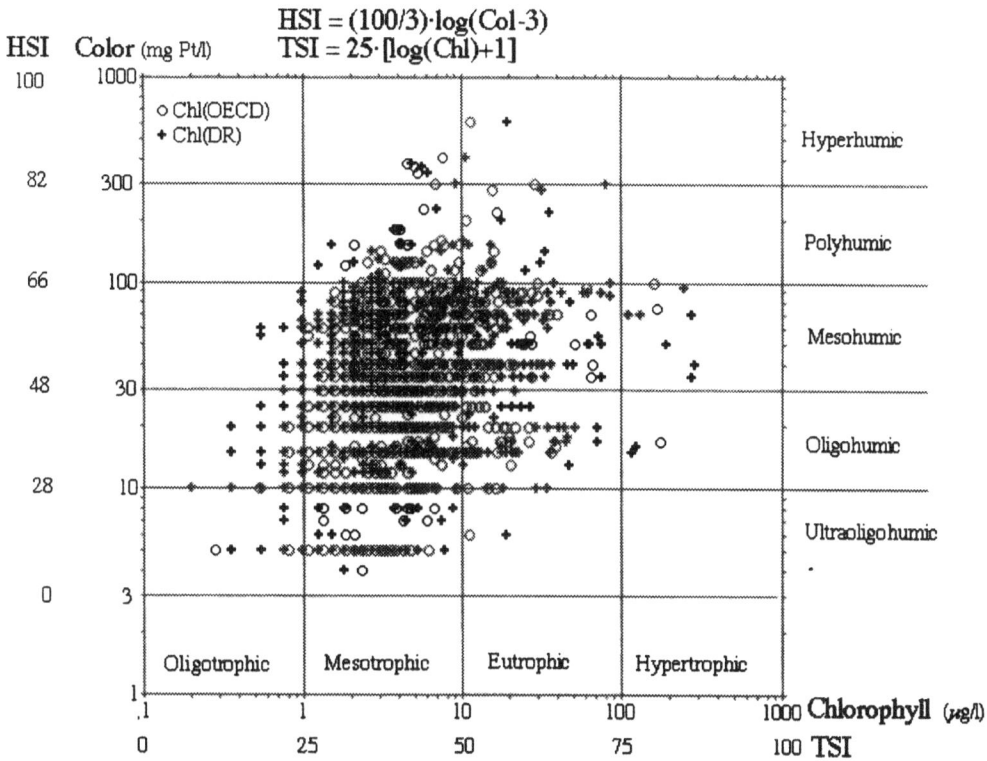

Fig. 2.24. The relationship between lake color and humic state on the y axis and lake chlorophyll and trophic state on the x axis and data from 936 lakes (from Håkanson and Boulion, 2002).

One can note that:

- There are very strong relationships between SPM (log transformed values) and TP (log(TP); $r^2 = 0.70$); and PP (log(PP); $r^2 = 0.76$).

- Also the relationships between SPM and dissolved P and particulate P (in mg/g) are highly significant ($r^2 = 0.19$ and 0.28, respectively).

- The very important relationship between the particulate fraction (PF, which is a fundamental importance in modelling fluxes of phosphorus; by definition, only the particulate fraction is subject to gravitational sedimentation and benthic fluxes), is weak ($r^2 = 0.045$) and not as significant (p = 0.01).

The statistical results in fig. 2.25 may be mechanistically understood since both phosphorus and SPM are transported to the lakes by rivers, particulate P largely behaves like SPM (sedimentation, mixing, resuspension, outflow, etc.), but the dissolved fraction does not behave in the same manner. It is the dissolved fraction of phosphorus that will trigger primary production.

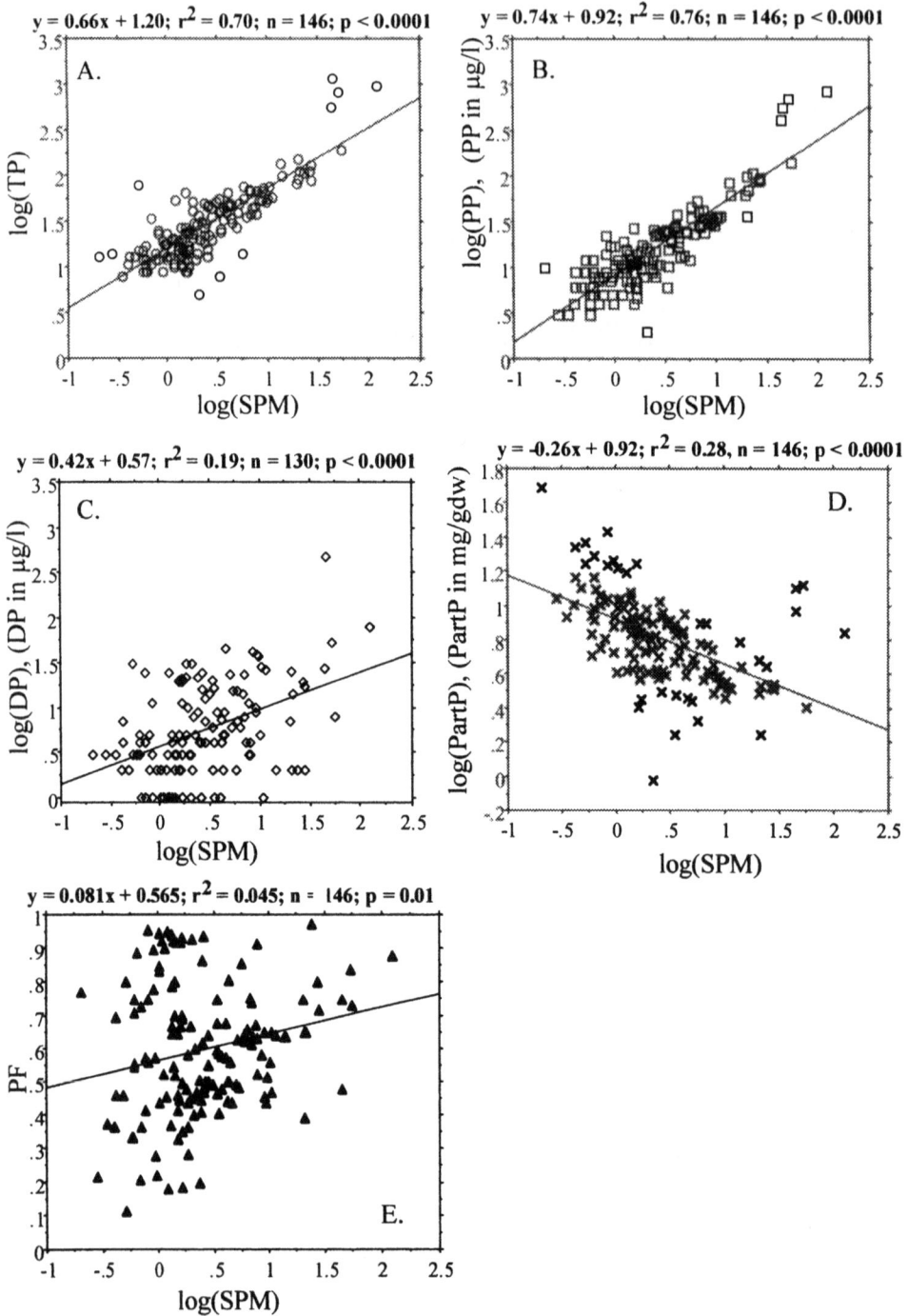

Fig. 2.25. Five regressions between SPM [log(SPM)] and (A) log(TP), (B) log(PP), where particulate P is given in μg/l, (C) DP, where dissolved P is given in μg/l, (D) particulate P in mg/g and (E) the particulate fraction (PF = PP/TP) using data from Johansson and Håkanson (2004).

Fig. 2.26 shows similar regressions between the two remaining limnological state variables, pH and color and SPM. There is a strong and highly significant relationship between SPM and pH – the higher SPM, the higher pH, and vice versa. High SPM means high TP, high production and high pH; low SPM means low TP, low production, low pH, higher aggregation and hence also even lower SPM. Using these data, there is no significant relationship between SPM and color (fig. 2.26D). From fig. 2.23, it is, however, evident that using a larger database, there is a strong (table 2.8) relationship between color and Secchi depth – the more colored substance suspended in the water, the lower the Secchi depth. One should also note that the results in fig. 2.26 are based not on mean lake conditions but on results from individual sampling events at given sites. Since SPM shows a very great variation (see chapter 3), a single SPM value may represent average lake conditions very poorly.

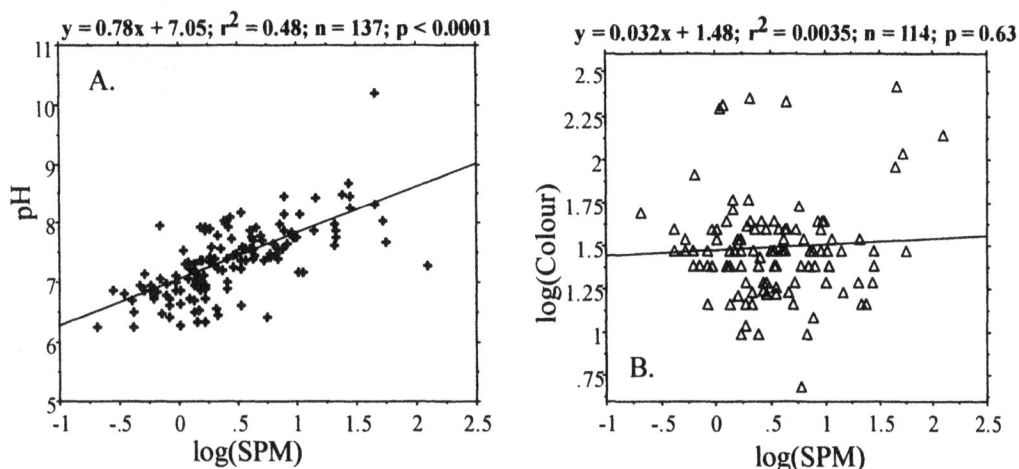

$y = 0.78x + 7.05; r^2 = 0.48; n = 137; p < 0.0001$

$y = 0.032x + 1.48; r^2 = 0.0035; n = 114; p = 0.63$

Fig. 2.26. Two regressions between SPM [log(SPM)] and (A) lake pH and (B) lake color (values in mg Pt/l) using data from Johansson and Håkanson (2004).

Table 2.8. Ladders based on stepwise multiple regressions. The target y variable is log(Sec); Sec = Secchi depth in m.

A. Data from 935 lakes (from Håkanson and Lindström, 1997).

Step	r^2	Variable	Model
1	0.58	log(TP)	$y=1.14-0.62 \cdot x1$
2	0.72	log(Col)	$y=1.41-0.49 \cdot x1-0.30 \cdot x2$

Range for TP (µg/l): 1 - 810

Range for Color (mg Pt/l): 0 – 600

B. Mean long-term averages for Secchi, TP and color; from 196 lakes (from Nürnberg and Shaw, 1998).

Step	r^2	Variable	Model
1	0.59	log(Col)	$y=1.09-0.51 \cdot x1$
4	0.73	log(TP)	$y=1.26-0.31 \cdot x1-0.36 \cdot x2$

Range for Color (mg Pt/l): 2 - 458

Range for TP (µg/l): 1-900

2.2.8. SPM versus latitude and altitude

Latitude and altitude are important integral parts of the catchment area submodel and the river submodel for SPM (see chapter 5). Latitude and altitude influence both SPM transport from land to water and the geometry of the river and this is one of the reasons why this section is included here. It is also interesting to see how altitude and latitude influence SPM in lakes. Lakes at high altitudes and latitudes should have lower water temperatures and hence also lower production than low altitude and low latitude lakes and the aim of this section is to use empirical data to see if this is true. The datasets come from Håkanson and Boulion (2002) and Johansson and Håkanson (2004).

Brylinsky and Mann (1973) gave one of the first analyses on these issues based on materials collected by the International Biological Program (IBP). Using data from a very large database with lakes located from 2°N (Africa) up to 80°40' N (the Land of Francis Joseph), Håkanson and Boulion (2002), showed (fig. 2.27) that the relationship between phytoplankton primary production (PrimP in g ww/m^2·yr) and latitude (Lat) gives an excellent fit if the proper transformations are being used:

$$\log(\text{PrimP}) = -1.86 \cdot (90/(90-\text{Lat}))^{0.5} + 5.8 \qquad (2.13)$$
$$(r^2 = 0.74; n = 63; p < 0.0001)$$

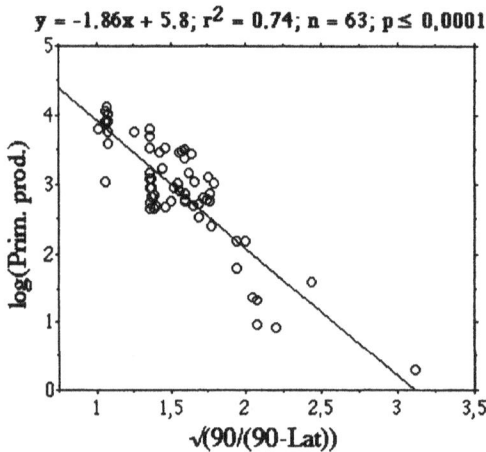

Fig. 2.27. The relationship between phytoplankton primary production (PrimP) and latitude; PrimP in g ww/m^2·yr; Lat in°N); data from Håkanson and Boulion (2002).

It is evident that latitude cannot be the only factor regulating PrimP, but it is interesting to note that as much as 74% of the variability among lakes in PrimP can be statistically explained by variations in latitude. Some of the scatter around the regression line in fig. 2.27 is evidently related to empirical uncertainties in the PrimP values. From this figure, latitude can be used as an excellent x variable to predict primary production. It is almost as useful in this context as the TP concentration, but very few persons would advocate the use of latitude as key criteria in trophic level classifications, as TP concentrations have been used. However, both are very important abiotic factors regulating lake production of suspended particles.

Fig. 2.28 gives four regressions relating altitude (√altitude) to Secchi depth (fig. A), TP concentrations (fig. B), pH (fig. C) and lake color (fig. D). These relationships are highly significant, except the one for color. This means that:

- High altitude (and latitude) lakes generally (but not always) can be expected to have a higher Secchi depth, a lower primary production and lower SPM values than lakes at lower altitudes – if all else is constant.

- High altitude (and latitude) lakes also generally have lower TP concentrations, and hence also lower primary production for this reason (and not just because these lakes also have lower temperatures; see the temperature submodel in appendix 9.4).

- Given the higher Secchi depths, the lower TP concentrations, the lower primary production and the lower SPM concentrations in high altitude lakes, it is logical that such lakes also generally have a lower pH than lakes at low altitudes and latitudes.

- It is also interesting to note that using this database including 936 lakes, there is no significant relationship between lake color and altitude.

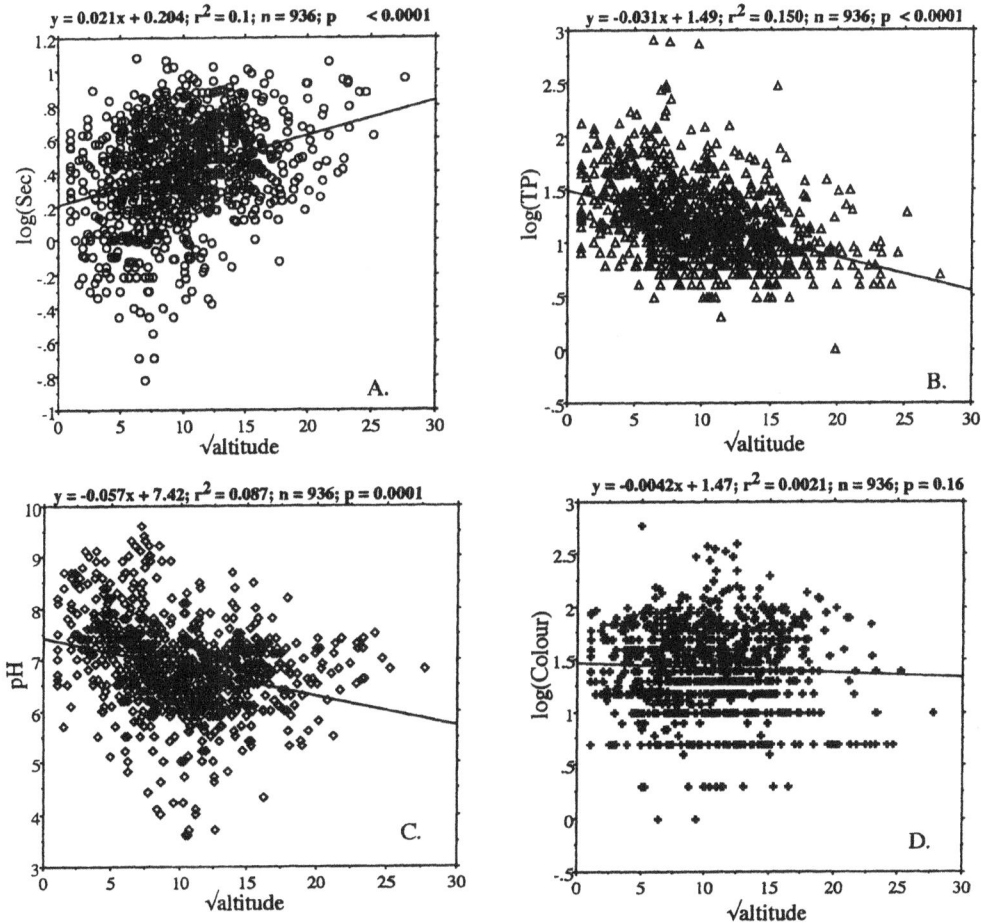

Fig. 2.28. Four regression between altitude (√altitude) and log(Sec), log(TP), pH and log(Color). Secchi depth in m, TP in µg/l, and color in mg Pt/l. Data from Håkanson and Boulion (2002).

3. Methodological Aspects Related to Variations in SPM

The variability, uncertainty and representativity of all variables is a major concern in predictive modelling. Variables may vary both within and among systems and the magnitude of these variations influences the predictive power of models. For example, SPM may vary much in a lake during a month. SPM may be rather uniformly distributed in the lake water or it may change with water depth. During windy conditions, SPM generally varies substantially both areally and vertically (see Hawley and Lesht, 1992; James and Barko, 1993; Bloesch, 1995; Weyhenmeyer, 1996). This means that there will be large uncertainties in single measurements. The same argument is valid for all water variables in all aquatic systems.

The coefficient of variation (CV) is defined by the ratio between the standard deviation (SD) and the mean value (MV), i.e., $CV = SD/MV$. Different substances have different characteristic CVs (table 3.1). In order to minimize model uncertainty, model variables with low CVs should preferably be used (Håkanson, 1999). This chapter will demonstrate why this is important not just for SPM in freshwater and marine systems, but for all variables in all types of ecosystems.

3.1. Background and aim

Variations and uncertainties in empirical data constrain our approaches to knowledge about ecosystems and our possibilities to make meaningful predictions. This methodological chapter addresses this question in discussing important concepts in predictive modelling in general, and in particular, in SPM modelling. The focus here is not on concentrations of SPM but rather on CV for SPM.

The first part of this chapter concerns the following question: How high is the highest possible predictive success for a given target variable (y), e.g., for y = SPM? It is evident that many factors are involved concerning sampling (such as the number of samples), analysis (the method and precision used in determining y), temporal and spatial variations, model structure (which model variables x_i are included), the reliability of the model variables and the statistical methods used to define predictive success, such as the r^2 value (= the coefficient of determination), as calculated from a regression between modelled values and empirical data of y. The r^2 value will be used here as a standard criterion of predictive success since this is a widely used concept in aquatic modelling (Håkanson and Peters, 1995). If a model is validated, i.e., tested against an independent set of data, the achieved r^2 value will not just depend on the uncertainty in the empirical y value (i.e., the uncertainty in the y direction), but also on the structuring of the model (i.e., which processes and model variables are accounted for and the empirical uncertainty of the model variables; the uncertainty in the x direction). This section will discuss three r^2 values:

1. the highest reference r^2 (r_r^2),

2. the empirically based highest r^2 (r_e^2) and

3. the achieved r^2 when modelled y values are compared to empirical y values (r^2).

To exemplify these concepts, this section will use empirical SPM data from appendix 9.1 and other literature sources.

CV values are probably the most commonly expressed statistic in contexts of ecosystem modelling (see Whicker, 1997). It is not likely that other statistical measures of uncertainty (see Gilbert, 1987) will substantially change the general conclusions about empirical uncertainties for the aquatic variables discussed.

Table 3.1. Coefficients of within site or within system variation (CV) and theoretically highest r^2 values (r_r^2) for variables from (A) sites in an open marine area (the Baltic Sea), (B) marine coastal areas in the Baltic), (C) rivers, and (D) lakes.

A. Marine open water areas (data from the Baltic Sea; from Håkanson and Eckhell, 2004)	CV	r_r^2
SPM	0.67	0.70
Temperature	0.40	0.89
Salinity	0.07	0.997

B. Marine coastal areas (data from Wallin et al., 1992 and Nordvarg, 2001)		
SPM	0.34	0.92
Secchi depth	0.19	0.98
Chlorophyll-a	0.25	0.96
Total N	0.13	0.99
Inorganic N	0.31	0.94
Total P	0.16	0.98
Inorganic P	0.28	0.95
O_2 concentration	0.26	0.96
O_2 saturation	0.26	0.96

C. Rivers (from Håkanson et al., 2004b)		
SPM	1.71	-0.93
Chlorophyll-a	1.15	0.13
Total phosphorus	0.51	0.83
Water discharge	1.42	-0.33
Particulate phosphorus	1.59	-0.67
Dissolved phosphorus	0.77	0.61
Chem. oxygen demand	0.65	0.72
Biological oxygen demand	0.54	0.81
Conductivity	0.44	0.87
Temperature	0.41	0.89
Particulate fraction of phosphorus	0.31	0.94
Dissolved oxygen	0.25	0.96
Redox potential	0.18	0.98
pH	0.03	1

D. Lakes (from Håkanson and Peters, 1995 and Håkanson, 1999)		
SPM	0.65	0.72
Secchi depth	0.15	0.99
Chlorophyll-a	0.25	0.96
Total phosphorus	0.35	0.92
Alkalinity	0.35	0.92
Fe concentration	0.25	0.96
K concentration	0.20	0.97
Color	0.20	0.97
Ca concentration	0.12	0.99
Hardness	0.12	0.99
Conductivity	0.10	0.99
pH	0.05	0.998

3.2. Highest r^2 of predictive models

Fig. 3.1 illustrates some fundamental concepts related to "the highest r^2." The data in the figure should emanate from several sampling occasions from a sampling site for defined time intervals (such as days, weeks, or months). The CV value for within site variability is always related to very complex climatological, biological, chemical, and physical conditions. It is often possible to define a characteristic CV value for a given variable, such as 0.65 for SPM in lakes. Table 3.1 gives a compilation of such CV values for many standard water variables from lakes, rivers, marine coastal areas, and open water areas. The data in this table will be used in this book and one can note that SPM varies very much. By definition, CV is also largely independent of n, the number of data used to determine the mean value and the standard deviation, if n is large enough (n > 4). CV should also be independent of the analytical method provided that the main difference between the methods can be expressed by a calculation constant ($y_1 = const \cdot y_2$). Table 3.1 illustrates:

- The characteristic inherent in CV for individual SPM data at river sites is very high indeed, 1.71. So, an important question is: How is it possible to reduce high CVs so that practically useful models may be derived, i.e., models giving r^2 values at least higher than 0.75?

- The highest reference r^2 (r_r^2) is lower than zero for SPM at many river sites. The achieved r^2 value can never be lower than zero but r_r^2 can. This will be explained in more detail in a coming section.

- The characteristic CV for SPM from individual marine open water sites is also very high, 0.67.

- The characteristic CV for SPM in lakes is 0.65, which is very high.

- The characteristic CV for SPM in marine coastal areas is 0.34, which is also a high value.

- Among all the studied water variables given in table 3.1, SPM has the highest inherent CV.

The uncertainty or variability associated with a given target variable is illustrated by the vertical uncertainty bands (error bars) in fig. 3.1. This uncertainty will evidently influence the result of regressions (e.g., the r^2 value). If CV for y is large, a model cannot be expected to predict y well. Fig. 3.1 illustrates a hypothetical model validation when modelled values of the target variable are put on the x axis. The uncertainty in the x direction is then related to the uncertainty associated with the model structure and the uncertainty of the model variables (x_i). Generally, the model uncertainty should be larger than the uncertainty in the y direction. This is also shown in fig. 3.1.

The achieved r^2 value depends on the range of data. In fig. 3.2, a regression is given between Secchi depth and SPM values. In this case, the data come from just one lake (Lake Erken, Sweden) and not from many lakes as in fig. 2.4. The data in fig. 3.2 provide a relatively narrow range and the r^2 values is just 0.29 as compared to 0.78 in fig. 2.4. The spread around the regression line and the mean difference [Diff = (x-y)/y] is, however, quite similar in the two examples. The lower r^2 value in fig. 3.2 is mainly explained by the narrow range in the data.

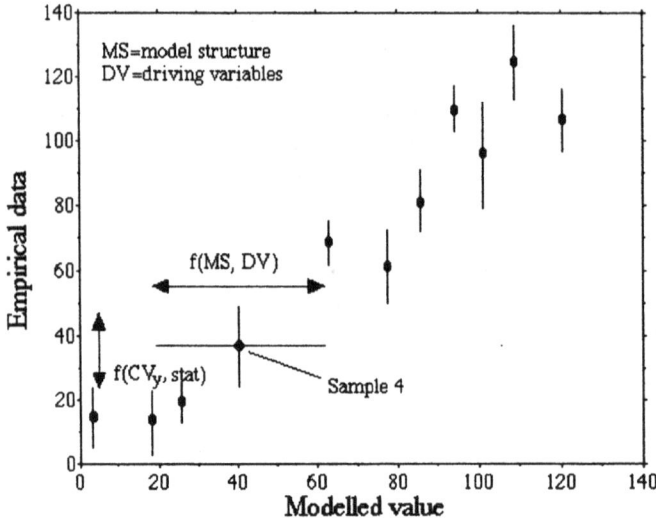

Fig. 3.1. Illustration of fundamental concepts related to the question of "the highest possible r^{2}" of models as determined from validations, i.e., a comparison between modelled values (on the x axis) and independent empirical data on the y axis. Note that the uncertainty in the y direction (e.g., given by the CV value) differs among variables and that the characteristic uncertainty in the y direction regulates the predictive power of the model.

3.2.1. Empirically based highest r^2, r_e^2

One way to estimate the highest possible r^2 of a predictive model is to compare two empirical samples (r_e^2) since models are not expected to predict better than empirical data. To illustrate this, SPM data from River Avon (the station at Evesham; see appendix 9.1) will be used. Fig. 3.3 shows that the r^2 value calculated from 32 samples taken within one week is 0.006 using actual data and 0.13 using logarithmic values. This includes analytical uncertainties but mainly depends on the very significant short-term (weekly) variability for SPM at this river site, which is also reflected in the very high characteristic CV of 1.71 for SPM in rivers. The higher the CV value, the more difficult it will be to establish representative and reliable empirical mean values of the given variable. Evidently, r_e^2 depends on the total number of samples in the regression and the number of samples for each individual value. The next r^2 value (r_r^2) has been defined as a standard reference r^2, which is meant to be independent of the sampling.

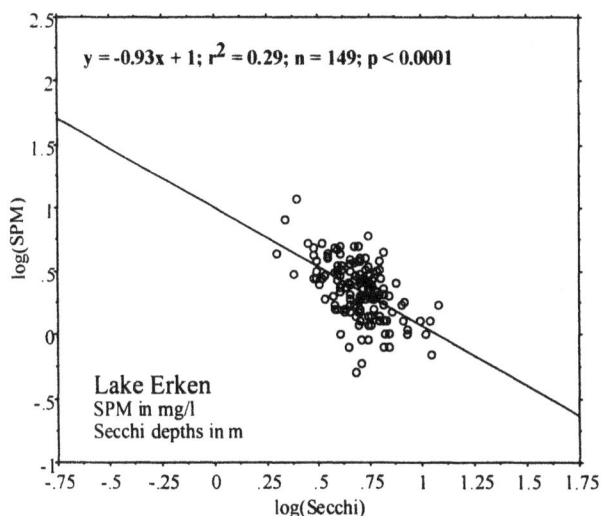

Fig. 3.2. Regression between Secchi depth and SPM data (logarithmic values) using data from Lake Erken, Sweden (from Malmaeus and Håkanson, 2003). This regression can be compared with a similar regression in fig. 2.4, which covers a much wider range in Secchi depths and SPM values.

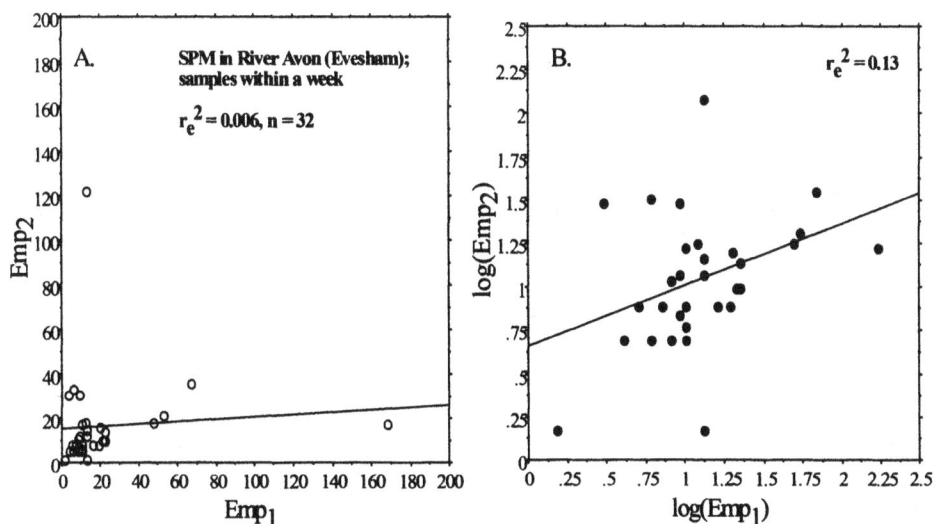

Fig. 3.3. Example illustrating determination of "empirically highest r^2" (r_e^2) for SPM from a regression between two parallel empirical samples taken within a week (Emp_1 vs Emp_2). The r_e^2 value is 0.006 for the given 32 data pairs for SPM in the River Avon (data from appendix 9.1).

3.2.2. Highest reference r^2, r_r^2

The highest reference r^2, r_r^2, has been presented and motivated by Håkanson (1999). From a statistical point of view, an equation has been derived that gives the highest reference r^2 value as a function of (1) the number of samples (n_i) for each y_i value in the regression; (2) the number of data points in the regression (N); (3) the standard deviations related to all individual data points;

(4) the standard deviation of all points in the regression and (5) the range of the y variable. The r_r^2 value is defined as:

$$r_r^2 = 1 - 0.66 \cdot CV_y^2 \qquad (3.1)$$

where CV_y is the characteristic within site variability for the given y variable. The equation is graphically shown in fig. 3.4. It is valid for actual (non transformed) y values. In practice, in ecosystem modelling, to obtain r^2 values higher than the r_r^2 values cannot be expected, so these values may be used as reference values for useful predictive models.

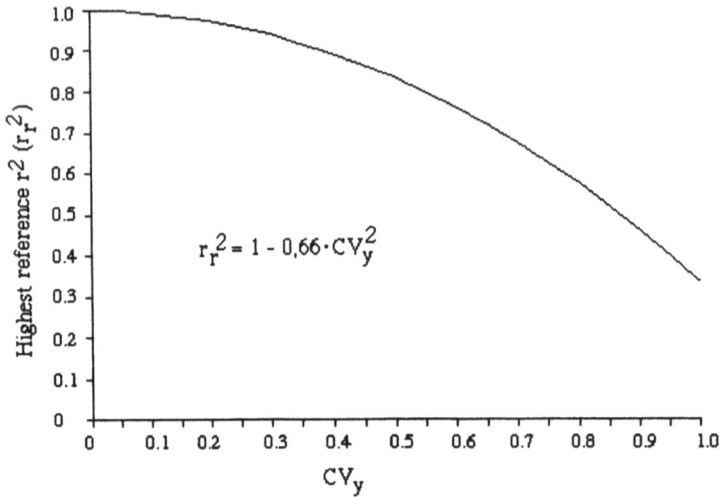

Fig. 3.4. The relationship between the "theoretically highest reference" r^2 (r_r^2) and the coefficient of variation for variability within (CV_y) ecosystems. Figure modified from Håkanson (1999).

Table 3.1 lists r_r^2 values for a number of water variables, some of these (e.g., salinity and pH) have low inherent CV values and high r_r^2 values, and SPM is the opposite extreme.

3.2.3. The sampling formula and uncertainties in empirical data

If the within site variability (CV_y or simply CV) is large, many samples must be taken to obtain a given level of certainty in the mean value. There is a general formula, derived from the basic definitions of the mean value, the standard deviation and the Student's t value, which expresses how many samples are required (n) in order to establish a mean value with a specified certainty (see Håkanson, 1999).

$$n = (1.96 \cdot CV/L)^2 + 1 \qquad (3.2)$$

where L is the level of error accepted in the mean value. For example, L = 0.1 implies 10% error so that the measured mean will be expected to lie within 10% of the true mean with the probability assumed in determining t. Since the mean value is often determined with a 95% certainty (p = 0.05), the t value is often 1.96 (or about 2).

The relationship between n, CV, and L is illustrated in fig. 3.5. If the CV is 0.65 (the characteristic CV for SPM in lakes), then 163 samples are required to establish a site-specific mean value provided that an error of L = 10% is acceptable. If a larger error is acceptable, e.g., L = 20%, fewer samples are required. If 4 samples are taken at a river site, the mean value may be estimated by an error of about 200%!

$$n = (1.96 \cdot CV/L)^2 + 1$$

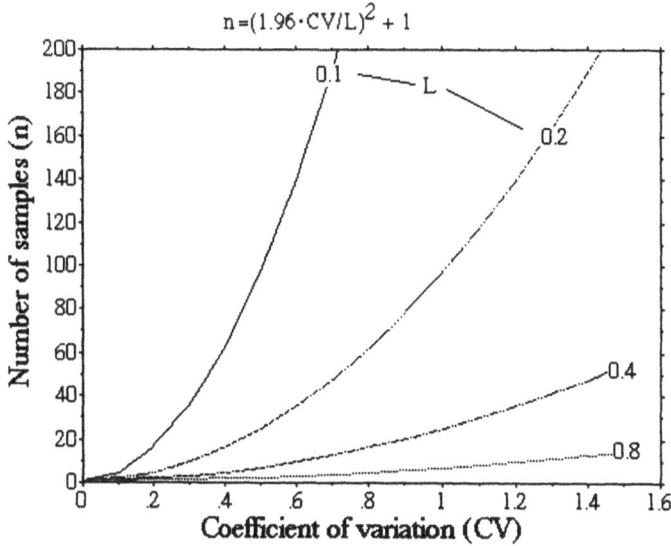

Fig. 3.5. Illustration of the sampling formula. The nomogram shows how many samples that must be analyzed (n) to establish a characteristic mean value (MV) with a given uncertainty or error (L) and a given confidence (95%).

This variability in SPM has profound implications for the modelling and understanding of SPM. An important question that will be addressed below is: How would CV depend on the length of the sampling period? Evidently, individual data from specific sites and sampling occasions will represent the characteristic/mean/median conditions very poorly when the CV value is high.

3.2.4. CV and sampling period

The data used in this example are measurements of SPM from River Avon (see appendix 9.1). Fig. 3.6A exemplifies all actual (n = 93) data for SPM from this river site and fig. 3.6B gives the same data represented as monthly values with uncertainty bands (the box-and-whisker plots that give median values, M_{50}, quartile values, M_{25} and M_{75}, and percentiles, M_{10} and M_{90}). An important question concerns the characteristic CV values based on daily, weekly, monthly, and annual samples for the variable, since these CV values regulate the predictive power (r_r^2; eq. 3.1) or the number of samples according to the sampling formula (eq. 3.2).

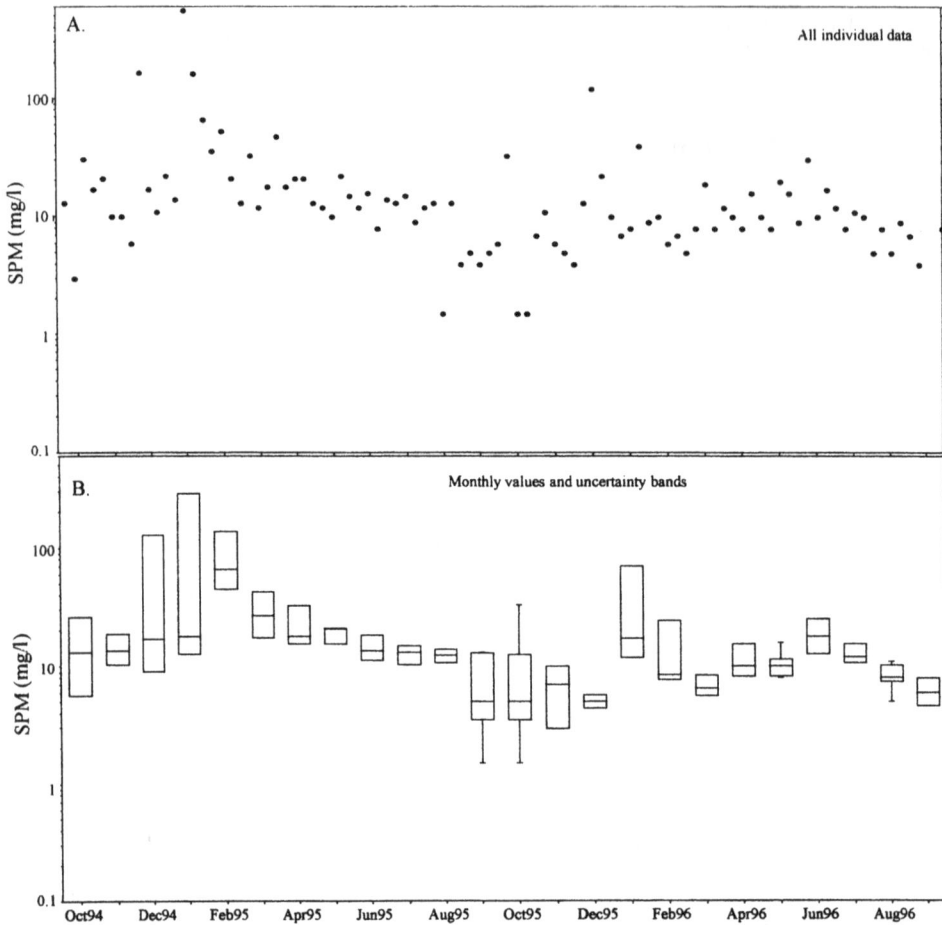

Fig. 3.6. A. All actual data (n = 92) from the sampling site in River Avon (Evesham station) for SPM 1994 to 1996. B. The same data but calculated as monthly values and the corresponding uncertainty bands (medians, M_{50}, quartiles, M_{25} and M_{75}, percentiles, M_{10} ands M_{90} and outliers).

Table 3.2 shows 90 individual SPM data from this river site for the period 1994 to 1996. This table is included here to stress some important concepts. The column marked "MV, n = 6" shows mean values of the actual data for n = 6. The CV for n = 6 is 1.43, which is lower than the CV for the actual data, CV = 2.55. The CV of 1.43 is significantly higher than the CV calculated when the actual data are randomly distributed in the given series; then the CV value is 1.00. It is a standard practice in many statistical contexts to approximate the CV for a mean value according to eq. 3.3, but then, as indicated in table 3.2, the mean value should come from a random sample from a normal frequency distribution. The point here is that there are often seasonal or long-term trends in water variables and when this is the case, this approach to reduce the CV cannot be used without due reservations:

$$CV_{MV} \approx CV_{ind}/\sqrt{n} \qquad\qquad (3.3)$$

Where CV_{MV} is the CV for the mean and CV_{ind} is the CV for the individual data; n is the number of data used to determine the mean value.

Table 3.2. Actual data on SPM concentrations (mg/l) River Avon (Evesham). Mean values (MV) from 6 samples, the actual data redistributed randomly, mean values of the redistributed data (n = 6). n = number of data, MV = mean value, SD = standard deviation, CV = coefficient of variation, and CV/$\sqrt{6}$.

Actual data	MV, n=6	Random order	MV, rand, n=6	Actual data	MV, n=6	Random order	MV, rand, n=6
13		15		1.5		12	
3		33		1.5		8	
31		16		7		11	
17		12		11		53	
21		13		6		11	
10	15.8	560	108.2	5	5.3	8	17.2
10		3		4		14	
6		67		13		12	
168		5		122		6	
17		4		22		33	
11		10		10		5	
22	39.0	48	22.8	7	29.7	1.5	11.9
14		4		8		10	
560		13		40		8	
165		18		9		16	
67		8		10		22	
36		17		6		13	
53	149.2	15	12.5	7	13.3	14	13.8
21		4		5		10	
13		10		8		13	
33		8		19		18	
12		122		8		168	
18		40		12		13	
48	24.2	31	35.8	10	10.3	1.5	37.3
18		10		8		17	
21		1.5		16		22	
21		12		10		5	
13		7		8		7	
12		7		20		20	
10	15.8	21	9.8	16	13.0	5	12.7
22		10		9		6	
15		12		31		11	
12		165		10		12	
16		10		17		36	
8		9		12		10	
14	14.5	8	35.7	8	14.5	17	15.3
13		6		11		21	
15		31		10		8	
9		6		5		9	
12		5		8		13	
13		19		5		21	
1.5	10.6	22	14.8	9	8	5	12.8
4		9					
5		9					
4		16					
5		10					
6		8					
33	9.5	21	12.2				

	Actual data	MV, n=6	Random order	MV, rand, n=6
n	90	15	90	15
MV	24.85	24.85	24.85	24.85
SD	63.29	35.50	63.29	24.94
CV	2.55	1.43	2.55	1.00
CV/$\sqrt{6}$	1.04			

Fig. 3.7A gives a compilation of CV values calculated for mean values from weekly, monthly and yearly values, as well as the CV value for all 459 individual data for chlorophyll-a in River Danube (Regensburg station; see Håkanson et al., 2003). The latter CV value is 0.96 and in fig. 3.7A, this value is compared to 19 similar values for chlorophyll for 19 sites in UK rivers (see appendix 9.1).

Fig. 3.7A illustrates:

- The mean CV for chlorophyll based on all data from the site in River Danube is close to the median value from 19 rivers sites in the UK.

- The daily, weekly, monthly, and yearly CVs increase in steps from 0.3, 0.35, 0.55 to 0.95, respectively for chlorophyll.

- Analogous results have been obtained also for green algae, diatoms, cryptophytes, and blue-greens (see Håkanson et al., 2003), and should apply for all water variables; fig. 3.7B gives data for SPM from River Avon.

3.3. Model testing and CV

Before it is meaningful to critically test a model, it should be calibrated. Calibration means that a given model set-up is tuned against empirical data so that the fit between modelled values and empirical data becomes as good as possible. All model variables in dynamic models (such as rates and distribution coefficients) could and should be tested. There are generally several combinations of values for the model variables that can give good predictions when calibrated against a set of empirical data from one system. All such combinations cannot be correct if one seeks the model constants that could be used as general default values in model simulations for many systems. This means that calibrations often involve many iterations. The idea with the calibration procedure is that for each round of iterations, the uncertainty in the values for the model variables should be reduced. When the model is properly calibrated, it should be validated, i.e., blind-tested against independent data. It is evident that it is preferable if the calibrations and the validations include reliable empirical data from as many systems as possible covering as wide a range as possible in model variable characteristics.

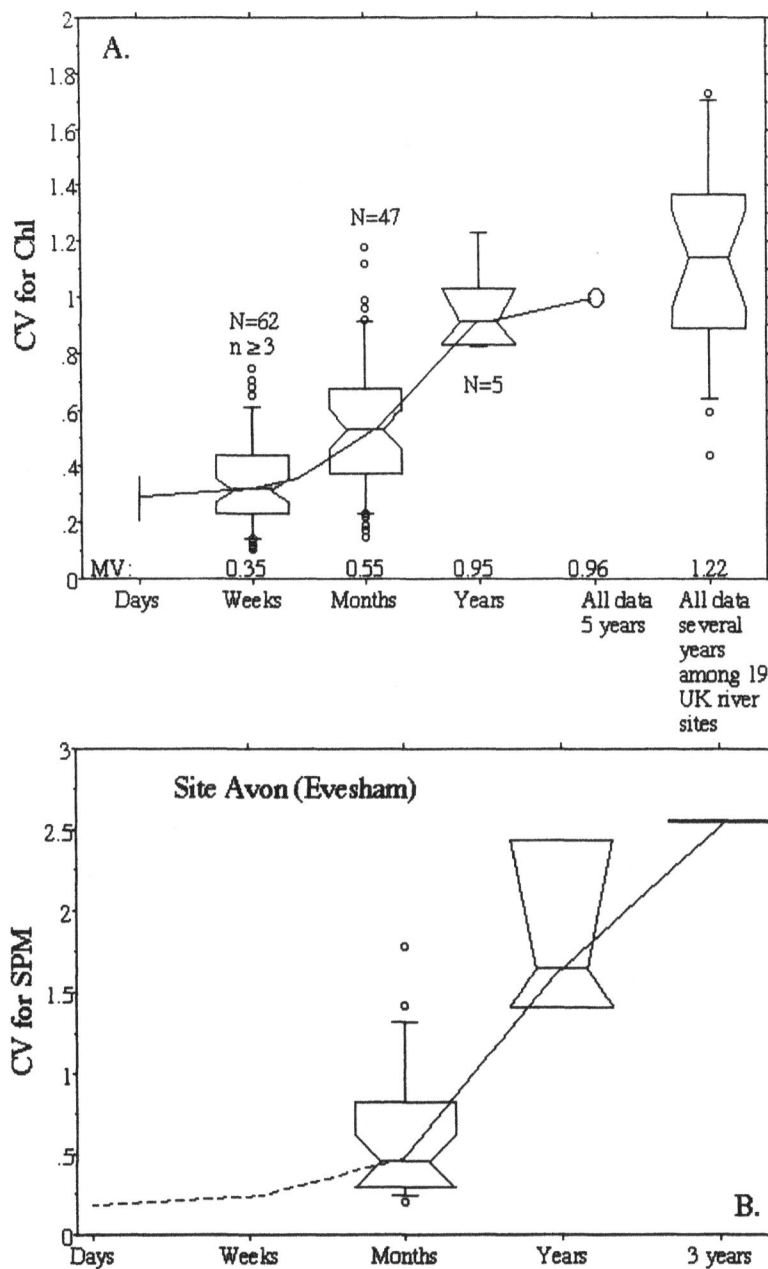

Fig. 3.7. Compilation of CV values based on mean values from different time intervals, daily, weekly, monthly (from individual months and monthly data from several years), yearly, and all data from 1995-1999 for chlorophyll from the River Danube (Regensburg station). The box-and-whisker plot to the right gives for comparative purposes CV values based on all data from several years from 19 UK rivers. The results for the UK river sites should be compared to the CV value of 0.96 calculated using all data from the Regensburg station (from Håkanson et al., 2003). B. Gives corresponding information for SPM in River Avon, UK.

3.3.1. Sensitivity tests

Sensitivity analysis involves the study of how an alteration of one model variable influences a given prediction, while everything else is kept constant. This analysis plays an important role in ecosystem modelling (Hinton, 1993; Hamby, 1995; IAEA, 1998). This section gives an example of how a sensitivity analysis can be performed and the following chapter will provide more examples.

Fig. 3.8 gives a sensitivity analysis where the sedimentation of SPM, as calculated by the dynamic coastal SPM model (see chapter 6), has been changed 100 times while all else in the model have been kept constant. The CV value has been set to 0.5 for sedimentation from surface water to ET areas (erosion and transport areas dominated by resuspension processes). From a normal frequency distribution with a mean value of 1 and a standard deviation of 0.5, 100 data have been drawn at random and used in the sensitivity test to produce the 100 curves for the target variable, modelled sedimentation on accumulation areas. It is evident from fig. 3.8, that the predictions of the y variable is sensitive to how sedimentation on ET areas (and hence also resuspension that depends on the amount of matter on ET areas) is calculated. The uncertainty in y is also shown by the box-and-whisker plot in fig. 3.8B. When steady-state conditions have been reached, the CV value for y is 0.32 related to the uncertainty in the given x variable (here sedimentation on ET areas).

The next step in a sensitivity analysis is often to repeat this type of calculation for all interesting model variables to try to produce a ranking of the factors influencing the target variable. The basic idea is to identify the most sensitive part of the model, i.e., the part that is most decisive for the model prediction. An example of such a comprehensive sensitivity analysis is given in fig. 3.9, again using the dynamic SPM model for coastal areas.

Fig. 3.9 gives the results when a uniform uncertainty (a CV of 0.5 and a normal frequency distribution around each mean value) has been used for all the fluxes in the coastal SPM model. The idea is to rank the importance of the various uncertainties for the predictions of sedimentation on A areas in this coastal area (Matvik, Sweden). From these presuppositions, the three most important uncertainties for this coastal area concern the SPM fluxes into and out of the surface water compartment (F_{outSW}, F_{inSW}, F_{SWET}, and F_{SWDW}). All other uncertainties in the fluxes are of less importance and the uncertainty in land uplift (F_{LU}) is of no significance since land uplift is zero in this region. It may be realistic to use a uniform uncertainty for fluxes, but not for the driving variables. In the next test, the aim is to see how characteristic uncertainties in the driving variables will influence the uncertainty in the y variable.

3.3.2. Uncertainty tests using Monte Carlo techniques

Two main approaches to uncertainty analysis exist, analytical methods (Cox and Baybutt, 1981; Beck and Van Straten, 1983; Worley, 1987) and statistical methods, like Monte Carlo techniques (Tiwari and Hobbie, 1976; Rose et al., 1989). In this section, Monte Carlo simulations will be discussed. Uncertainty tests using Monte Carlo techniques may be done in several ways, using uniform CV values, or more realistically, using characteristic CV values (e.g., from table 3.1). For practically useful predictive models based on several uncertain model variables (rates, etc.), the uncertainty in the prediction of the target variable (y) depends on such uncertainties. The cumulative uncertainty from many uncertain x variables may be calculated by Monte Carlo simulations.

Monte Carlo simulations is a technique to forecast the entire range of likely observations in a given situation; it can also give confidence limits to describe the likelihood of a given event. Uncertainty analysis (which is a term for this procedure) is the same as conducting sensitivity analysis for all given model variables at the same time. A typical uncertainty analysis is carried out in two steps. First, all the model variables are included with defined uncertainties and the resulting uncertainty for the target variable calculated. Then, the model variables are omitted from the analysis one at the time. The procedure is illustrated in fig. 3.10.

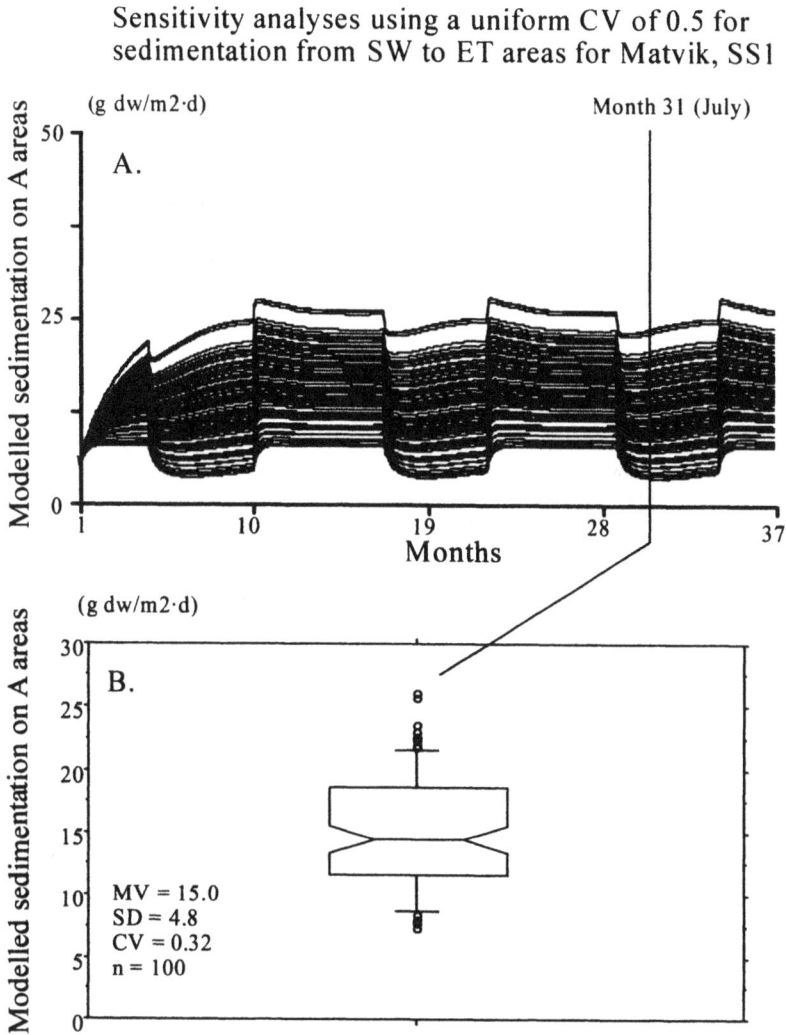

Fig. 3.8. A. Results from a sensitivity test (100 runs) using the dynamic coastal model for SPM (see chapter 6) and data for the coastal area of Matvik, Sweden. The uncertainty for one of the fluxes, sedimentation from surface water to ET areas, is defined by a CV of 0.5 and by a normal frequency distribution around the mean, and all else have been kept constant. The idea is to study how this uncertainty influences the uncertainty in the given target variable, sedimentation on accumulation areas. For the following tests, data from July (month 31) have been selected, as indicated in the figure.
B. Gives the corresponding box-and-whisker plot and statistics (CV for y = 0.32). Figure from Håkanson et al. (2004a).

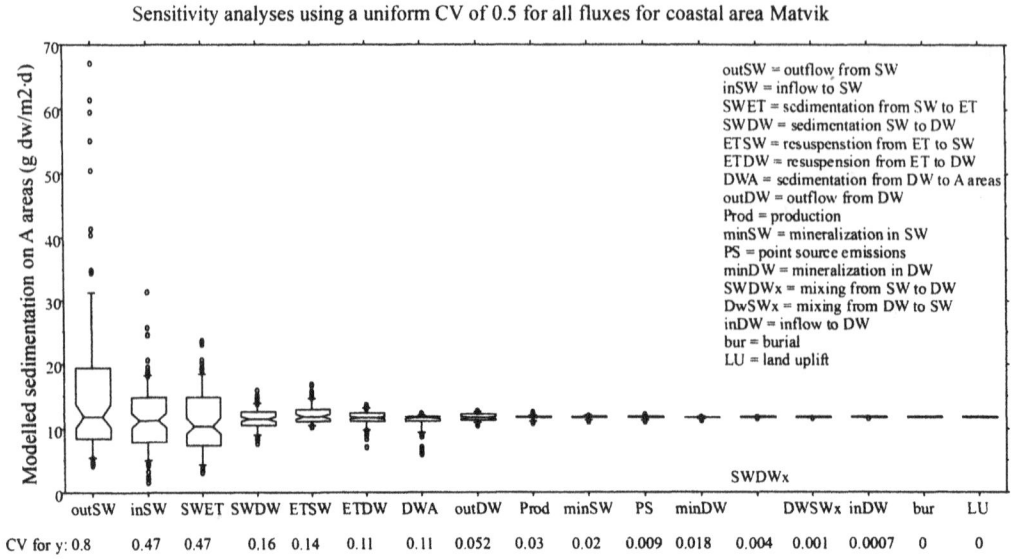

Fig. 3.9. Sensitivity tests where all fluxes in the coastal SPM model are accounted for, one by one, and all else kept constant. In this example, a uniform uncertainty has been used for all the fluxes (a CV of 0.5 and a normal frequency distribution around the mean value). The idea is to rank the importance of the fluxes in relation to the prediction of the target variable, sedimentation on A areas, under these presuppositions for this coastal area (Matvik, Sweden). The figure gives the box-and-whisker plots (median, 25 and 75 quartiles, 10 and 90 percentiles and outliers) as well as the CV for the y variable (e.g., 0.8 related to the uncertainty in the flux of SPM from surface water in July). From Håkanson et al. (2004a).

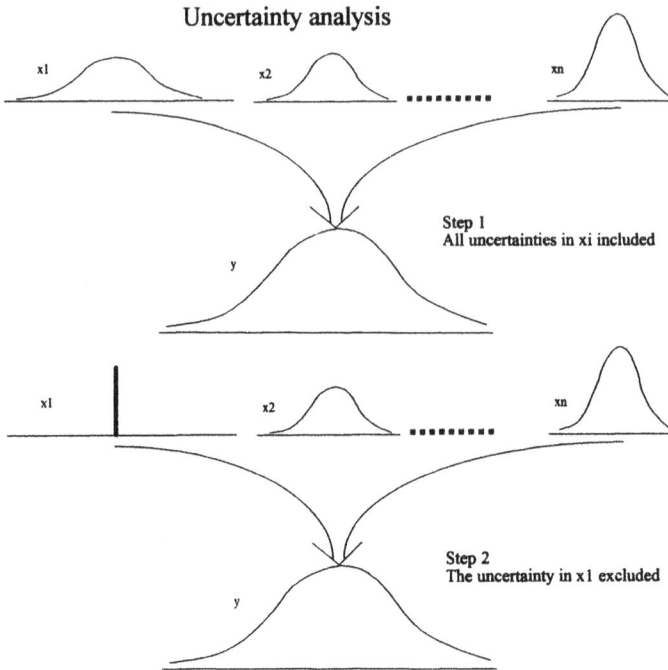

Fig. 3.10. Illustration of the principles of uncertainty analysis using Monte Carlo simulations.

There are major differences among model variables in inherent CV (table 3.1). Morphometric parameters can often be determined very accurately (Pilesjö et al., 1991), some model variables, like rates and distribution coefficients, can, on the other hand, not be empirically determined for real water systems, but have to be estimated from laboratory tests or theoretical derivations. This means that the values used for such model variables are often very uncertain. Table 3.3 gives a compilation of typical, characteristics CV values for different types of lake variables used in aquatic modelling. Note that in rivers, the CV for a given variable should be a factor of 1.5 to 2 higher than in lakes and coastal areas. From table 3.3, model variables like rates and distribution coefficients generally can be given CV values of 0.5. High CV values also appear for many sedimentological variables. In the following uncertainty tests, the CV values given in table 3.3 are used.

Table 3.3. Compilation of characteristic CV values for different types of lake variables (from Håkanson, 1999).

Lake variables	CV	Sedimentological variables	CV
Lake area (Area)	0.01	Percent ET areas (ET)	0.05
Mean depth (D_m)	0.01	Mean water content for E areas	0.30
Maximum depth (D_{max})	0.01	Mean water content for T areas	0.20
Volume (Vol)	0.01	Mean water content for A areas	0.05
Theoretical water retention time (T_w)	0.10	Mean bulk density for E areas	0.10
		Mean bulk density for T areas	0.10
Climatological variables		Mean bulk density for A areas	0.02
Annual runoff rates	0.10	Mean organic content for E areas	0.50
Annual precipitation	0.10	Mean organic content for T areas	0.50
Temperatures	0.20	Mean organic content for A areas	0.10
		Fall velocities	0.50
		Age of A sediments	0.50
		Age of ET sediments	0.50
		Diffusion rates	0.50

The tests have been done for the following variables:

1. The concentration of SPM in the surface water outside the coast calculated from empirical data on Secchi depth, Sec_{SeaSW}. This is, as the previous sensitivity analyses indicated, a very important variable. The characteristic CV for Sec_{SeaSW} is set to 0.25.

2. The concentration of SPM in the deep water outside the coast (Sec_{SeaDW}). This is also a rather uncertain value and the characteristic CV is also set to 0.25.

3. The value used for the ET areas (ET). This is an important distribution coefficient in mass-balance modelling and the ET value in these simulations is predicted by an empirical submodel. Those predictions are occasionally very uncertain. The characteristic CV is set to 0.5.

4. The theoretical surface water retention time (T_{SW}). This value is important because it regulates the fluxes of SPM to and from the sea and/or adjacent coastal areas. T_{SW} is calculated from an empirical model that has given an r^2 value of 0.95 (see chapter 6), but occasionally, the predicted T_{SW} value is uncertain and CV is set to 0.25.

5. The theoretical deep water retention time (T_{DW}). T_{DW} is also calculated from an empirical model, which gave an r^2 value of 0.79. CV is set to 0.35.

6. Land uplift (LU). It is evidently very difficult to set a reliable CV for the influence of land uplift. The CV for LU is likely very high and set to 0.5.

7. The chlorophyll-a concentration regulating primary production. The CV is set to 0.25 (Wallin et al., 1992)

8. Salinity (Sal). There are comparatively reliable data on the salinity and the CV is set to 0.15.

9. The feed conversion ratio (FCR) influencing the point source emissions of SPM from a fish cage farm in this coastal area. CV is set to 0.1.

10. Surface water temperature (SWT); SWT is predicted from a modified version of a well-tested model and the CV value is set to 0.1.

11. Finally, the coastal area. This value can be determined quite well but there may be uncertainties related to where the boundary line defining the coastal area is drawn. This CV is set to 0.05.

The aim now is to produce a ranking of these uncertainties for the target variable, sedimentation of SPM on A areas (F_{DWA}).

Also the following test uses data for coastal area Matvik, Sweden (see chapter 6). The results are given in fig. 3.11. Note that:

• The total calculated CV for F_{DWA} according to this testing procedure is 0.83. The empirical CV for sedimentation in sediment traps is 0.5 (Wallin et al., 1992). This means that the assumptions concerning the CV values for the model variables should be reasonable. The CVs for the model variables should be set according to the precautionary principles so that the calculated CV is not smaller than the empirical CV.

• The most important factor is the uncertainty associated with the value used for the ET areas. If this uncertainty is omitted, CV for F_{DWA} decreases the most, from 0.83 to 0.36. This means that future model development should concentrate on getting more reliable data and/or submodels for the ET areas.

• The model is not so well balanced since the model predictions depend much on one single uncertainty. In a balanced model, no part of the model dominates the calculated uncertainty in the target variables.

Uncertainty analyses using characteristic CVs for the driving variables for coastal area Matvik

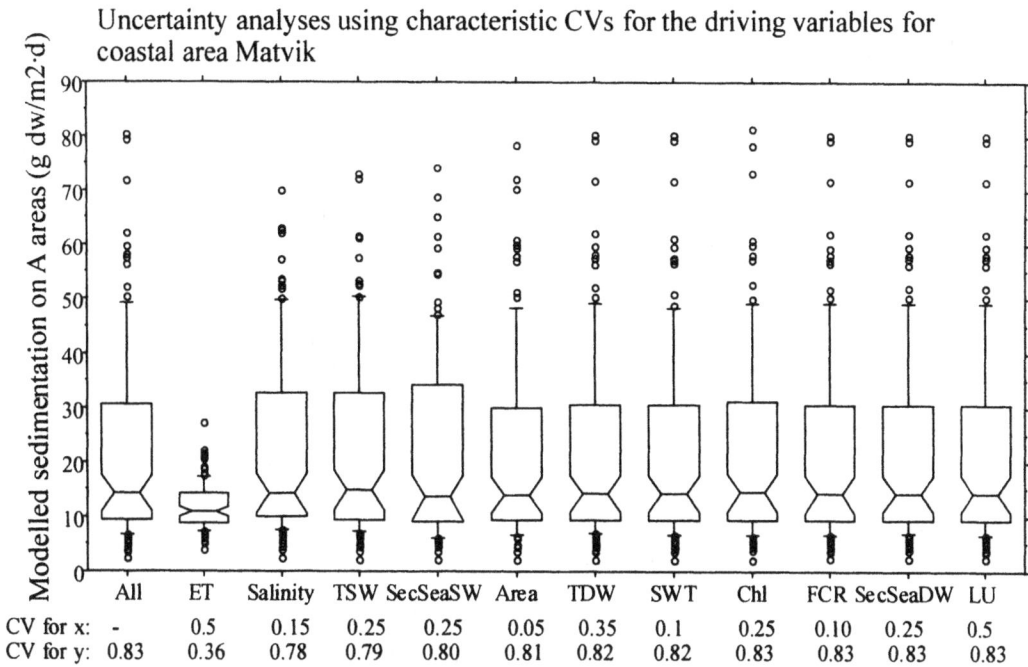

	All	ET	Salinity	TSW	SecSeaSW	Area	TDW	SWT	Chl	FCR	SecSeaDW	LU
CV for x:	-	0.5	0.15	0.25	0.25	0.05	0.35	0.1	0.25	0.10	0.25	0.5
CV for y:	0.83	0.36	0.78	0.79	0.80	0.81	0.82	0.82	0.83	0.83	0.83	0.83

Fig. 3.11. Uncertainty analyses using Monte Carlo techniques using data for area Matvik, Sweden, for the target variable, sedimentation on A areas. The figure gives characteristic CVs for the eleven selected driving variables (e.g., 0.5 for ET, 0.15 for salinity and 0.05 for coastal area), calculated CVs for sedimentation (e.g., 0.83 when all these uncertainties are accounted for at the same time and 0.36 when the uncertainty for ET is neglected and all other uncertainties are accounted for) and a ranking based on the calculated CVs of how uncertainties in these driving variables influence the uncertainty for sedimentation. From Håkanson et al. (2004a).

3.4. K_d and SPM

In aquatic ecosystems, substances in the water column can, as pointed out in chapter 2, be divided into two main parts, the dissolved phase and the particulate phase, relating to their fates and transport routes (pelagic versus benthic). The distribution (= partition = partitioning) coefficient of substances depends on the association to suspended particulate matter (SPM). Predictive mass-balance models should consider this particle association since it is a key factor regulating flows of substances in aquatic ecosystems. Particulate bound substances are, by definition, subject to gravitational sedimentation. Hence, they are to a high degree retained within the system and affect benthic habitats.

One well-known and general approach to describe the affinity of substances to carrier particles is by means of the partition coefficient, K_d (l/mg). K_d is generally defined as the ratio of filter-retained to filter-passing concentrations calculated as:

$$Kd = (C_{part}/SPM)/C_{diss}   (3.3)$$

where SPM is in mg/l, C_{diss} is the dissolved (filter-passing) concentration (mg/l) and C_{part} is the particulate concentration (mg/l).

In environmental investigations, total concentrations of trace substances are often found to increase with increased SPM (Cuthbert and Kalff, 1993; Balogh et al., 1997; Hurley et al., 1998a). For streams and rivers (Balogh et al., 1997; Kronvang et al., 1997; Solo-Gabriele and Perkins, 1997; Hurley et al., 1998a) and sediment resuspension (Slotton and Reuter, 1995), this is most often due to a marked increase in the particulate concentration of the given substance, C_{part}.

Contradictory to this, K_d is generally found to decrease with increasing SPM. For trace metals and hydrophobic organic pollutants, the relationship with decreasing K_d with increasing SPM is well-known (e.g., O'Connor and Connelly, 1980; Duursma and Bewers, 1986; Honeyman and Santschi, 1988; Benoit et al., 1994; Benoit and Rozan, 1999; Turner et al., 1999). It is sometimes called "the particle concentration effect" and several mechanistic explanations have been suggested, e.g., sorption to colloids, filtration artefacts, particle particle interactions, kinetics, qualitative variations in surface chemistry, irreversible adsorption, or incomplete desorption (Benoit, 1995). It should be noted that the particle concentration effect has been called 'strange' and the slope between $\log(K_d)$ and $\log(SPM)$ of -1, irrespective of substance, has been called 'striking results' (quoted from Morel and Gschwend, 1987). Slopes between $\log(K_d)$ and $\log(SPM)$ for trace metals found in environmental studies are generally in the range -1.0 to -0.5 (Honeyman and Santschi, 1988) and for hydrophobic organic pollutants in the range -1.5 to -0.42 (from the compilation by Turner et al., 1999). Johansson et al. (2001) have demonstrated that the same dependency may be explained by spurious correlations. Spurious self-correlation is, in fact, a mathematical inevitability due to the formulation of the K_d value as a ratio. One should generally be careful when a ratio is used in a regression versus one of the arguments of the ratio (Kenney, 1982; Jackson et al., 1990; Krambeck, 1995; Berges, 1997). The theory of such spurious correlations was developed by Pearson (1897) and Reed (1921); see also Kenney (1982). In the following, this will be examined and illustrated with a few tests.

According to Johansson et al. (2001) the particulate fraction (PF, dimensionless; $PF = C_{part}/(C_{part}+C_{diss})$ is a much better alternative to K_d in mass-balance models. There are three reasons for this conclusion:

1. PF is the variable that is actually necessary in mass-balance models since it, directly and not indirectly as K_d, describes and distributes mass flows of substances, which after all, is the goal of mass-balance models. To achieve the same mass flow distribution, the K_d value must be recalculated into a PF value.

2. As stressed, the CV values vary between different substances and influence the redictive power of models. Table 3.4 gives a compilation of mean values and CV values for many substances and different chemical forms of the substances (Ag, Cd, Co, Cr, Cu, Mn, Ni, P, Pb, and Zn) in many freshwater and marine systems. Since SPM is included in the calculations of K_d (eq. 3.3), but not of PF, the variability in SPM influences the CV for K_d. This is a major reason why CV for K_d is significantly larger than CV for PF, fig. 3.12 which gives frequency distributions of CV values for C_{part}, C_{diss}, SPM, PF and K_d using the data given in table 3.4. One can note that the CVs for K_d generally are much (2-3 times) higher than for PF. This means that in mass-balance models, PF is preferable to K_d to distinguish between the dissolved and the particulate phases due to lower within system variability.

3. Due to the definition of K_d as a ratio, spurious correlations may contribute to observed correlations between K_d values and environmental variables (Kenney, 1982; Jackson et al., 1990; Krambeck, 1995). This means that regression models to varying degrees may overestimate correlations with variables that are included or closely related to any of the variables included in the K_d ratio and this will be discussed in the next section.

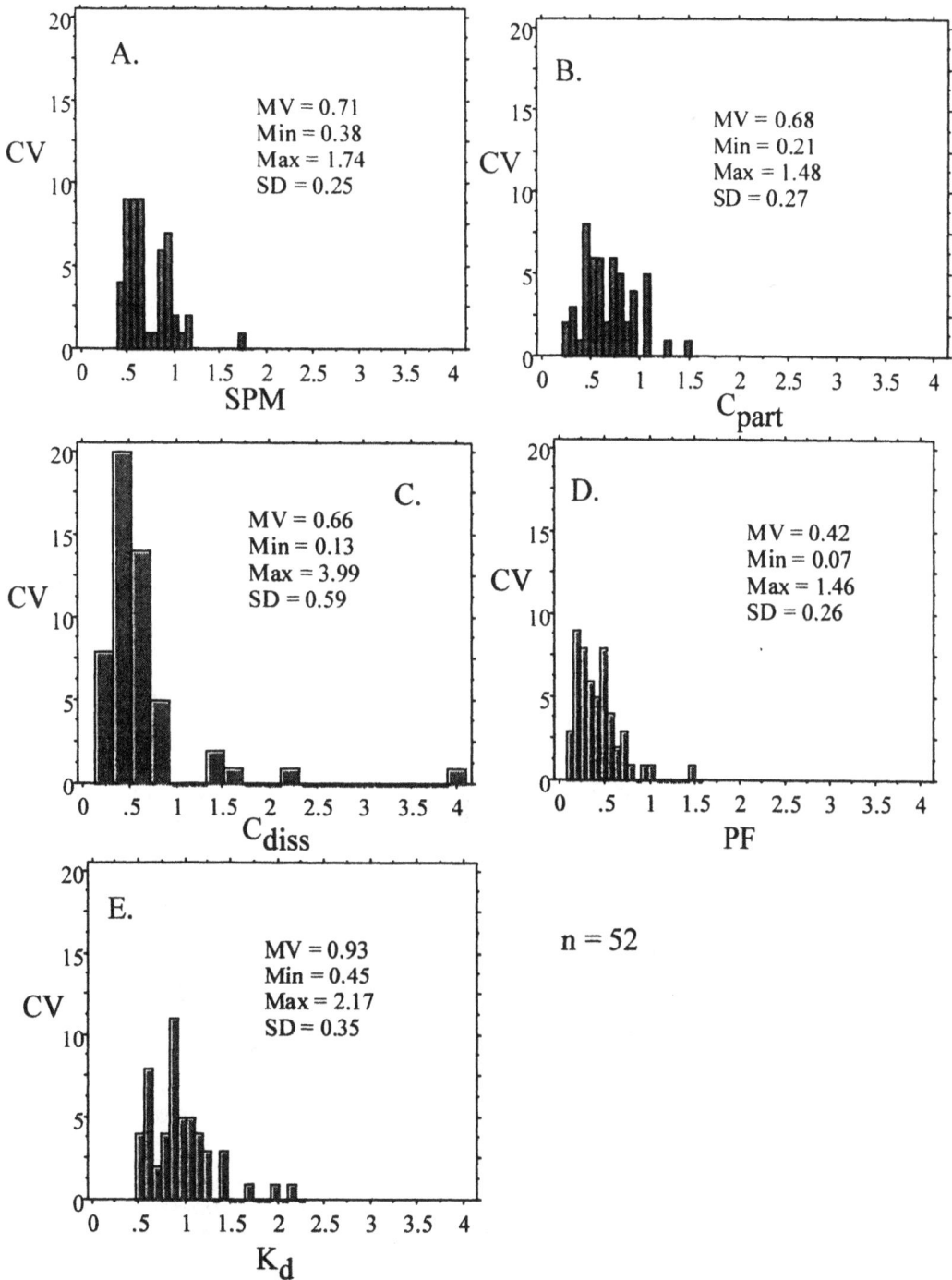

Fig. 3.12. Frequency distributions for SPM, the particulate concentration (C_{part}), the dissolved concentration (C_{diss}), the particulate fraction (PF), and K_d using the data (n = 52) for the substances given in table 3.4.

Table 3.4. A compilation of mean values and coefficients of variation (CV) for particulate concentrations, Cpart (†), dissolved concentrations, Cdiss (†), SPM (mg/l), the particulate fraction, PF (-), and Kd (l/kg·) for various substances, n = number of data. From Johansson et al. (2001).

Substance	n	Mean					CV					System	Ref*
		C_{part}	C_{diss}	SPM	PF	K_d (10^6)	C_{part}	C_{diss}	SPM	PF	K_d		
Ag^{VIII}	10	5.53	6.21	10.38	0.46	0.14	0.93	0.68	0.69	0.62	0.86	Sabine estuary	1
Ag^{VIII}	5	9.28	5.68	19.70	0.52	0.15	1.26	0.59	0.92	0.37	1.10	Galveston estuary	1
Ag^{VIII}	6	4.00	4.53	17.40	0.58	0.24	0.55	1.45	0.95	0.46	0.82	Colorado estuary	1
Ag^{VIII}	6	1.95	2.45	29.32	0.45	0.050	0.31	0.30	0.98	0.16	0.85	Lavaca estuary	1
Ag^{VIII}	8	5.86	4.58	90.73	0.54	0.019	0.57	0.49	0.38	0.40	0.80	San Antonio est	1
Ag^{VIII}	6	3.75	4.55	26.03	0.46	0.062	0.49	0.52	0.83	0.41	0.75	Corpus Christi est	1
Ag^{IX}	44	118	30.2	10.56	0.74	0.59	0.96	1.53	0.88	0.24	0.86	Quinnipiac River	2
Ag^{IX}	5	6.40	3.80	10.64	0.63	0.19	0.50	0.54	0.47	0.07	0.45	Tivoli South Bay	2
Cd^{V}	9	6.81	12.33	1.27	0.32	0.85	0.90	0.26	0.64	0.72	1.07	Lake Sammamish	3
Cd^{IX}	46	155	68.6	9.96	0.64	0.43	1.08	0.89	0.93	0.29	0.95	Quinnipiac River	2
Cd^{IX}	5	18.8	12.80	10.64	0.61	0.18	0.55	0.86	0.47	0.19	0.59	Tivoli South Bay	2
Cd^{I}	37	0.001	0.013	0.17	0.08	0.63	0.85	0.32	0.49	1.03	1.24	The Baltic Sea	4
Co^{IV}	34	0.27	0.49	1.26	0.61	3.70	0.54	2.21	0.60	0.46	1.45	Lake Sammamish	3
Cr^{II}	40	0.22	1.01	0.99	0.18	0.26	0.65	0.20	0.61	0.59	0.67	Lake Sammamish	3
Cu^{IV}	46	1.08	5.82	1.15	0.17	0.19	0.59	0.24	0.46	0.74	0.62	Lake Sammamish	3
Cu^{VII}	13	0.16	0.97	10.51	0.17	0.025	0.69	0.56	0.57	0.74	0.78	Sabine estuary	1
Cu^{VII}	16	0.24	0.94	12.96	0.21	0.034	0.74	0.33	0.91	0.63	0.93	Galveston estuary	1
Cu^{VII}	12	0.13	0.90	13.47	0.12	0.018	0.87	0.23	0.92	0.56	0.88	Colorado estuary	1
Cu^{VII}	13	0.14	0.79	19.58	0.16	0.022	0.81	0.52	1.07	0.77	1.24	Lavaca estuary	1
Cu^{VII}	14	0.21	1.08	66.23	0.16	0.003	0.71	0.40	0.59	0.52	0.50	San Antonio est	1
Cu^{VII}	13	0.23	0.86	24.55	0.23	0.015	0.41	0.49	0.56	0.40	0.56	Corpus Christi est	1
Cu^{I}	37	0.013	0.387	0.17	0.06	0.43	0.74	0.42	0.49	1.46	1.39	The Baltic Sea	4
Mn^{I}	18	8.41	67.6	0.15	0.53	20.8	1.48	3.99	0.67	0.51	1.19	The Baltic Sea	4
Ni^{IV}	40	0.52	4.88	1.53	0.09	0.09	0.93	0.25	0.62	0.92	1.14	Lake Sammamish	3
P^{I}	17	5.35	3.71	1.13	0.62	2.49	0.27	0.68	0.39	0.28	1.04	Lake Njupfatet	7
P^{I}	11	26.7	28.31	6.35	0.52	0.26	0.28	0.44	0.83	0.20	0.45	Lake Björkaren	7
P^{I}	19	47.5	28.32	12.02	0.60	0.22	0.71	0.49	1.00	0.19	0.68	Lake Kundby	7
P^{I}	27	7.33	5.19	1.07	0.58	2.71	0.45	0.56	0.51	0.35	1.17	Lake Siggefora	7
P^{I}	63	11.1	13.72	2.39	0.47	0.62	0.42	0.53	0.47	0.20	1.38	Lake Erken	7
P^{I}	18	7.39	12.22	1.71	0.37	0.49	0.33	0.13	0.82	0.18	0.57	Lake Örträsket	7
Pb^{II}	66	0.25	0.07	1.28	0.71	4.97	0.56	0.62	0.60	0.17	0.82	Lake Sammamish	5
Pb^{VII}	16	0.16	0.09	11.36	0.60	0.26	0.80	0.73	0.55	0.32	1.06	Sabine estuary	1
Pb^{VII}	16	0.20	0.07	12.93	0.65	0.32	0.81	0.41	0.91	0.32	1.01	Galveston est	1
Pb^{VII}	12	0.17	0.07	10.48	0.69	0.75	0.66	0.49	0.49	0.24	2.17	Colorado estuary	1
Pb^{VII}	13	0.11	0.07	18.62	0.56	0.13	1.07	0.46	1.14	0.30	0.82	Lavaca estuary	1
Pb^{VII}	15	0.15	0.10	63.31	0.62	0.047	0.42	0.69	0.62	0.25	0.85	San Antonio est	1
Pb^{VII}	13	0.15	0.08	24.55	0.55	0.10	1.10	0.63	0.56	0.49	0.88	Corpus Christi est	1
Pb^{IX}	51	1091	155.1	10.35	0.80	1.15	0.79	0.63	0.89	0.25	0.91	Quinnipiac River	2
Pb^{IX}	13	446	96.2	7.15	0.81	1.02	0.47	0.50	0.60	0.12	0.73	Tivoli South Bay	2
Pb^{IX}	21	17.7	12.1	0.48	0.60	11.77	0.57	0.91	1.74	0.32	0.90	Bear Brook	2
Pb^{I}	37	0.008	0.020	0.17	0.33	5.07	0.49	0.70	0.49	0.50	1.03	The Baltic Sea	4
^{210}Pb^{III}	61	2.46	0.68	1.43	0.75	6.08	1.04	1.33	0.59	0.26	1.11	Lake Sammamish	5
^{210}Pb^{VI}	20	3.68	1.89	0.68	0.65	3.69	0.42	0.47	0.43	0.21	0.58	Lake Crystal	6
^{210}Po^{III}	59	0.89	0.54	1.44	0.64	2.13	0.44	0.63	0.59	0.24	1.64	Lake Sammamish	5
^{210}Po^{VI}	20	3.07	1.08	0.68	0.75	5.56	0.21	0.42	0.43	0.09	0.48	Lake Crystal	6
Zn^{II}	8	1.06	3.35	1.25	0.26	0.51	0.46	0.36	0.78	0.56	0.99	Lake Sammamish	3
Zn^{VII}	14	0.66	1.16	11.75	0.34	0.055	0.80	0.43	0.55	0.49	0.60	Sabine estuary	1
Zn^{VII}	14	1.59	1.85	13.65	0.43	0.11	1.07	0.63	0.91	0.45	0.98	Galveston estuary	1
Zn^{VII}	13	0.78	0.98	13.46	0.45	0.098	0.44	0.38	0.88	0.35	0.57	Colorado estuary	1
Zn^{VII}	13	1.22	1.72	18.62	0.44	0.086	0.60	0.74	1.14	0.42	0.87	Lavaca estuary	1
Zn^{VII}	15	1.38	0.83	63.31	0.60	0.034	0.60	0.34	0.62	0.17	0.58	San Antonio est	1
Zn^{VII}	13	4.27	3.15	24.55	0.53	0.14	0.74	0.51	0.56	0.46	1.96	Corpus Christi est	1

† Concentrations in: I = µg/l; II = nM; III = dpm 100 per l; IV = nmol/kg; V = pmol/kg; VI = pCi/100l; VII = ppb; VIII = ppt; IX = ng/l.
* 1 = Data from Benoit et al., 1994; 2 = Data from Benoit, 1995; 3 = Data from Balistrieri et al., 1992; 4 = Data from Pohl and Hennings, 1999; 5 = Data from Balistrieri et al., 1995; 6 = Data from Talbot and Andren, 1984; 7 = Håkanson and Johansson, 2004.

3.4.1. Spurious correlations - K_d

There are two important areas involving SPM and spurious correlations, one concerns SPM and the distribution of substances into dissolved and particulate phases in the water; the other concerns river transport of SPM. These two cases will be addressed in the following section.

Spurious correlations is a fundamental statistical problem in situations where the y variable is a function of x. To illustrate this, a random parameter test has been conducted where 100 random numbers have been generated for several variables that have then been regressed pair-wise and as ratios in the same manner as is done for real data to calculate K_d. So, SPM_{rand} represents randomly generated data that will be used in the same manner as real data for SPM are used; random values corresponding to total concentrations ($C_{totrand}$) and particulate concentrations ($C_{partrand}$) have also been produced. From these randomly produced data, PF (here called PF_{rand}) and K_d (here called K_{drand}) are calculated. Fig. 3.13A illustrates the regression between $\log(PF_{rand})$ and $\log(SPM_{rand})$. The spread around the regression line is, as it should be, totally random ($r^2 = 0.018$). However, there is a statistically significant ($p < 0.0001$) relationship between $\log(PF_{rand})$ and $\log(C_{partrand})$; $r^2 = 0.50$ because by definition $PF_{rand} = C_{partrand}/C_{totrand}$. The same situation is evident for $\log(K_{drand})$ (see fig. 3.13D). There is a significant ($p < 0.0001$) relationship between the randomly produced values $\log(K_{drand})$ and $\log(C_{partrand})$ because K_{drand} is by definition equal to $(C_{partrand}/SPM_{rand})/(C_{totrand}-C_{partrand})$. By pure chance in this random parameter test, there is also a strong correlation between $\log(C_{partrand})$ and $\log(C_{totrand})$, $r^2 = 0.35$ (fig. 3.13F).

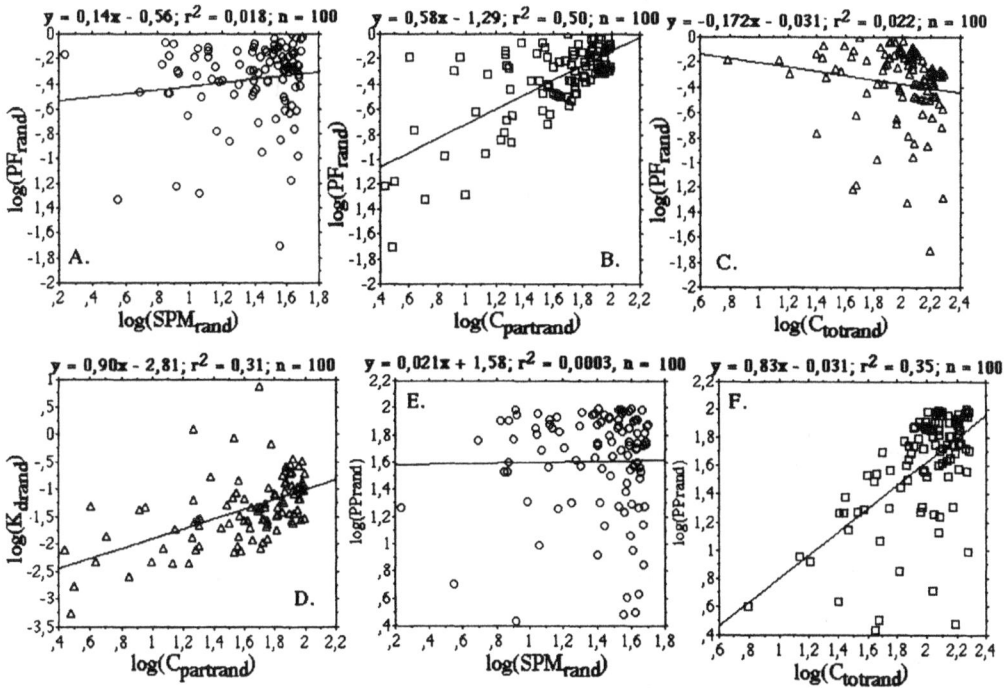

Fig. 3.13. Test to illustrate that randomly produced parameters can give significant correlations when ratios are being regressed against parameters used in the ratios. Note that all these variables on the y and x axes have been produced by a random number generator. The abbreviations are: SPM = suspended particulate matter, C_{part} = particulate concentration, C_{tot} = total concentration, PF = particulate fraction and K_d.

From these random parameter tests, it can be concluded that there are several statistical problems in predicting ratios like PF, and the x variables in regression models for PF should preferably not include C_{part} and C_{tot}, which by definition are used in defining PF.

3.4.2. River transport of SPM and spurious correlations

It is a common practice in physical geography and hydrology to correlate sediment transport (e.g., in g/s; i.e., Q·SPM) on the y axis with water discharge (Q in m^3/s) on the x axis. This is spurious and the following test will demonstrate why. First, a random number generator has been used to produce 1000 random data called Q_{rand} and 1000 random data called SPM_{rand}. The correlation between these data are shown in fig. 3.14C; the r^2 value is 0.001. If the sediment transport is given on the y axis as $log(Q_{rand}·SPM_{rand})$ and $log(Q_{rand})$ on the x axis, which is a "standard" practice in many contexts (see Jansson, 1982), there is a highly significant correlation ($r^2 = 0.49$, p < 0.0001), but it is spurious and a mathematical consequence of regressing a product against a component in the product. If 10 classes like the one shown in fig. 3.14A are used (to illustrate results from 10 different measurement stations), the results are shown in fig. 3.14B. The r^2 value is 0.973, a "perfect" correlation, but entirely spurious.

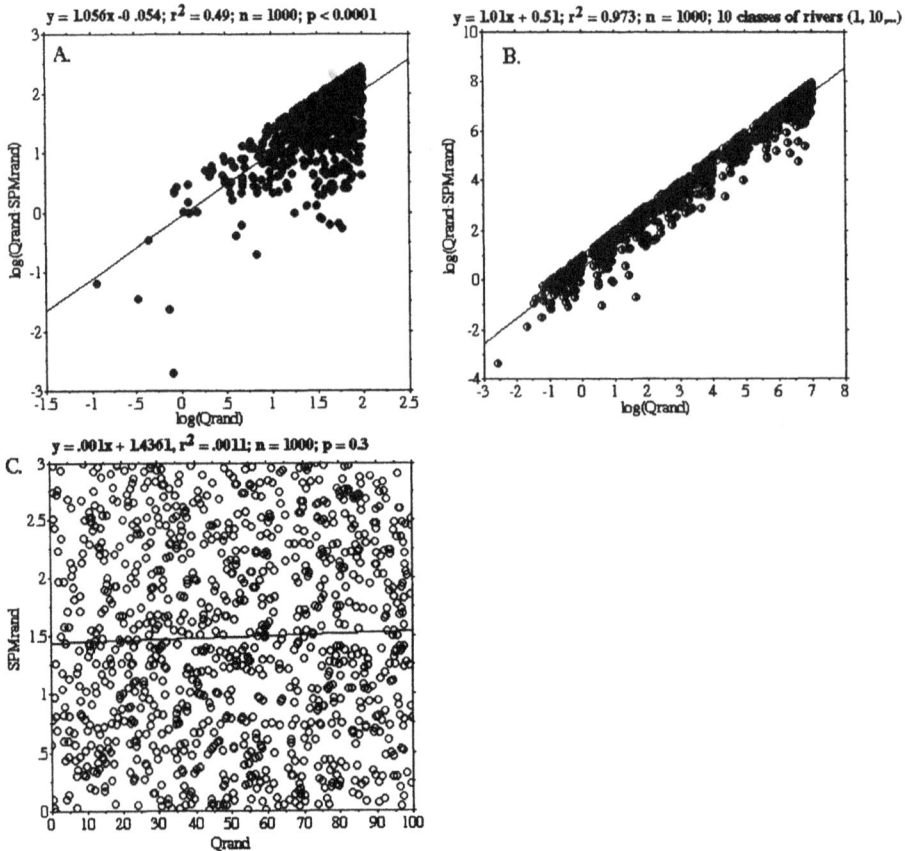

Fig. 3.14. Illustration of spurious correlations. 1000 data on "water discharge" (Q_{rand}) and "suspended particulate matter" (SPM_{rand}) have been generated and the correlation is shown in fig. C. Fig. A gives the regression between "sediment transport" ($Q_{rand}·SPM_{rand}$) and "water discharge" (Q) for the data in fig. C, and fig. B gives a regression when 10 datasets like the one in fig. A have been used.

4. SPM in Lakes

4.1. Introduction and aim

The basic aim of this chapter is to present statistical/empirical models and a dynamic model for SPM in lakes. Although SPM is an important variable in aquatic management and science, it is not often measured in regular monitoring programs. This section first discusses an empirical model yielding reliable estimations of SPM in lakes (the model has been presented by Lindström et al., 1999). This model is meant to be used in ecological models for target variables such as primary production or in models predicting concentrations of toxins in fish.

Fig. 2.3 gave some important principles concerning variability of water variables, such as SPM, within and among lakes. Fig. 2.3A illustrates how changing weather conditions create variations within a lake over a year, and/or for given time periods. This type of variability is not handled by the following empirical model for SPM. There must, however, exist factors that can explain the variability in mean SPM levels among lakes, factors which the following empirical model aims to identify, quantify, and rank. Due to the large within lake variability for SPM in lakes (characteristics CV = 0.65), it is not possible to derive a predictive model yielding higher r^2 than about 0.7-0.75 (table 3.1) for models based on individual data. Here, however, the aim is not to predict SPM in individual samples but rather to predict mean SPM values for entire lakes. So, as discussed, one should expect to get a higher r^2 value than 0.70 to 0.75 with the empirical model discussed here.

Fig. 4.1 exemplifies the temporal within lake variability for SPM for two subbasins of Lake Mälaren, Sweden. The within year variations are large (as well as the among year variability). The dynamic model for SPM discussed in section 4.3 is meant to explain also variations within lakes.

4.2. Empirical models for SPM

The empirical model discussed here for mean annual concentrations of SPM is, like all ecosystem models, only valid in a defined domain. This section will also discuss criteria defining the model domain.

4.2.1. Methods and data

Linear and stepwise multiple regressions have been used to quantify and rank how the tested x variables influence the variability in the target y variable. The structuring of the data and the statistical procedures are discussed in more detail by Håkanson and Peters (1995).

An extensive dataset from several investigations has been used (see appendix 9.1 and table 9.1). The lakes vary in size from 0.042 to 1147 km^2, in mean depths from 2.1 to 90 m, in theoretical lake water retention times from 0.0017 to 137 years, and in mean TP concentrations from 5 to 60 µg/l.

4.2.2. Statistical modelling

To obtain the best possible normal frequency distributions for the following regression analyses, the transformations recommended by Håkanson and Lindström (1997) have been used. Fig. 4.2 exemplifies the frequency distributions for the actual data for TP and SPM, as well as the logarithmic transformations. Both TP and SPM are positively skewed and the log transformation

SPM, site Görväln S, 10 m

Legend:
- 1992 (1.9)
- 1993 (1.8)
- 1994 (1.3)
- 1995 (2.4)

SPM, site Galten, 10 m

Legend:
- 1992 (15.3)
- 1993 (13.4)
- 1994 (14.8)
- 1995 (15.6)

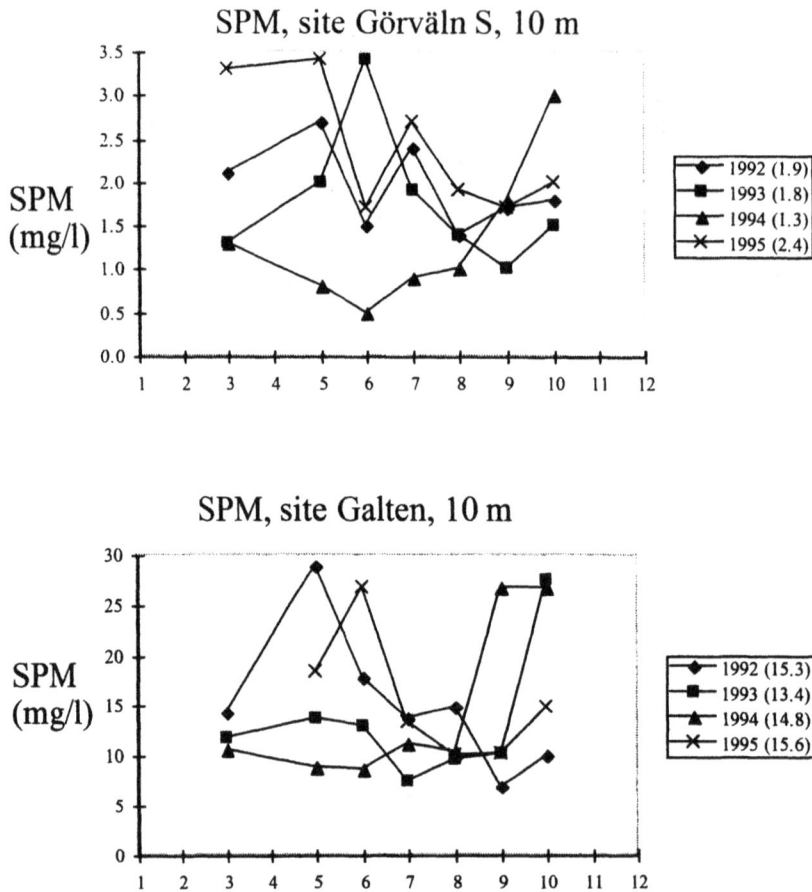

Fig. 4.1. Illustration of SPM variations at two stations in Lake Mälaren, Sweden, the Görväln site of 0-10 m water depth (upper figure) and the Galten site (lower figure). The within year variations depend on, e.g., water temperature, nutrient concentrations and light conditions, controlling lake bioproduction. There is also a difference in the average SPM level between the stations (yearly averages in brackets), which this empirical SPM model aims to quantify. From Lindström et al. (1999).

provides more normal distributions, which should be valid not just for the given sample of lakes but also for a much wider population of lakes (see Håkanson and Peters, 1995).

A correlation matrix, table 4.1, shows the linear correlation coefficients (r) between all the tested x variables (lake area, mean depth, D_m, maximum depth, D_{max}, theoretical lake water retention time, T, water discharge, Q, lake pH, the lake form factor, Vd, and the dynamic ratio, DR) assumed to influence the target y variable. It can be noted that log(TP) and log(DR) show the highest correlations to log(SPM), r values of 0.86 and 0.79, respectively (DR is the dynamic ratio = $\sqrt{Area/D_m}$).

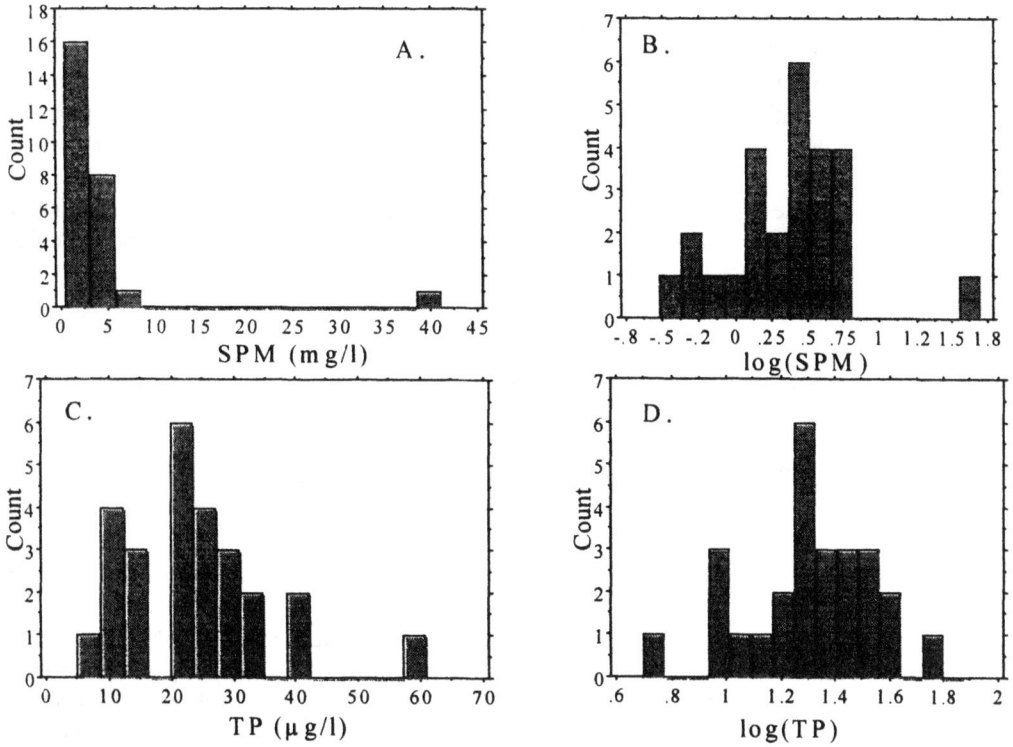

Fig. 4.2. Frequency distributions for SPM and log(SPM) and for TP and log(TP). The actual data are positively skewed for both SPM and TP; and the log transformations give more normal frequency distributions.

Table 4.1. Correlation matrix showing the pair-wise linear correlation coefficients, r, among the tested x variables and the target y variable, log(SPM). n = 26, r values > 0.39 show significant correlation at the 95% level (p = 0.05). From Lindström et al. (1999).

	log(SPM)	log(Area)	log(D_m)	log(D_{max})	log(TP)	log(T)	log(Q)	pH	Vd	log(DR)
log(SPM)	1.00									
log(Area)	0.35	1.00								
log(D_m)	-0.42	0.53	1.00							
log(D_{max})	-0.47	0.53	0.95	1.00						
log(TP)	0.86	0.23	-0.44	-0.43	1.00					
log(T)	-0.44	0.24	0.60	0.47	-0.64	1.00				
log(Q)	0.49	0.75	0.25	0.34	0.55	-0.43	1.00			
pH	0.43	0.49	0.19	0.03	0.16	0.26	0.23	1.00		
Vd	0.20	-0.10	-0.06	-0.37	0.02	0.31	-0.34	0.51	1.00	
log(DR)	0.79	0.60	-0.36	-0.31	0.67	-0.30	0.58	0.36	-0.05	1.00

Some of the x variables are functionally and statistically related, e.g., the mean depth (D_m) and the maximum depth (D_{max}), which correlate with r = 0.95 (table 4.1). Since "everything depends on everything else" in complex ecosystems such as lakes, it is essential to identify different cluster groups that show a minimum of interdependence and select representatives from such groups for the model variables. Table 4.1 is an aid to identify correlations among variables, and hence also clusters.

Stepwise multiple regression analysis has been used to examine the order in which the x variables influence the y variable, how much each step contributes to the r^2 value and the influence on the level of significance (expressed by the F value). The steps form a ladder to higher degrees of statistical explanation (table 4.2).

Table 4.2. Results of the stepwise multiple regression analysis. n = 26. From Lindström et al. (1999).

Step	r^2	F>	Model Variable	Model
1	0.74	67	TP	log(SPM) = 1.56·log(TP) - 1.69
2	0.83	11	pH	log(SPM) = 1.47·log(TP) + 0.18·pH - 2.85
3	0.87	6	DR	log(SPM) = 1.148·log(TP) + 0.137·pH + 0.286·log(DR) - 1.985

The first and most important variable to statistically explain the variability in mean SPM values among these lakes is the total phosphorus concentration (TP). This represents the bioproduction part of SPM and is quite logical (Vollenweider, 1968; Dillon and Rigler, 1974; OECD, 1982; Peters, 1986); $r^2 = 0.74$. Lake pH enters at step 2 with a positive sign and raises r^2 to 0.83. The correlation matrix (table 4.1) shows that pH is mostly correlated to log(Area) and then to log(SPM). pH adds information and predictive power by accounting for both allochthonous and autochthonous processes. pH affects aggregation of particles that affects sedimentation of SPM. pH is also positively related to bioproduction (Håkanson and Peters, 1995). Finally, the dynamic ratio (DR) raises the r^2 value to 0.87 with a positive correlation. This is mechanistically understandable since DR is designed to represent resuspension. Large and shallow lakes have high DR values and large sediment areas exposed to wind/wave energy (fig. 2.18).

Thus, the model is built on the x variables TP, pH and DR. According to the correlation matrix, the strongest dependency among these variables is between TP and DR (r = 0.67). This shows that internal loading/resuspension of TP influences water concentrations of TP. TP and DR will both be accepted in the model, because they represents two different processes: Production and resuspension.

Fig. 4.3 shows that the empirical SPM model is well balanced in the sense that the residuals are normally distributed.

The ranges of the model variables are given in table 4.3 and the model should not be used for lakes outside the domain defined by these ranges.

The two regressions in fig. 4.4 show the relationships between modelled and empirical actual data and log data, respectively. The uneven distribution of the SPM is evident from the regression in fig. 4.4A using actual data, where the extreme value for Lake IJsselmeer totally controls the slope of the regression line and the r^2 value.

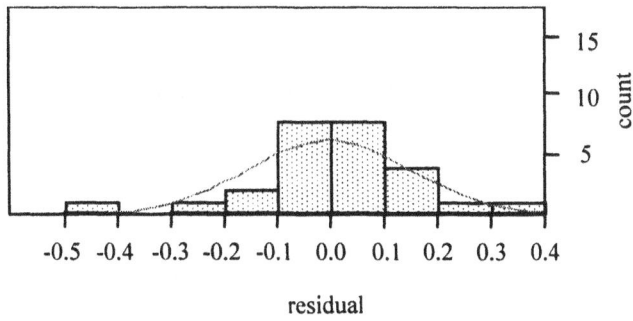

Fig. 4.3. Histogram of the residual distribution of the empirical SPM model, with a fitted normal curve for comparison.

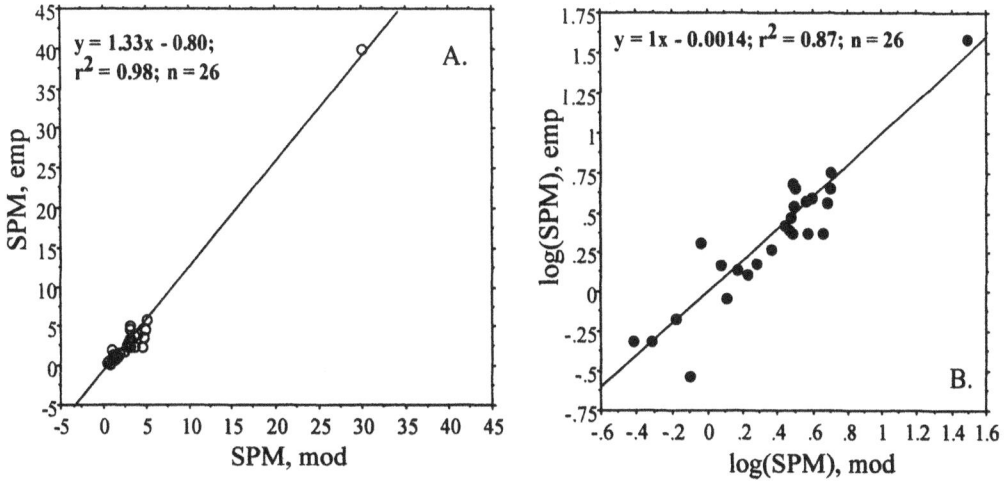

Fig. 4.4. Regressions showing the relationships between measured data and SPM values using the empirical model, (A) for actual data (note the high r^2 value related to one outlier, Lake IJsselmeere) and (B) for logarithmic data.

Table 4.3. The ranges of the model variables as criteria for the model domain.

	Unit	Minimum value	Maximum value
TP	[µg/l]	5	60
pH	[-]	5.10	8.50
DR	[-]	0.07	7.88

To conduct a stability test, 10 lakes were first randomly excluded, and the best model calculated with the remaining 16 lakes. The model variables and the slope coefficients may then be compared to the corresponding values in the original model and, if this is repeated, the stability of the model and its x variables can be evaluated (see Håkanson and Peters, 1995).

Table 4.4. Stability test of the empirical regression model for SPM. For each of the 15 cases, 10 lakes have randomly been excluded; data for 16 lakes have been used in the regressions. The first row shows the original coefficients of the model variables. For some combinations, new variables enter [here Vd and log(Q); Vd = the form factor, DR = the dynamic ratio, Q = the tributary water discharge]. From Lindström et al. (1999).

No.	\multicolumn{10}{c} 10 randomly excluded lakes.										r^2	log(TP)	pH	log(DR)	Vd	log(Q)
	1	2	3	4	5	6	7	8	9	10	0.87	1.15	0.14	0.29		
1	4	8	9	11	12	15	16	17	18	20	0.84	1.46		0.39		
2	1	4	8	9	12	14	17	19	23	25	0.91	1.42		0.40		
3	1	5	7	8	9	12	16	17	22	23	0.90	0.87		0.49		
4	2	8	9	14	17	18	19	22	23	24	0.85	1.00		0.46	0.26	
5	3	5	7	8	9	13	14	19	21	23	0.95	0.84		0.52	0.24	
6	4	5	7	11	12	14	16	22	23	25	0.90	0.69	0.18	0.36		
7	1	3	6	10	11	15	19	20	22	24	0.84	1.53	0.24			
8	1	3	8	9	10	11	12	15	17	24	0.91	1.06		0.49	0.24	
9	3	8	9	11	12	13	17	19	20	24	0.92	1.07		0.49	0.23	
10	1	5	8	11	15	17	20	21	23	25	0.97	1.14	0.21	0.40		-0.10
11	1	7	8	11	12	17	21	23	25	26	0.90	0.79		0.57	0.33	
12	1	3	8	9	10	12	14	17	20	23	0.86	1.60	0.23			
13	3	4	5	8	10	11	15	21	23	26	0.87	0.94		0.67		-0.11
14	2	3	8	9	10	19	20	21	23	24	0.89	0.92		0.42	0.28	
15	1	3	11	12	13	19	20	21	23	25	0.89	1.14	0.21	0.31		
										Min	0.84	0.69	0.18	0.31	0.23	-0.11
										Mean	0.89	1.10	0.21	0.46	0.26	-0.10
										Max	0.97	1.60	0.24	0.67	0.33	-0.10
										CV (%)	4.20	25.80	10.37	20.43	13.80	6.85

Table 4.4 shows the results of such a stability test. TP is the most stable model variable, followed by DR. The slope coefficient for TP varies around the value for the best model; the slope coefficient for DR is often higher than that of the best model. The volume development, Vd ($=3 \cdot D_m/D_{max}$), enters in 6 of the 15 tests as the third model variable. This is one time more than pH, the third variable in the best model. The range in which the slope coefficients vary gives a confidence interval for the given slope coefficient, which can be used in Monte Carlo simulations.

In a stability test, lakes may also be excluded not randomly but in a structured manner. Table 4.5 gives results of such a structured stability test, first when excluding the lake with the highest SPM concentration, Lake IJsselmeer. Test 2 shows the results based only on nonSwedish lakes, while test 3 shows the opposite. Test 4 controls if there is anything particular with the lakes in the River Kolbäcksån water system (Sweden) by excluding these; No. 5 is only based on these lakes. Tests 6 to 12 show the effects of extreme size (area and T), form (DR), acidity (pH), and phosphorus (TP).

Table 4.5. Structured stability tests of the empirical SPM model. The most extreme lakes from different aspects are excluded to systematically test the model stability to outliers. From Lindström et al. (1999).

No.	Excluded lakes.	r^2	log(TP)	pH	log(DR)	Area	Vd
		0.87	1.15	0.14	0.29		
1	Extreme SPM (Lake IJsselmeer)	0.77	1.34	0.14			
2	Swedish lakes, n=7 lakes	0.95	2.01				0.64
3	Non-Swedish lakes, n=19 lakes	0.80	0.67		0.43	-0.20	
4	"Kolbäcksån", n=12	0.85	0.92		0.55		
5	All but "Kolbäcksån", n=14	0.79	1.29				
6	3 largest, area (Ijssel, Bracc, Zürich)	0.74	1.28	0.15			
7	3 most acid (Iso V, Devoke, Örtäsk)	0.85	1.18	0.17	0.30		
8	4 least acid (Flats, Ijssel, Bracc, Devoke)	0.76	1.19		0.33		
9	3 largets, T (Iso V, Bracc, Erken)	0.87	1.15	0.22	0.28		
10	3 highest DR (IJssel, Åmmän, Virsbos)	0.78	1.33	0.14			
11	3 most TP (IJssel, Östers, Freden)	0.74	1.34	0.14			
12	3 least TP (Bracc, Devoke, Siggefora)	0.89	1.59		0.33		
	Min	0.74	0.67	0.14	0.28	-0.20	0.64
	Mean	0.82	1.23	0.17	0.38	-0.20	0.64
	Max	0.95	2.01	0.22	0.55	-0.20	0.64
	CV (%)	8.07	29.51	20.94	29.66		

From these tests, it can be concluded that the model is not sensitive to extreme values. It can also be noted from tables 4.4 and 4.5, that TP is an extremely stable model variable and that pH and DR are less stable model variables, but do add predictive power.

The plots in fig. 4.5 visualise the effects of varying two model variables in the range of the model, and keeping the third model variable fixed. The effects of pH and DR on SPM are small. The "synergetic" effect of the three variables, due to the logarithmic model transformations, is clearly seen, which underlines the dominance of TP.

4.2.3. Conclusions

The empirical lake model for SPM in table 4.2 is meant to be a practically useful tool to predict lake characteristic values of SPM in contexts of mass-balance modelling of water pollutants. The degree of statistical explanation is $r^2 = 0.87$, which is close to the highest possible for a predictive lake model for SPM and it is high enough to give useful predictions in individual lakes ($r^2 \geq 0.75$, see Praire, 1996). The domain of the model is wide and the model is stable due to the relative independence between the x variables and their wide ranges. However, it is a statistical (static) model which does not give information on fluxes. The next section will discuss a dynamic lake model for SPM.

The most important model variable (explaining 74% of the mean SPM variability among these lakes) is the TP concentration, indicating that the autochthonous production is the most important process determining SPM in lakes. Lake pH (i.e., a collective variable for aggregation and allochthonous and autochthonous processes influencing SPM) and the dynamic ratio (i.e., resuspension) both add significantly to the predictive power of the model. The unexplained residual, 13%, may be related to analytical and sampling uncertainty in the empirical data.

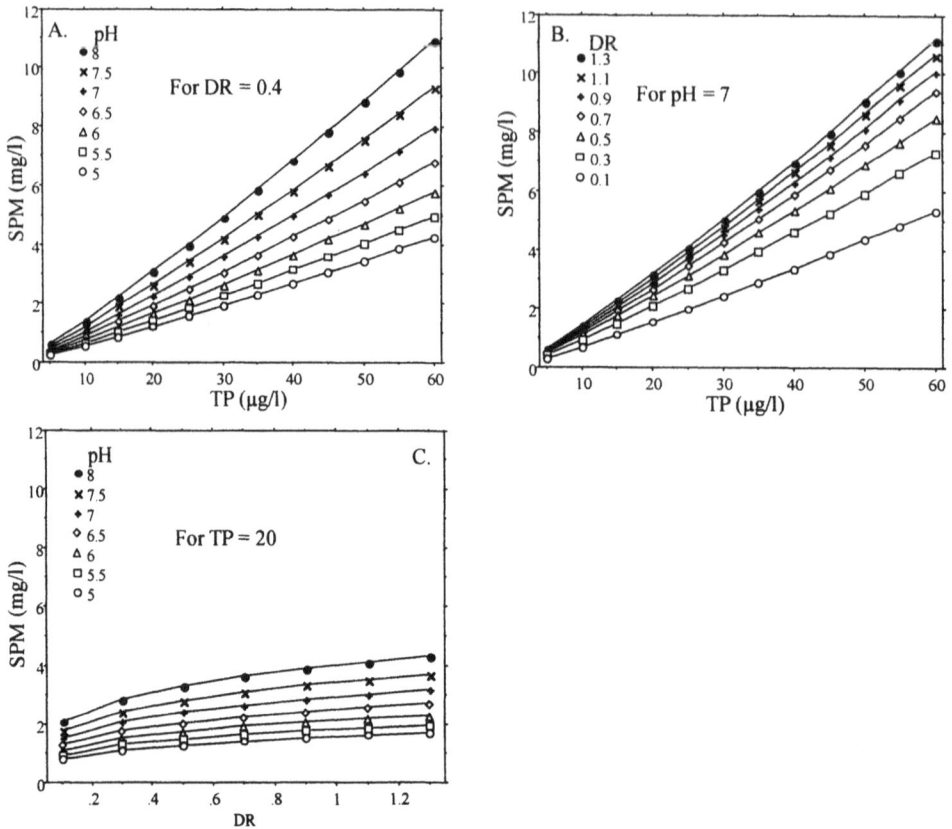

Fig. 4.5. Three-dimensional diagrams illustrating how SPM is related to the three model variables (TP, pH and DR).

4.3. Dynamic modelling of SPM in lakes

SPM concentrations have been dynamically modelled at specific sites using wind speed and current velocity as driving variables (e.g., Aalderink et al., 1984; Luettich et al., 1990; Somlyódy and Koncsos, 1991). This section will focus on a general dynamic model for predicting mean SPM concentrations in lakes but not for individual sites and not using data on winds and/or currents. The aim is instead to model on a monthly basis from few and readily accessible driving variables.

4.3.1. Background and aim

It is not possible to understand how lakes function without a basic knowledge of the transport processes that exist in all aquatic systems and apply to all substances. Fig. 4.6 gives a compilation of such processes. The aim of this section is to present and motivate equations for these transport processes for SPM in lakes. It should be stressed that these processes have the same names for all lakes, rivers, and marine areas and for all substances, so sedimentation is the flux from water to sediments, resuspension is the advective flux from sediments back to water, primary production creates living suspended particles in the water, mixing appears in stratified systems and is the transport between deep water and surface water, mineralization is the bacterial decomposition of organic matter, burial is the transport of matter from the lake biosphere to the geosphere often of substances from the technosphere, and outflow is the flux out of the system of water and everything dissolved and suspended in water.

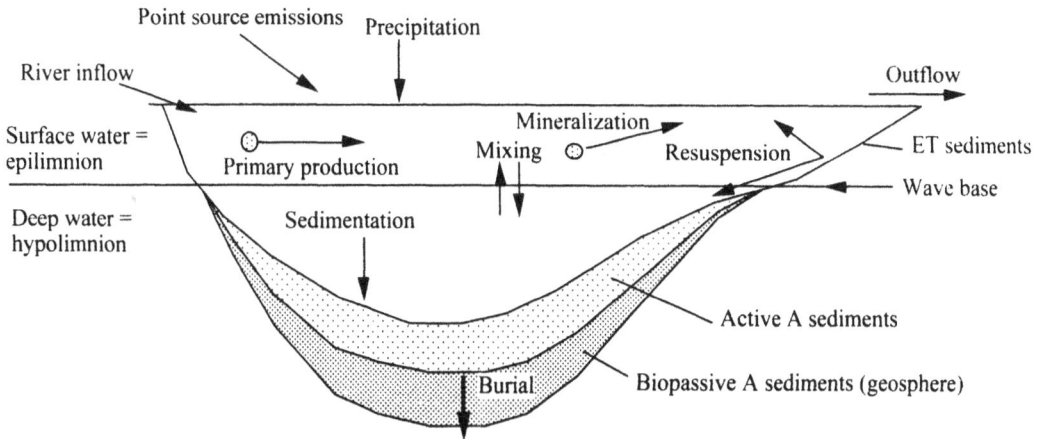

Fig. 4.6. Illustration of general transport processes of SPM to, within and from lakes.

Mass-balance models are since long fundamental tools in aquatic sciences to gain knowledge about processes and fluxes, in water management to identify sources of contamination and relate natural fluxes to anthropogenic fluxes, and hence also to get realistic expectations of remedial measures. The approaches presented by Vollenweider (1968, 1976, 1990) have been followed by many others (Chapra and Reckhow, 1979, 1983; Chapra, 1980; OECD, 1982; Håkanson, 1999). The mass-balance approach to eutrophication has played a paramount role in lake management.

During the last 10 years, there has been something of a "revolution" in predictive aquatic ecosystem modelling. The major reason for this development is the Chernobyl accident. To follow the pulse of radionuclides through ecosystem pathways has meant that important transport routes have been revealed and the algorithms to quantify them developed and tested (Håkanson, 2000). It is important to stress that many of those structures and equations are valid not just for radionuclides, but for most types of contaminants, e.g., for metals, nutrients and organics - and for SPM - in most types of aquatic environments.

The predictive power of any mass-balance model for any given substance depends on the structure of the model and the numerical values and equations applied in the model. Generally, there are many ways to structure mass-balance models depending on the target y variable to be predicted and the scale of the modelling, i.e., if calculations are done on a short-term basis (minutes to days) or long-term basis (years to centuries) or if site-typical information (using partial differential equations) or area typical (whole lake) approaches are being used (ordinary differential equations). This section addresses an SPM model based on monthly calculations to obtain seasonal variations. It uses ordinary differential equations. That is, the model is based on the ecosystem scale, a scale of fundamental importance in water management.

Since this dynamic model mainly concerns SPM in the water, the particulate fraction, PF, is set to one and diffusion from accumulation area sediments has been omitted from the general mass-balance model (LakeMab; see Håkanson, 2000). There are several simplifications of that kind in this model concerning:

- Winds and waves. Since this is modelling per month, the basic assumption is that winds can blow from all directions and with a variety of speeds, so the algorithm for wind influences (see Malmaeus and Håkanson, 2003) has been omitted to gain simplicity.

- Light. The primary production is higher on a light day than on a cloudy day but since this is SPM modelling on a monthly basis, the algorithm for light has also been omitted from the lake production submodel, and for the same reason as just mentioned for winds. On a monthly basis there is generally a marked covariation between temperature and light and the submodel for production includes variations in water temperatures.

- Burial, bioturbation and compaction. These processes depend on sedimentation and sedimentation depends on SPM concentrations in water, so there is a close link between SPM in water and burial. But burial, bioturbation and sediment compaction generally influence SPM concentrations in water only marginally or not at all, so to gain simplicity, the submodel for burial (see Håkanson, 2003b) is not included in the following model.

This dynamic model for SPM is meant to apply for lakes, rivers, and coastal areas. It is based on the previous SPM models for coastal areas (Håkanson et al., 2004a) and for lakes (Malmaeus and Håkanson, 2003), but there are some new calculation routines, which will be described in the following text. The new parts mainly concern the algorithms for sedimentation, mixing, and mineralization.

4.3.2. Lake modelling – key concepts

4.3.2.1. Basic structure

The model is meant to account for all important processes regulating the transport of SPM to, within, and from a given system - no more, no less. "Everything should be as simple as possible, but not simpler," according to Albert Einstein, and this modelling approach tries to follow that advice.

Fig. 4.7 gives the basic structure of the dynamic model, which will be elaborated in more detail in the following section. There are some key structural components for all models of the type given in fig. 4.7. When there is a partitioning of a flow from one compartment to two or more compartments, this is handled by a distribution (= partitioning = partition) coefficient. This could be a default value, a value derived from a simple equation or from an extensive submodel. There are three such distribution coefficients (DCs) in this model:

1. The DC regulating sedimentation either to areas of erosion and transport (ET areas) above the wave base (= the critical depth, D_{crit}) or to the deep water areas beneath the wave base.

2. The DC describing resuspension flux from ET areas back either to the surface water compartment water or to the deep water areas.

3. The DC describing how much of the suspended particles in the water that has been resuspended and how much that has never been deposited and resuspended.

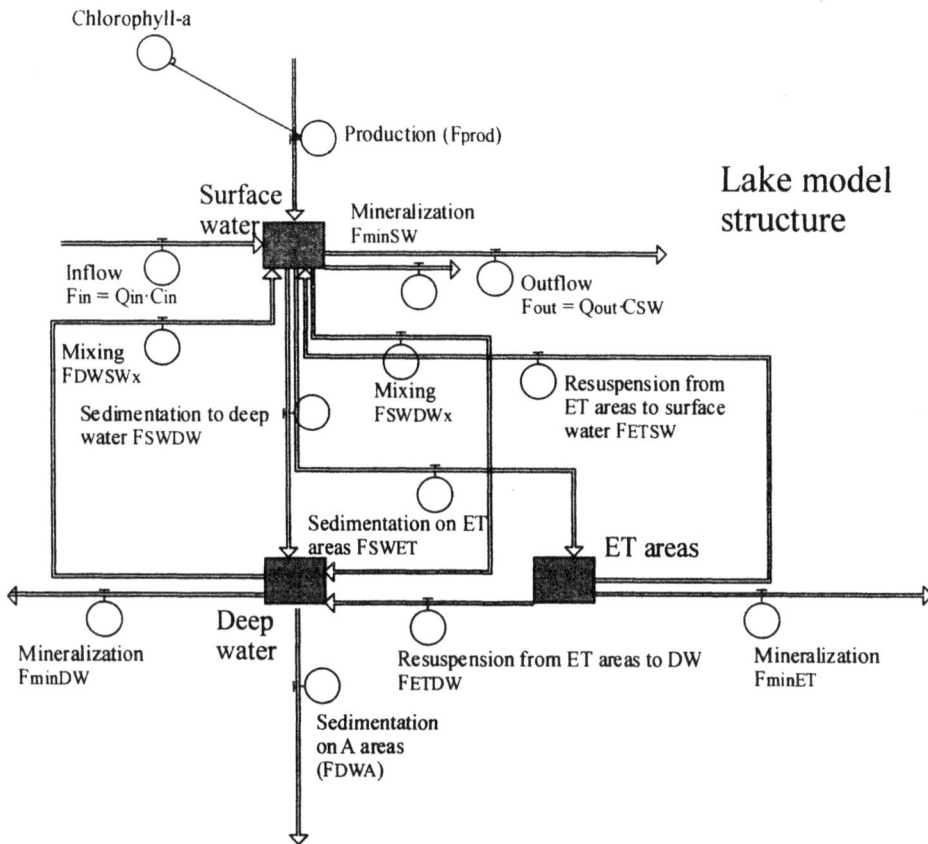

Fig. 4.7. Illustration of the basic structure of the mass-balance model (LakeMab) for SPM, which includes three compartments, the surface water (SW) and the deep water (DW), areas of fine sediment erosion plus transport (ET areas) and a flux to accumulation areas where fine sediments are continuously being deposited (A areas).

There are three compartments: (i) surface water (SW), (ii) deep water (DW) and (iii) areas where processes of erosion and transport dominate the bottom dynamic conditions (ET areas). Also note the abbreviations (table 4.6 gives a compilation of the nomenclature used throughout this book): F for fluxes (g/month), C for concentrations (g/m^3), R for rates (1/month), M for masses (= amounts in g). Fluxes from one compartment to another are denoted, e.g., F_{SWDW} for sedimentation from surface water (SW) to deep water (DW). The resuspended matter can be transported either back to the surface water (F_{ETSW}) or to the deep water (F_{ETDW}). How much that will go in either direction is regulated by a distribution coefficient calculated from the form of the lake. In the following, submodels will be presented concerning:

1. Volumes, i.e., surface water and deep water volumes.

2. Inflow, i.e., the allochthonous (tributary) inflow of SPM to a lake (F_{in}). For simplicity, this modelling does not include SPM from atmospheric input, which is often relatively small.

3. Lake production (F_{prod}) of living and dead suspended particles from both primary (phytoplankton, benthic algae and macrophytes) and secondary sources (e.g., zooplankton and faeces from fish) and bacterioplankton.

Table 4.6.   Abbreviations and dimensions of variables used in the dynamic SPM model for lakes.

Morphometric variables

Size measures

D = Depth (m);

D_A = Mean depth of A areas (m)

D_{crit} = Critical depth (m)
D_{ET} = Mean depth of ET areas (m)
D_m = Mean depth (m)
D_{max} = Maximum depth (m)
D_{SW} = Surface water depth (m)
A = Lake area (m²)
A_{DA} = Drainage area (km²)
V = Lake volume (m³)
V_{SW} = Surface water volume (m³)
V_{DW} = Deep water volume (m³)

Form parameters

Vd = Volume development, ($3 \cdot D_m/D_{max}$, dim.less.)
DR = Dynamic ratio ($DR=\sqrt{A}/D_m$)

Other parameters

C_{TP} = Concentration of TP (µg/l)
DC_{DW}= Resuspended fraction of SPM in deep water (dim.less)
DC_{rs} = Resuspended fraction of SPM in surface water SPM (dim.less)
ET   = Fraction of ET areas (dim.less)
DWT= Deep water temperature (°C)
Lat = Latitude (°N)
Prec = Annual precipitation (mm)
Q_{in} = River discharge (m³/month)
Q_{yr} = Mean yearly discharge (m³/yr)
Sec = Secchi depth (m)
SPM = Concentration of SPM (mg/l)
SWT = surface water temperature (°C)
T_{ET} = Age of ET sediments (months)
T_{Sw} = Theoretical surface water retention time (y)
v   = Fall velocity for SPM (m/month)
Y_Q = Moderator for river discharge
Y_{SPM} = Moderator for SPM influences on v
Y_{sal} = Moderator for salinity influences on v

Mass-balance for substances

M   = Mass (= amount, g)
M_{SW} = Mass in surface water (g)
F   = Flow (= flux, g/month)
F_{SWDW}=Sedimentation from surface water to deep water
R   = Rate (1/month)

Compartments

DW = Deep water (= hypolimnetic water)
SW = Surface water (= epilimnetic water)
ET   = Areas of fine sediment erosion and transport

4. Sedimentation. The transport from surface water to deep water (F_{SWDW}), from deep water to A areas (F_{DWA}), and from surface water to ET areas (F_{SWET}).

5. Internal loading from the ET areas (F_{ETSW} and F_{ETDW}). The resuspension (or advection) rate is given by the age of the ET sediments (T_{ET}). This retention time also includes a consideration to the fact that the substance can be entrapped by macrophytes (Håkanson and Boulion, 2002). Diffusion is omitted in this modelling for SPM.

6. Mineralization (F_{min}), i.e., the bacterial decomposition of organic matter in the surface water (F_{minSW}), the deep water (F_{minDW}) or in ET sediments (F_{minET}), but not in A area sediments since this would not affect SPM concentrations in water.

7. Mixing, i.e., the transport from deep water to surface water (F_{DWSWx}) or from surface water to deep water (F_{SWDWx}).

8. Outflow (F_{out}), i.e., the "size of the exit gate" from the system.

This modelling uses the definitions given for the two sediment compartments, ET areas and A areas in chapter 2 (see, e.g., fig. 2.18). The approach to define the two water compartments, surface water (SW) and deep water (DW), will be discussed in the next section.

4.3.2.2. Stratification and mixing

The water depth that differentiates the surface water (epilimnetic) part of the lake from the deep water (hypolimnetic) is called the "critical" water depth (= the wave base), D_{crit}. It could, potentially, be related to (i) water temperature conditions and the thermocline, (ii) vertical concentration gradients of dissolved or suspended particles, (iii) wind/wave influences and wave characteristics and (iv) sedimentological conditions associated with resuspension and internal loading. Evidently, there are several ways to characterize vertical heterogeneities in lakes. From a mass-balance perspective, it is necessary to work with clearly defined depths, volumes and substance concentrations. This section will briefly discuss benefits and drawbacks with the four mentioned approaches and gives a general method, which is also used in this dynamic model. If the critical water depth is defined in a relevant way from a mass-balance perspective, this will also open possibilities to calculate outflow in a simple, operational, and mechanistic manner since the outflow is given by:

$$F_{out} = Q_{out} \cdot C_{SW} \qquad (4.1)$$

where

F_{out} = Outflow of SPM in g per month.
Q_{out} = Water transport out of the lake in m^3 per month (see 4.3.3.4.)
C_{SW} = The concentration of the SPM in the surface water in g/m^3 or mg/l.

The concentration of SPM in the surface water, C_{SW}, is defined from the following ratio:

$$C_{SW} = M_{SW}/V_{SW} \qquad (4.2)$$

where

M_{SW} = The mass or amount of SPM in the surface water compartment (g).
V_{SW} = The volume of the surface water compartment (m^3).

The volume of the surface water compartment, in turn, is a function of the lake area (Area in m^2), the critical depth separating surface water from deep water (D_{crit} in m) and the form of the lake (as this can be given by the form factor = the volume development = Vd; see later).

The definition of the critical water depth is also important in quantifying sedimentation, because sedimentation of SPM from surface water to the deep water, and from deep water to the A area, is related to the settling velocity of the suspended particles (v in m/month) and the mean depth of the two respective compartments, D_{SW} and D_{DW}. Basically, sedimentation (F_{SWDW}, i.e., sedimentation from surface water to deep water) is given by:

$$F_{SWDW} = M_{SW} \cdot (v/D_{SW}) \cdot PF \qquad (4.3)$$

Where PF is the particulate fraction, the fraction subject to gravitational sedimentation. PF = 1 for SPM. The average depth of the surface water compartment is given by:

$$D_{SW} = D_{crit}/2 \qquad (4.4)$$

And the average depth of the deep water compartment by:

$$D_{DW} = (D_{max} - D_{crit})/2 \qquad (4.5)$$

where D_{max} is the maximum depth. Algorithms quantifying sedimentation based on D_{crit} and the lake form will be presented.

Defining the surface water and the deep water compartments also means that one can define upward and downward mixing, and algorithms will be given for this that are meant to apply for both lakes and marine coastal areas. Also mineralization, i.e., the loss of matter due to substrate decomposition by bacteria, will be quantified.

Algorithms to calculate the bottom areas where resuspension is likely to occur (the areas of fine sediment erosion and transport, the ET areas) will be presented. These are the areas above the critical depth, D_{crit}, and a general algorithm to estimate the ET areas will be given. This also means that the areas of continuous accumulation of fine particles (following Stokes' law), the accumulation areas (A areas), where by definition there should be no resuspension are defined.

4.3.2.2.1. The "critical" depth from water temperature

To use the thermocline as the key criteria to differentiate between the surface water and the deep water is the most classical approach (Wetzel, 2001). But that does not necessarily imply that it is operational in contexts of mass-balance calculations. This is demonstrated in fig. 4.8 using data from Lake Erken, Sweden. For single sites after longer periods without strong winds in the summer, one can certainly see clear thermoclines. However, for many sites over longer periods of time (weeks and months), it is generally very difficult to define the critical water depth requested in mass-balance calculations - any water depth between 0.1 m and 18 m in Lake Erken, which has a maximum depth of 21 m, could be used! The data shown in fig. 4.8 for Lake Erken are included here to emphasize that for this lake, and for water systems in general, the theromocline is a very poor alternative for the requested critical water depth. Water temperatures do not provide consistent information to define the boundary between the surface water and deep water compartments. Most limnologists and marine ecologists can probably accept this conclusion because temperature profiles in most (but maybe not all) lakes and coastal areas look like those shown in fig. 4.8.

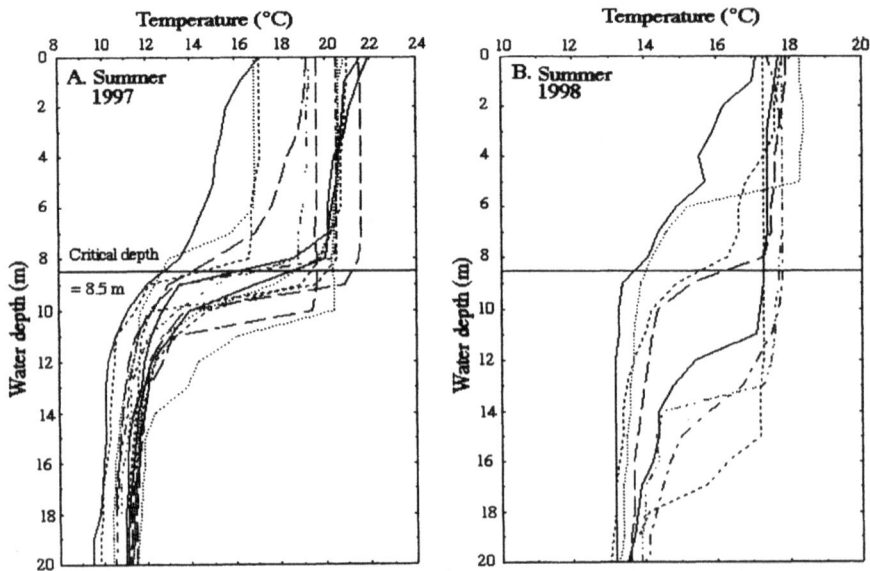

Fig. 4.8. Temperature data from Lake Erken, Sweden (data for the summers of 1997 and 1998) and illustration of the critical depth (= the wave base, D_{crit}) in this lake. Each line represents the temperature profile at the monitoring station at a given sampling occasion (from Håkanson et al., 2004c).

4.3.2.2.2. The "critical" depth from concentration profiles

Fig. 4.9 shows vertical variations at six sites on algal biomass in Lake Erken. From this figure, the same conclusion can be drawn as from fig. 4.8. That conclusion would also be valid for most substances (and not just for algal biomass) in most lakes and coastal areas. So, for aquatic systems in general, concentration gradients would not provide very useful data to define the requested critical water depth.

4.3.2.2.3. The "critical" depth from the wave equation

The wave equation (see, e.g., Smith and Sinclair, 1972) is:

$$g \cdot H/w^2 = 0.0026 \cdot (g \cdot EF/w^2)^{0.47} \qquad (4.6)$$

where g = the acceleration due to gravity (m/s^2); H = the wave height (m); w = the wind speed (m/s); and EF = the effective fetch (m).

The wave base, which would correspond to the requested critical water depth, is often set to 1/3 of the wave length rather than the wave height. In fact, whatever criteria that would be used from the wave theory (even if accounting for the duration of the wind), it would not give a value that could be used in a simple and rational manner in the requested mass-balance model. Instead, the wave theory gives a whole array of wave bases related to different wind situations. During a period of one week or one month, winds can blow from many directions and with many velocities.

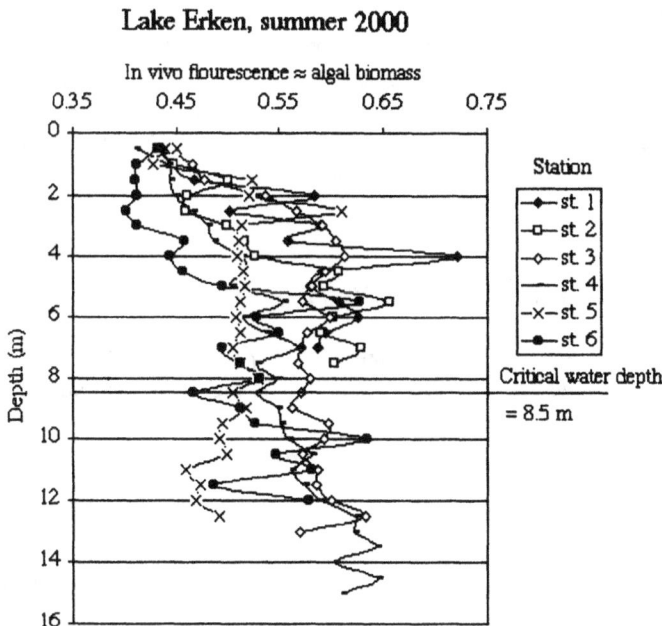

Fig. 4.9. Data on algal biomass in Lake Erken, Sweden (summer of 2000) from 6 stations and illustration of the critical depth (D$_{crit}$) in this lake. All the profiles are sampled during the same two day period in July (from Håkanson et al., 2004c).

4.3.2.2.4. The "critical" depth from the ETA diagram

This approach focuses on the behavior of the cohesive fine materials settling according to Stokes' law. The approach used here (from Håkanson and Jansson, 1983) to estimate the critical water depth, D_{crit}, was shown in fig. 2.18. D_{crit} (in m) is calculated from lake area (note that the area should be in km^2 and not in m^2 in eq. 4.7), which is related to the effective fetch and how winds and waves influences the bottom dynamic conditions:

$$D_{crit} = (45.7 \cdot \sqrt{Area})/(21.4 + \sqrt{Area}) \tag{4.7}$$

This approach gives one value of the critical depth for any given lake (8.5 m in Lake Erken, fig. 4.8). Eq. 4.7 will be applied in this dynamic SPM model for lakes and it will also be used in a modified version for marine coastal areas.

4.3.2.3. Mixing

The mixing between surface water and deep water depends on the stratification, which in turn depends on many climatological factors (prevailing winds, season of the year, etc.). The following submodel for mixing is simple. It first gives the monthly mixing rate (R_{mix}; 1/month) as a function of the absolute difference between mean monthly surface and deep water temperatures.

$$\text{If ABS(SWT - DWT)} < 4 \text{ (°C) then } R_{mix} = 1 \text{ else } R_{mix} = 1/\text{ABS(SWT - DWT)} \tag{4.8}$$

That is, if the absolute difference between epilimnetic (SWT) and hypolimnetic temperatures (DWT) is smaller than 4°C, it is assumed that the lake is not stratified and R_{mix} is set to 1. It has also been tested if the mixing rate, R_{mix}, during homothermal conditions should be larger than one for very large and shallow lakes with high dynamic ratios (DR) and lower for small and deep lakes, but that does not seem to improve model predictions. Complete mixing, i.e., setting the mixing rate to one (rather than to a higher number) generally seems to be sufficient.

In calculating mixing, the same amount of water should be transported from surface water to deep water and from deep water to surface water. The downward mixing from surface water to deep water is given by:

$$F_{SWDWx} = M_{SW} \cdot R_{mix} = (M_{SW}/V_{SW}) \cdot (V_{SW}/T_{SW}) = C_{SW} \cdot Q_{mix} \tag{4.9}$$

Where the concentration of SPM in surface water (C_{SW}) is equal to M_{SW}/V_{SW} and the water flux related to mixing (Q_{mix}) is equal to V_{SW}/T_{SW}, where T_{SW} is the theoretical surface water retention time.

This means that mixing from deep water to surface water will cause a transport given by:

$$F_{DWSWx} = M_{DW} \cdot R_{mix} \cdot (V_{SW}/V_{DW}) \tag{4.10}$$

Where $M_{DW} = V_{DW} \cdot C_{DW}$. If the wave base ($D_{crit}$) is very close to the maximum depth (D_{max}), and hence the deep water volume is very small and the ratio V_{SW}/V_{DW} very large, the flux from deep water to surface water can be so large that it will become difficult to get stable solutions using Euler's or Runge-Kutta's calculation routines. This means that the following boundary condition is used for mixing:

$$\text{If } R_{mix} \cdot (V_{SW}/V_{DW}) > 500 \text{ then } 500 \text{ else } R_{mix} \cdot (V_{SW}/V_{DW}) \tag{4.11}$$

Data on surface and deep water temperatures (in eq. 4.8) may be calculated from the temperature submodel presented by Ottosson and Abrahamsson (1998; see appendix 9.4). It is well known that water (and air) temperatures are governed by many complicated climatological relationships. This temperature submodel has proven to give good predictions of mean monthly temperatures for European lakes. The temperature submodel requires latitude (°N), altitude (m.a.s.l.), continentality (km from ocean), lake mean depth (D_m, m), and lake volume (= $D_m \cdot A$, m^3) as driving variables.

Evidently, the modelled temperatures could be replaced by measured values, but in the following calculations, temperature data have generally been calculated from the temperature submodel, except for Lake Kinneret, which lies outside the domain of the temperature submodel. For this lake, empirical temperature data have been used.

The lake is not likely stratified if the dynamic ratio (DR) is higher than 3.8 (fig. 4.10). Then, the lake is probably not dimictic, but polymictic. A boundary condition for this is added to the model: If DR > 3.8 then $R_{mix} = 1$.

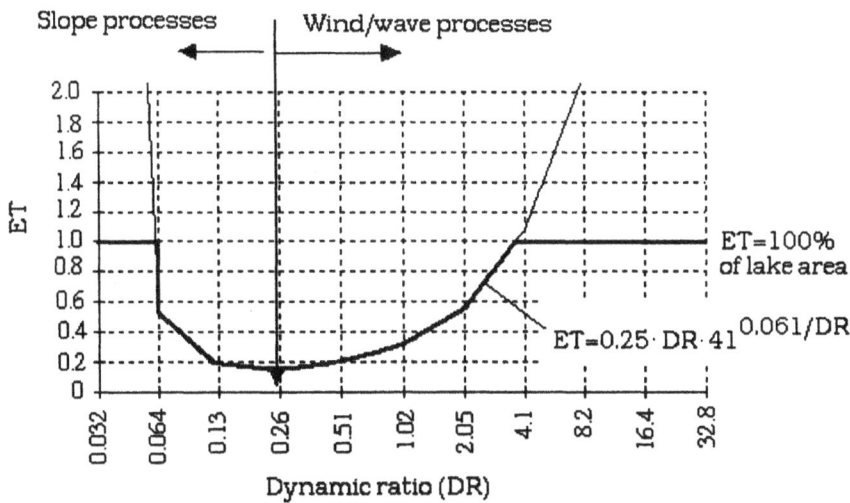

Fig. 4.10. A graphical illustration of the relation between the bottom areas dominated by processes of erosion and transportation expressed as a fraction of lake area (ET) and the dynamic ratio (DR = √Area/D_m; Area in km^2; D_m = mean depth in m). Resuspension from wind and wave action dominates in lakes with high dynamic ratios (DR > 0.25). Turbidity currents related to slope processes are more important in lakes with low DR (modified from Håkanson and Jansson, 1983).

4.3.2.4. Submodels for volumes

The volume of the surface water is calculated in the following way (fig. 4.11):

$$V_{SW} = V - V_{DW} = Area \cdot D_m - Area_A \cdot (D_{max} - D_{crit})/3 \qquad (4.12)$$

Where $Area_A$ is the area below the wave base (the accumulation areas). $Area_A$ (in m^2) is calculated from an equation describing the hypsographic curve (from Håkanson, 1999) and given by:

$$Area_A = (Area \cdot ((D_{max} - D_{crit})/(D_{max} + D_{crit} \cdot EXP(3 - Vd^{1.5})))^{(0.5/Vd)}) \qquad (4.13)$$

To calculate $Area_A$, and hence also $Area_{ET} = Area - Area_A$, data are needed on the maximum depth (D_{max}), the critical depth (D_{crit}), and the form factor (= volume development), Vd, which is a standard expression also defined in fig. 4.11 ($Vd = 3 \cdot D_m/D_{max}$, where D_m = the mean depth).

From this general background, the next part puts these concepts into the framework of mass-balance modelling.

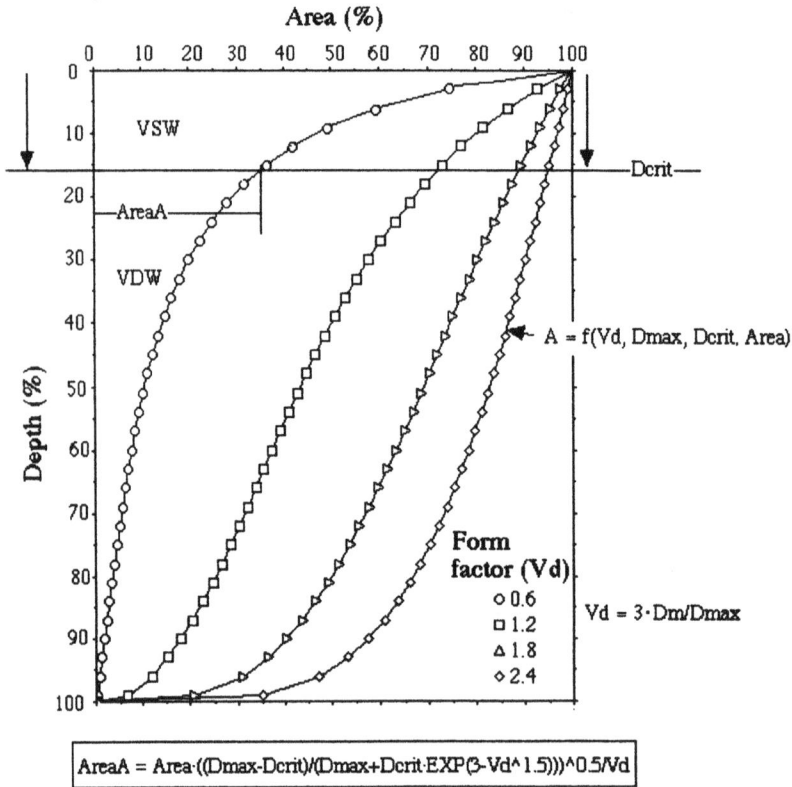

AreaA = Area·((Dmax-Dcrit)/(Dmax+Dcrit·EXP(3-Vd^1.5)))^0.5/Vd

Fig. 4.11. Illustration of relative hyposographic curves (= depth area curves) for four lakes with different forms (and form factor = volume development = Vd). The form influences the ET areas above the critical water depth (D_{crit}), and the A areas below the critical water depth. The form of the lake is also important in defining the surface water volume (V_{SW}) and the deep water volume (V_{DW}). Shallow lakes with a small Vd have relatively large areas above the wave base, where processes of wind/wave-induced resuspension will influence the bottom dynamic conditions. Deep, U-formed lakes generally have smaller areas above the wave base (the ET areas). A is the lake area. Modified from Håkanson (2000).

4.3.2.5. Basic mass-balance modelling for lakes

A simple mass-balance model for a lake is depicted in fig. 4.12. A typical mass-balance model envisions the lake as a "reactor tank" in the sense that the lake mixes completely during an interval of time dt. The flow of SPM or a given contaminant to and from such a lake may be described by the following ordinary differential equation:

$$V \cdot dC/dt = Q \cdot C_{in} - Q \cdot C - M_W \cdot R_{sed} + M_S \cdot R_{res} \qquad (4.14)$$

where

V   = lake volume; usually m^3 or km^3;

dC/dt   = the change in concentration of SPM, dC, per unit of time (dt; e.g., in $g/(m^3 \cdot month)$);

C   = the concentration of the substance in the lake water; units usually g/m^3; $C = M_W/V$;

C_{in} = the concentration of the substance in the tributary; Cin has the same dimension as C;

Q   = the tributary water discharge to the lake; expressed as $m^3/month$;

R_{sed} = the sedimentation rate of a given substance in the lake; like all rates, R_{sed} has the dimension 1/time and its unit is usually 1/month or 1/yr;

M_W = the mass (= amount) of the substance in the lake water ($M_W = C \cdot V$); units often in g;

M_S = the mass (= amount) of the substance on the lake bed; units in g;

R_{res} = the internal loading rate; or resuspension rate; units usually in 1/month or 1/yr.

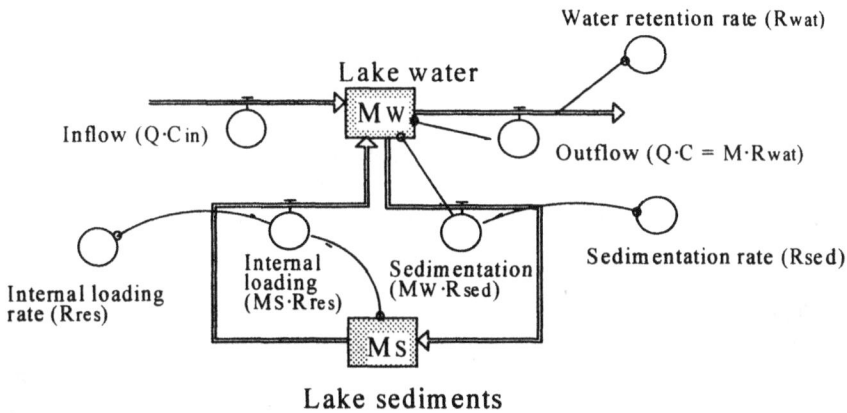

$$V \cdot dC/dt = Q \cdot Cin - Q \cdot C - MW \cdot Rsed + MS \cdot Rres$$
dC/dt = change in conc. in lake water
C = MW/V; V = lake volume
MW = Mass in water
MS = Mass in sediments
Tw = V/Q; Q = water discharge
Cin = Tributary concentration

Fig. 4.12. The basic mass-balance equation for a lake with internal loading.

The simplified model given in fig. 4.12 is only meant to provide a framework for the following sections.

The lake water retention time (T, in days, months or years) is a fundamental concept in limnology. It is defined by the ratio between the lake volume and the mean annual water discharge (Q):

$$T = V/Q \tag{4.15}$$

Q can be derived from time series of measurements; or be given as a mean monthly value; or as a mean yearly value; or estimated from the submodel given in appendix 9.3. If annual data for entire lakes are used, T is referred to as the theoretical lake water retention time.

By setting $dC/dt = 0$, the steady-state solution to eq. 4.14 is given and one can solve for C, the equilibrium concentration of the substance of interest:

$$C = (Q \cdot C_{in} + M_S \cdot R_{res})/(Q + V \cdot R_{res}) \qquad (4.16)$$

When Vollenweider (1968) presented his first loading model for lake eutrophication, this meant a break-through not "just" for lake management but also for lake modelling. Vollenweider simplified the mass-balance model first by omitting seasonal variations and instead gave the annual budget, he omitted different nutrients and different forms of the nutrients, and instead made the calculations for total phosphorus, he disregarded internal loading (the $M_S \cdot R_{res}$ term in eq. 4.14) and simplified the sedimentation term ($M_S \cdot R_{sed}$, which he approximated to $T^{1/2}$). This gave the famous Vollenweider model:

$$C = C_{in}/(1 + T^{1/2}) \qquad (4.17)$$

It is evident that substances with large sedimentation rates (R_{sed} values) settle rapidly, near the point of discharge, and that substances with small R_{sed} values may be distributed over larger areas. It is important to determine or predict the R_{sed} value. R_{sed} is related to the settling velocity, v (v = $z \cdot R_{sed}$, where z is the distance through which the particle sinks in the given time interval). The settling velocity, v, is generally given in cm/s, m/month, or m/yr. Given R_{sed} or v, one can model or predict where high and low concentrations are likely to appear in water and sediments using partial differential equations. This is the key to predicting sites where high and low ecological effects may appear, e.g., related to reductions of zoobenthos biomass. It is important to note both the similarity and difference between the settling velocity (v in cm/s) and the sedimentation rate (R_{sed} in 1/s).

Since Vollenweider omitted internal loading from the basic mass-balance equation, and since the Vollenweider model is so frequently used in practical lake management, one can ask: Is it motivated to complicate matters and account for internal loading? When or why is this important?

Fig. 4.12 gives a compilation of all major fluxes of phosphorus in a real lake situation using data from Lake S. Bullaren, Sweden (from Håkanson, 1995; lake area = 8.3 km²; mean depth = 10.1 m; dynamic ratio = 0.29; TP concentration = 35 µg/l; the lake had a fish farm producing 500 tons/yr of rainbow trout). The typical tributary inflow was about 4000 kg TP/yr and the total inflow from the fish farm about 1900 kg TP/yr. These were the two most important primary TP fluxes to this lake; all other fluxes depend on these two fluxes, including sedimentation (7320 kg) and internal loading (3000 kg). So, in this lake, it is no doubt that internal loading is important. In fact, it is important in most lakes and not just for phosphorus as in this example but for all substances with a PF value higher than zero; and SPM has a PF value of one. In shallow lakes, wind/wave-influences (resuspension or advective processes) dominate the internal loading. In deep lakes, slope processes (turbidity currents, slumping, etc.) are important; and, for some substances like phosphorus and radiocesium, diffusion may also cause a considerable internal loading.

Thus, the key problem in many contexts of water science and management is to quantify internal loading and relate internal loading to all other major fluxes to, within, and from a lake.

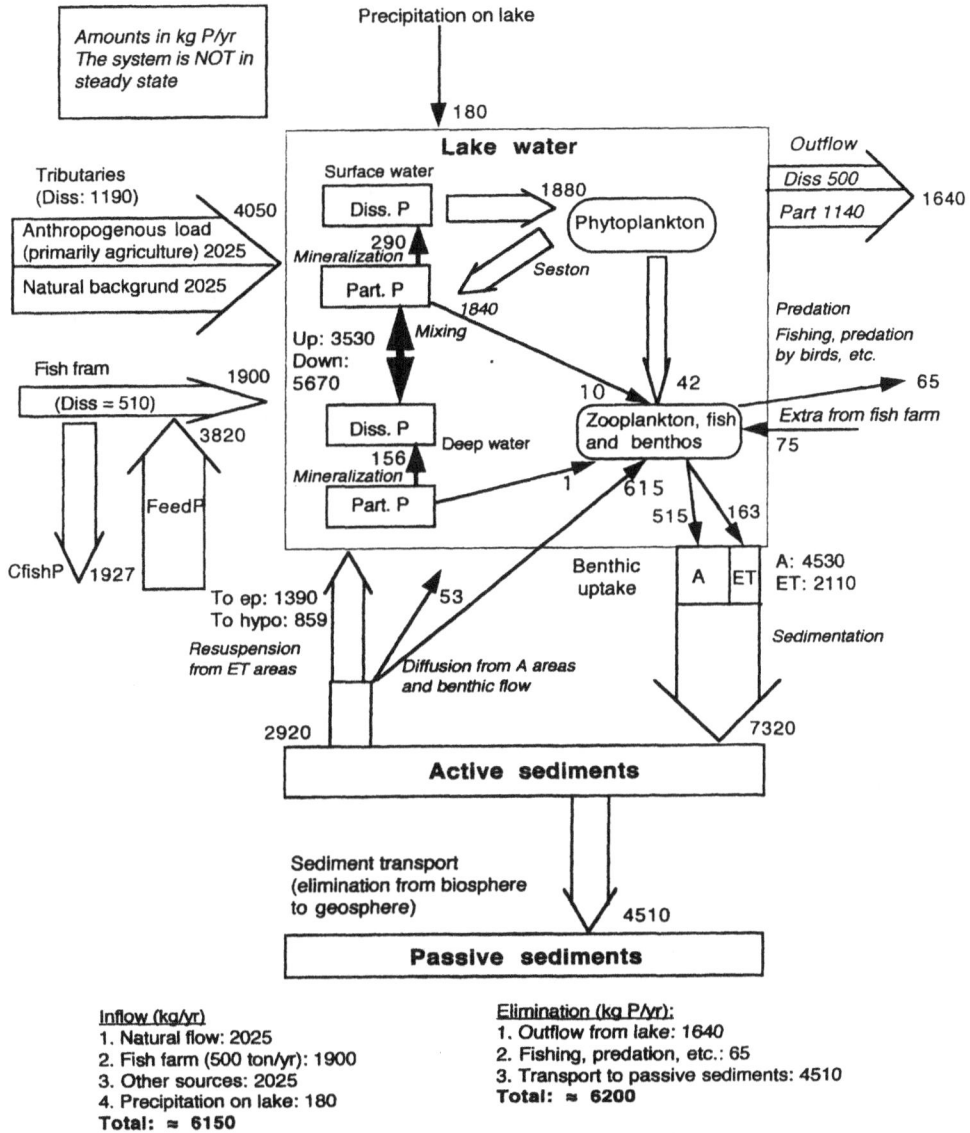

Fig. 4.13. A phosphorus budget for Lake S. Bullaren (with a fish farm producing 500 tons fish/yr). All major abiotic and biotic fluxes are calculated with this mass-balance model including the fluxes related to internal loading (from Håkanson, 1995)

4.3.3. Comprehensive mass-balance modelling

The aim of the following section is to present and motivate a more comprehensive dynamic mass-balance model for SPM in lakes.

4.3.3.1. River inflow to lake

The inflow (F_{in} in g SPM per month) to a lake from tributary rivers is generally calculated from water discharge (Q) times the concentration in the tributary (C_{in}), i.e.:

$$F_{in} = Q \cdot C_{in} \tag{4.18}$$

In this modelling, the inflow may be calculated from the river model (chapter 5) or obtained from empirical data on Q and C_{in}.

Some rather specialized studies, utilizing GIS processing of catchment features such as soil properties, slope, and drainage density (e.g., Verstraeten and Poesen, 2001) have been presented that predicts SPM transport in rivers, but not very well. The link between soil loss and sediment transport is problematic, as is the spatial and temporal lumping of catchment characteristics (Walling, 1983; Atkinson, 1995). Chapter 5 will discuss this in more detail.

If reliable empirical SPM data are not available from the tributaries to a lake (which is generally the case), the modelling discussed here provides an alternative approach to estimate the river inflow of SPM from the river concentration of total phosphorus (C_{TPIn} in g TP/month), which would generally be more accessible than SPM data. This approach simply uses the phosphorus content in phytoplankton, 2.03% (Håkanson and Jansson, 1983) to estimate the SPM concentration, i.e.:

$$F_{in} = Q_{in} \cdot C_{TPIn}/0.0203 \tag{4.19}$$

where Q_{in} is river discharge in m^3/month.

Q_{in} may be obtained from measurements or predicted by a submodel presented by Abrahamsson and Håkanson (1998) (see appendix 9.3). Then, the mean yearly discharge (Q_{yr}) is given by (from Håkanson and Peters, 1995):

$$Q_{yr} = A_{DA} \cdot (Prec/650) \cdot 0.01 \cdot 60 \cdot 60 \cdot 24 \cdot 365 \tag{4.20}$$

where A_{DA} is the catchment area (km^2).

In eq. 20, the annual precipitation (Prec, mm/yr) is related to a standard annual precipitation (650 mm/yr) and multiplied by a general specific runoff rate (0.01 $m^3/km^2 \cdot s$) and conversion factors into m^3/yr. The requested mean monthly discharge (Q_{in}) is predicted by means of a dimensionless moderator (Y_Q, basically Q_{month}/Q_{year}; see appendix 9.3) so that:

$$Q_{in} = Y_Q \cdot Q_{yr}/12 \tag{4.21}$$

Using this approach, Q_{in} can be estimated from data on latitude (Lat, °N) and altitude (Alt, meter above sea level, m.a.s.l.). Norm values meant to describe monthly variability in flow patterns are related to differences in latitude, altitude and mean annual discharge are used in this submodel. These norm values (Lx, Li, Ax, Ai, Qx, and Qi) are defined in appendix 9.3.

4.3.3.2. Lake bioproduction

Calculating lake bioproduction is a focal issue in limnology and many authors have discussed primary production in lakes. Generally, mean summer chlorophyll-a concentrations are predicted from water temperature, light conditions and nutrient status (e.g., Dillon and Rigler, 1974; Smith, 1979; Riley and Prepas, 1985; Evans et al., 1996). The equation to quantify the amount of SPM generated on a monthly basis in a given lake used here comes from Håkanson and Boulion (2002). In this approach, total SPM production is calculated from chlorophyll-a, accordingly:

$$F_{prod} = (30.6 \cdot Chl^{0.927}) \cdot 0.45 \cdot 30 \cdot Area \cdot Sec \cdot 0.001 \cdot ((SWT+0.1)/9) \cdot (BM_{PL}/BM_{PH}) \tag{4.22}$$

Chl = the mean monthly chlorophyll concentration (μg/l); the expression ($30.6 \cdot Chl^{0.927}$) transforms Chl into phytoplankton production (in μg C/l·d). The factor 0.45 is a standard transformation factor to change g C to g dw. Multiplication with 0.001, 30 days, lake area (Area) and the mean monthly value of the effective depth of the photic zone (= the Secchi depth = Sec in m) gives the biomass of phytoplankton produced per month (g dw per month). Note that for lakes, chlorophyll concentrations may be predicted from TP concentrations (table 2.4). For very humic lakes or for marine coastal areas, e.g., in the Bothnian Sea, the Black Sea or the Mediterranean, chlorophyll concentrations are generally better predicted from total nitrogen concentrations (which will be discussed in chapter 6).

BM_{PL}/BM_{PH} = the ratio between the biomass of all sorts of plankton (phytoplankton, bacterioplankton zooplankton, etc.; BM_{PL}) to the calculated biomass of phytoplankton (BM_{PH}) comes from the foodweb model (LakeWeb, see appendix 9.2). This foodweb model has been used to get a simple general calculation constant for this ratio along a trophic state gradient. Those results are shown in fig. 4.14 and this ratio is on average about 2.5 along the entire gradient. It indicates that the total biomass of bacterioplankton plus zooplankton plus phytoplankton is a factor of about 2.5 higher than the phytoplankton biomass. Fig. 4.14 shows that there is a marked temporal variability around this mean value from seasonal changes in surface water temperatures. This is quantified by the dimensionless moderator for SWT in eq. 4.22.

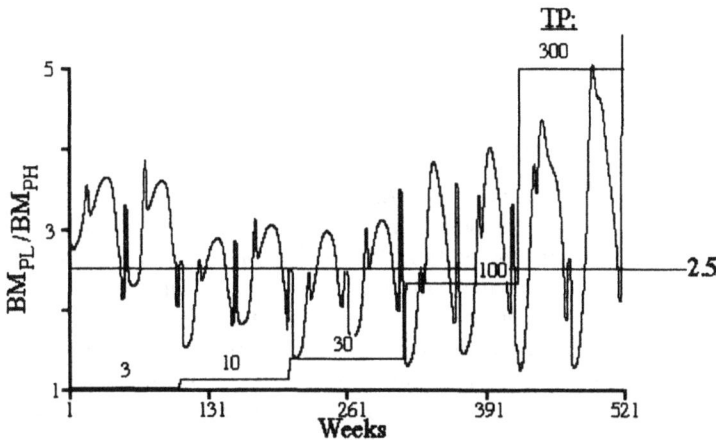

Fig. 4.14. Calculations along a trophic state gradient (TP concentrations set to 3, 10, 30, 100, and 300 μg/l in two-year steps) using the LakeWeb model (from Håkanson and Boulion, 2002) to see if it is possible to find a simple, general calculation constant between the biomass of all sorts of plankton (phytoplankton, bacterioplankton and zooplankton, BM_{PL}) and the biomass of phytoplankton (BM_{PH}).

SWT = Mean monthly surface water temperatures (°C), e.g., calculated from the temperature submodel given in appendix 9.4. By dividing SWT with a reference temperature of 9°C (related to the duration of the growing season), this approach accounts for seasonal variations in SWT in a dimensionless manner. The moderator is (SWT+0.1)/9. The constant 0.1 is used since SWT may approach 0°C during the winter and since there is also production under the ice.

Sec = Secchi depth (in m) as measured using Secchi discs or calculated from empirically or dynamically modelled values of SPM. This modelling will generally use values of Secchi depth calculated from the regression between Secchi depth and SPM discussed in chapter 2.

4.3.3.3. Internal processes

4.3.3.3.1. Sedimentation

Sedimentation of SPM in this mass-balance model depends on:

1. A default settling velocity, v_{def}, which is substance-specific and related to the type of carrier particle, e.g., 12 m/yr for illite clays, the carrier particles for radiocesium (Håkanson, 2000), 72 m/yr for planktonic materials and the particulate fraction applicable for phosphorus in lakes (Håkanson and Boulion, 2002). Here, 72 m/yr will also be used as a general default value for the complex mixture of substances making up SPM in the water in lakes, rivers, and coastal areas. The default settling velocity is changed into a rate (1/month) by division with the mean depth of the surface water areas (D_{SW}) for sedimentation in these areas and by the mean depth of the deep water areas (D_{DW}) for sedimentation in deep water areas.

2. The SPM concentration will also influence the settling velocity - the greater the aggregation of suspended particles, the bigger the flocs and the faster the settling velocity (Kranck, 1973, 1979; Lick et al., 1992). This is expressed by a dimensionless moderator (Y_{SPM}; see eq. 4.27).

3. The salinity of the water will influence the settling velocity: The higher the salinity, the greater the aggregation, the bigger the flocs, the faster the settling velocity (Kranck, 1973, 1979). This is expressed by a dimensionless moderator for salinity (Y_{sal}) operating on the default settling velocity (see eq. 4.28).

4. Burban et al. (1989, 1990) have demonstrated that changes in turbulence are very important for the fall velocity of suspended particles. Generally, there is more turbulence, which keeps the particles suspended, and hence causes lower settling rates, in the surface water than in the calmer deep water compartment. The turbulence in the surface water compartment is also generally greater in large and shallow lakes with high dynamic ratios (DR) compared to small and deep lakes. In this modelling, two new dimensionless moderators (Y_{DW} and Y_{DR}; see equations 4.29 and 4.34) related to the theoretical lake water retention time and the dynamic ratio have been used to quantify how turbulence is likely to influence the settling velocity in the surface water and deep water compartments.

5. The settling velocity also depends on the amount of resuspended matter. The resuspended particles have already been aggregated and have generally been influenced by benthic activities, which will create a "gluing effect." They have also a comparatively short distance to fall after being resuspended (see Håkanson and Jansson, 1983). The longer the particles have stayed on the bottom, the larger the potential gluing effect and the faster the settling velocity if the particles are resuspended. The default value for the age of the particles on ET areas in lakes is one year; the corresponding default age of the particles in marine coastal areas is 1 month. Marine coastal areas are generally very dynamic with typical theoretical surface water retentions times of about 5 days; a typical theoretical water retention time for a lake is 12 months (see Håkanson, 2000). The resuspended fraction is calculated in the model and the resuspended particles settle faster. This is expressed by another new dimensionless moderator (Y_{res}; see eq. 4.31).

Sedimentation from the surface water compartment to the ET areas (F_{SWET}) is given by:

$$F_{SWET} = M_{SW} \cdot R_{SW} \cdot ET \qquad (4.23)$$

where

M_{SW} = the mass of SPM in the surface water compartment (g);
R_{SW} = the sedimentation rate (1/month);
ET = the fraction of ET areas (ET = $Area_{ET}$/Area), which is either measured from field work or estimated by a submodel given in the next section.

Sedimentation from surface water to deep water areas is:

$$F_{SWDW} = M_{SW} \cdot R_{SW} \cdot (1 - ET) \tag{4.24}$$

and sedimentation from deep water areas on A areas:

$$F_{DWA} = M_{DW} \cdot R_{DW} \tag{4.25}$$

The basic sedimentation rates for surface and deep water areas may be written as $R_{SW} = v_{SW}/D_{SW}$ and $R_{DW} = v_{DW}/D_{DW}$, respectively. The mean depths of the surface and deep water areas, D_{SW} and D_{DW}, are calculated from equations 4.4 and 4.5. The monthly settling velocity for the surface water compartment (v_{SW}) is calculated from the default settling velocity of v_{def} m/yr accordingly:

$$v_{SW} = (v_{def}/12) \cdot Y_{SPMSW} \cdot Y_{salSW} \cdot Y_{DR} \cdot ((1 - DC_{resSW}) + Y_{res} \cdot DC_{resSW}) \tag{4.26}$$

where

v_{def} = 72 m/yr;
Y_{SPMSW} = the dimensionless moderator expressing how changes in SPM concentrations in the surface water compartment (C_{SW} in mg/l) influence the settling velocity;
Y_{salSW} = the dimensionless moderator expressing how changes in salinity (in %) in the surface water compartment influence the settling velocity;
Y_{DR} = the dimensionless moderator expressing how changes in dynamic ratio (turbulence) influence the settling velocity;
Y_{res} = the dimensionless moderator expressing how changes in the age of the resuspendable ET sediments influence the settling velocity;
DC_{resSW} = the resuspended fraction in the surface water compartment.

Y_{SPMSW} is given by:

$$Y_{SPMSW} = (1 + 0.75 \cdot (C_{SW}/50 - 1)) \tag{4.27}$$

This dimensionless moderator quantifies how changes in SPM influence the settling velocity of the suspended particles. The amplitude value is calibrated in such a manner that a change in SPM by a factor of 10, e.g., from 2 mg/l (which is a typical value for oligotrophic lakes) to 20 mg/l (which is typical for eutrophic lakes), will cause a change in the settling velocity by a factor of 2. The borderline value for the moderator is 50 mg/l, since it is unlikely that lakes will have higher mean SPM values than that. The default settling velocity differs for different types of carrier particles. For example, fish faeces fall very quickly, $v \approx$ 10,000 m/yr, dead plankton (seston) falls slower, $v \approx$ 20-200 m/yr, humic substances even slower, $v \approx$ 10-100 m/yr and certain clay minerals, like illite, which carries radiocesium, has a fall velocity of about 10-15 m/yr (from Håkanson, 1999). In this modelling, SPM has a default settling velocity of 72 m/yr in lakes with SPM values of 50 mg/l, and in lakes with lower SPM concentrations the fall velocity is lower, as expressed by eq. 4.27.

In traditional mass-balance models, an amount (kg) is multiplied by a rate (1/month) to get a flux (i.e., amount·rate or amount·rate·1). In this modelling, one multiplies kg·(1/month)·Y (= amount·rate·mod), where Y is a dimensionless moderator quantifying how an environmental variable (like SPM) influences the given flux (e.g., sedimentation). Instead of building a large mechanistic submodel for how environmental factors influence given rates, this technique uses a simple, general algorithm for the dimensionless moderator. Empirical data can be used for the calibration of the moderator. The dimensionless moderator defined by eq. 4.27 uses a borderline value, i.e., a realistic maximum value of SPM = 50, to define when the moderator, Y_{SPM}, attains the value of 1. For all SPM values smaller than the borderline value, Y_{SPM} is smaller than unity. One can also build normal value moderators in such a way that the Y_{SPM} is 1 for the "normal" value and higher or lower than 1 for SPM values higher and lower than the defined normal value (e.g., SPM = 5). The amplitude value regulates the change in Y_{SPM} when the actual SPM value differs from the borderline value and/or the normal value.

The dimensionless moderator for salinity (Y_{sal} or Y_{salSW} or Y_{salDW}) is given by:

$$\text{If salinity} <1\% \text{ then } Y_{sal} = 1 \text{ else } Y_{sal} = (1 + 1 \cdot (Sal/1-1)) = 1 \cdot Sal/1 = Sal \qquad (4.28)$$

The norm value of the moderator is 1% and the amplitude value is set to 1. This means that if the salinity changes from 5 to 10%, the moderator (Y_{sal}) changes from 5 to 10 and the settling velocity increases by a factor of 2. This moderator attains a value higher than one for saline lakes and marine coastal areas.

The dimensionless moderator for the dynamic ratio (the potential turbulence), Y_{DR}, is given by:

$$\text{If DR} < 0.26 \text{ then } Y_{DR} = 1 \text{ else } Y_{DR} = 0.26/DR \qquad (4.29)$$

Lakes with a DR value of 0.26 (fig. 4.10) are likely to have a minimum of ET areas (15% of the lake area) and the higher the DR value, the larger the lake area relative to the mean depth and the higher the potential turbulence and the lower the settling velocity.

The resuspended fraction of SPM in the surface water compartment is calculated by means of the distribution coefficient (DC_{resSW}), which is defined by the ratio between resuspension from ET areas to surface water relative to all fluxes (except mixing) to the surface water compartment:

$$DC_{resSW} = F_{ETSW}/(F_{ETSW} + F_{in} + F_{prod}) \qquad (4.30)$$
F_{ETSW} = Resuspension from ET areas to surface water areas (g SPM/month).
F_{in} = Inflow of SPM to the lake (g SPM/month).
F_{prod} = Lake production of SPM (g SPM/month).

DC_{resSW} is calculated automatically in the model.

The new dimensionless moderator expressing how much faster resuspended particles settle compared to primary particles is given by:

$$Y_{res} = (((T_{ET}/1)+1)^{0.5} \qquad (4.31)$$

Where T_{ET} is the mean retention time (the mean age) of the particles on ET areas in months. The default T_{ET} value for lakes is 1 year so $Y_{res} = \sqrt{(12+1)} = 4.46$ for lakes, which means that resuspended particles settle 4.46 times faster than suspended primary materials.

The corresponding equation for the settling velocity in deep water areas (v_{DW}) is given by:

$$v_{DW} = (v_{def}/12) \cdot Y_{SPMDW} \cdot Y_{salDW} \cdot Y_{DR} \cdot Y_{DW} \cdot ((1-DC_{resDW})+Y_{res} \cdot DC_{resDW}) \qquad (4.32)$$

where

v_{def} $= 72$ m/yr;

Y_{SPMDW} = the dimensionless moderator expressing how changes in SPM concentrations in the deep water compartment (C_{DW} in mg/l) influence the settling velocity;

Y_{salDW} = the dimensionless moderator expressing how changes in salinity (in %) in the deep water compartment influence the settling velocity;

Y_{DR} = the dimensionless moderator expressing how variations among systems in the dynamic ratio and turbulence (DR) are likely to influence the settling velocity for SPM;

Y_{DW} = the dimensionless moderator expressing how hydrodynamic properties, as expressed by the theoretical water retention time for the entire lake, T, influence turbulence in the deep water areas and the settling velocity;

DC_{resDW} = the resuspended fraction in the deep water compartment;

Y_{res} = the dimensionless moderator expressing how changes in the age of the ET sediments influence the settling velocity.

Y_{SPMDW} is given by in the same manner as Y_{SPMSW} as:

$$Y_{SPMDW} = (1 + 0.75 \cdot (C_{DW}/50 - 1)) \tag{4.33}$$

The dimensionless moderator for salinity (Y_{sal} or Y_{salSW} or Y_{salDW}) is the same as in eq. 4.28. Y_{DW} is calculated from:

$$Y_{DW} = (T/1+1)^{0.5} \tag{4.34}$$

Where T is the theoretical water retention time for the given system in months. The resuspended fraction of SPM in the deep water compartment is given by:

$$DC_{resDW} = F_{ETDW}/(F_{ETDW} + F_{SWDW}) \tag{4.35}$$

F_{ETDW} = Resuspension from ET areas to deep water areas (g SPM/month).

F_{SWDW} = Sedimentation, i.e., transport from surface water to deep water areas (g SPM/month).

DC_{resDW} is calculated automatically in the model.

The same dimensionless moderator (eq. 4.31) expressing how much faster resuspended particles settle compared to primary particles is used for the deep water compartment as for the surface water compartment.

4.3.3.3.1.1. Determination of ET areas

The following approach to estimate the fraction of ET areas may be used if reliable empirical data on ET are missing. This method calculates ET areas from the wave base (= the critical depth, fig. 3.20) and the form of the lake or the coastal area. The wave base (WB) or the critical depth (D_{crit}) is set equal to the depth separating T areas and A areas. It was given by eq. 4.7. An evident boundary condition for eq. 4.7 is that if $D_{crit} > D_{max}$ then $D_{crit} = D_{max}$.

The area below the wave base (Area$_A$) is calculated from the hypsographic form of the lake, which in turn is calculated from the form factor (= volume development, Vd), see eq. 4.13. There are two boundary conditions for ET (= the fraction of ET areas in the lake):

If ET > 0.99 then ET = 0.99
If ET < 0.15 then ET = 0.15

ET areas are generally larger than 15% (ET = 0.15; fig. 4.10) of the lake area since there is always a shore zone dominated by wind/wave activities. For practical and functional reasons, one can generally also find sheltered areas and deep holes with more or less continuous sedimentation, i.e., areas that actually function as A areas, so the upper boundary limit for ET is set at ET = 0.99.

The value for the ET areas is used as a distribution coefficient in this model. It regulates sedimentation of SPM either to A areas or to ET areas. These equations and presupposition for the algorithms to calculate ET are summarized in fig. 4.15.

4.3.3.3.2. Resuspension

By definition, the materials settling on ET areas will not stay permanently where they were deposited but will be resuspended by wind/wave activity. If the age of the material (T_{ET}) is set to a very long period, e.g., 10 years, these areas will function as accumulation areas; if, on the other hand, the age is set to 1 week or less, they will act as erosion areas. Often, it is assumed that the mean age of these deposits is about 1 year (T_{ET}) for lakes (see Håkanson, 1999).

Resuspension back into surface water, F_{ETSW}, i.e., mostly wind/wave-driven advective fluxes, is given by:

$$F_{ETSW} = (M_{ET} \cdot (1 - Vd/3))/T_{ET} \qquad\qquad (4.36)$$

Resuspension to deep water areas, F_{ETSW}, by:

$$F_{ETDW} = (M_{ET} \cdot (Vd/3))/T_{ET} \qquad\qquad (4.37)$$

where

M_{ET}	= the total amount of resuspendable matter on ET areas (g);
Vd	= the form factor; note that Vd/3 is used as a distribution coefficient to regulate how much of the resuspended material from ET areas that will go the surface water or to the deep water compartment. If the lake is U-shaped, Vd is about 3 (i.e., $D_{max} \approx D_m$) and all resuspended matter from ET areas will flow to the deep water areas. If, on the other hand, the lake is shallow and Vd is small, most resuspended matter will flow to the surface water compartment.
T_{ET}	= the age of SPM on ET areas (default value for lakes = 12 months).

The ET submodel

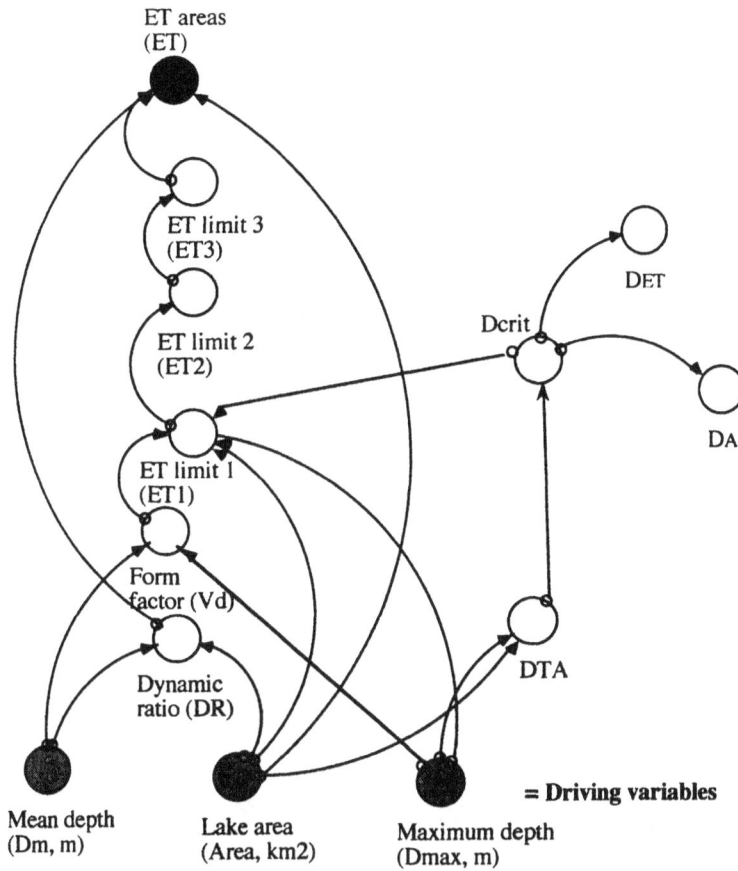

Equations:

$DR = \sqrt{Area}/Dm$

If $(45.7 \cdot \sqrt{Area})/(21.4 + \sqrt{Area}) > Dmax$ then $DTA = Dmax$ else

$Dcrit = WB = (45.7 \cdot \sqrt{Area})/(21.4 + \sqrt{Area})$

$Vd = 3 \cdot Dm/Dmax$

$ET1 = 1 - (Area \cdot 10^6) \cdot 0.01 \cdot (100 \cdot EXP(-3 + Vd^{1.5}) \cdot (Dmax - Dcrit)/(Dcrit + Dmax \cdot EXP(-3 + Vd^{1.5})))^{(0.5/Vd)} \cdot 10^{(2 - 1/Vd)})/(Area \cdot 10^6)$

If $ET1 > 0.99$ then $ET2 = 0.99$ else $ET2 = ET1$

If $ET2 < 0.15$ then $ET3 = 0.15$ else $ET3 = ET2$

If $ETemp > 0$ then $ET4 = ETemp$ else $ET4 = ET3$

If $DTA < 0.5$ then $DTA2 = 0.5$ else $DTA2 = DTA$

$DA = Dcrit + (Dmax - Dcrit)/2$

$DET = Dcrit/2$

Fig. 4.15. The submodel to calculate ET and the average depths of the surface water and deep water compartments, D_{SW} and D_{DW}.

4.3.3.3.3. Mineralization

Mineralization means net losses of SPM from mainly bacterial decomposition of the organic fraction of SPM. The value used for the mineralization rate, R_{min}, regulates the total amount of

SPM being lost each month. The mineralization rate operates in this model on SPM in the surface water and deep water compartments, and in the sediment compartment for ET areas. Mineralization is further assumed to be proportional to temperature (SWT = surface water temperature in°C and DWT = deep water temperatures in°C; Håkanson et al., 2000). The loss of SPM from mineralization in surface water is:

$$F_{minSW} = M_{SW} \cdot R_{min} \cdot Y_{ET} \cdot (SWT/9)^{1.2} \qquad (4.38)$$

where 9°C is a reference temperature related to duration of the growing season (see Håkanson and Boulion, 2002); the default value for the mineralization rate (R_{min}) is set to 0.125 (per month); and the mass (= amount) of SPM in the surface water (M_{SW}) is calculated automatically by the model. The radio SWT/9 is used as a simple dimensionless moderator and the exponent 1.2 stresses the nonlinear temperature dependence of bacterial decomposition (see Törnblom and Rydin, 1998). Y_{ET} is a dimensionless moderator quantifying in a simple manner a more complicated phenomena related to the fact (fig. 4.16 for illustration) that resuspended particles are older and more likely to have been mineralized and have a lower organic content than primary particles. Fig. 4.16A shows data on:

1. the organic content (= loss on ignition) of materials collected in sediment traps placed in surface water (IG_{SW} in % ww) from 17 lakes (data from Håkanson and Peters, 1995, and references in Swedish cited in that publication);

2. the organic content of materials collected in sediment traps placed in the deep water areas of these lakes; and

3. the organic content of surficial (0-1 cm) sediment samples from A areas collected from at least three sites in each of these lakes.

The box-and-whisker plots demonstrate that the organic content is generally highest in the material collected in the surface water traps (dominated by primary materials) and lowest in the sediment samples, where the sediments have been decomposed (mineralized) to a larger extent. This is logical and expected and it demonstrates that bacterial decomposition is important in understanding changes in SPM values and that resuspended matter should be expected to have a lower organic content than primary materials. Fig. 4.16B shows a regression between the organic content of surficial A sediments and the fraction of ET areas in 38 lakes of very different characteristics (areas from 0.04 to 5650 km^2; mean depths from 1.3 to 39 m, TP values from 5 to 44 µg/l; data from Håkanson and Boulion, 2002). There is a statistically significant correlation between ET and IG_{sed} – the higher the ET value, the lower the organic content. The information in fig. 4.16 lies behind the dimensionless moderator Y_{ET}, which is meant to quantify that the mineralization rate should be higher for lakes dominated by primary materials and lower for lakes dominated by resuspension and SPM which has already been mineralized. In this modelling ET, varies between 0.15 and 0.99, and Y_{ET} is defined by:

$$Y_{ET} = (0.99/ET) \qquad (4.39)$$

This means that Y_{ET} = 1 for lakes totally dominated by resuspension (ET = 0.99) and Y_{ET} = 6.6 in lakes with a minimum of resuspension (ET = 0.15). For such lakes, the mineralization rate is also 6.6 times higher.

The mineralization loss from the deep water compartment is then:

$$F_{minDW} = M_{DW} \cdot R_{min} \cdot Y_{ET} \cdot (DWT/9)^{1.2} \qquad (4.40)$$

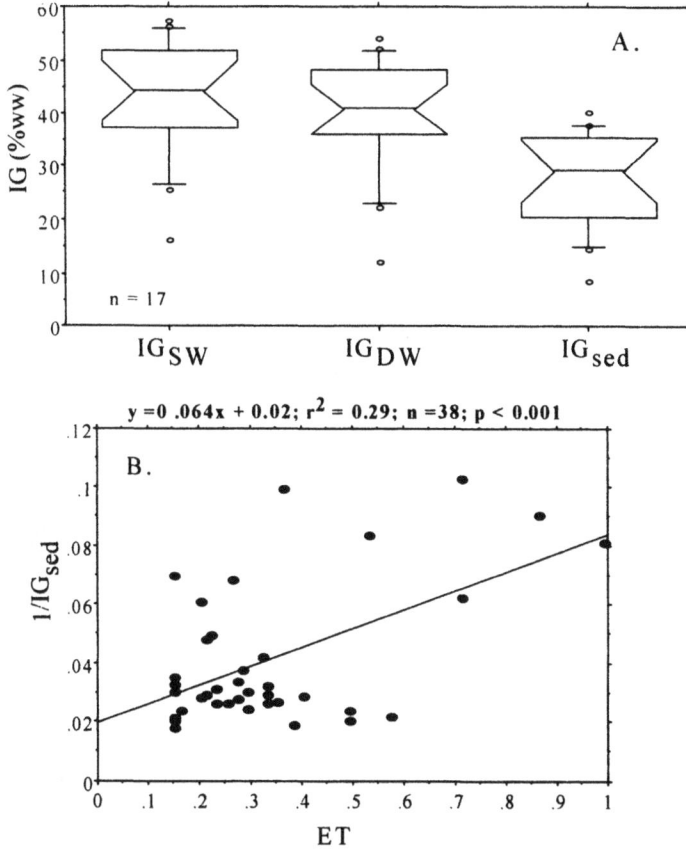

Fig. 4.16. A. The relationship between the organic content (loss on ignition) of materials from surface water sediment traps, deep water sediment traps and surficial sediment samples from accumulation areas (data from 17 lakes from Håkanson and Peters, 1995).
B. Regression between empirical data on organic content (loss on ignition) in surficial sediment samples from A areas and the fraction of ET areas using data from 38 lakes of different limnological characteristics (data from Håkanson and Boulion, 2002).

4.3.3.4. Outflow

The outflow (fig. 4.17) is given by eq. 4.41. This submodel for outflow also accounts for the balance between the precipitation of water and the evaporation loss of water from the lake surface in a simple manner by two dimensionless moderators, Y_{SWT} and Y_{prec}.

$$F_{out} = Q_{out} \cdot C_{SW} = Q_{in} \cdot Y_{SWT} \cdot Y_{prec} \cdot C_{SW} \qquad (4.41)$$

Q_{out} = The water transport out of the lake (m³/month).
Q_{in} = The water transport to the lake (m³/month).
C_{SW} = The SPM concentration in the surface water compartment (g/m³).

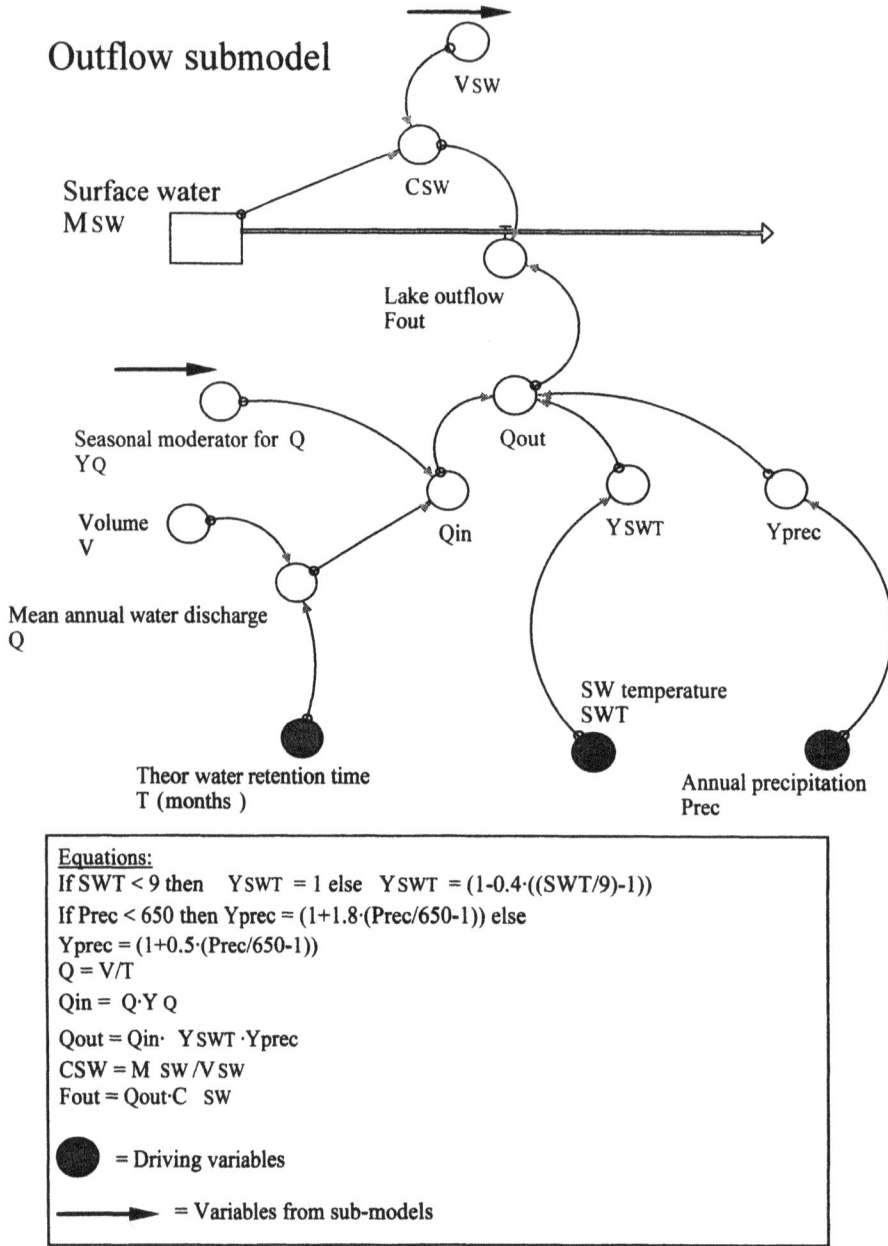

Fig. 4.17. Illustration of the submodel for outflow. The given equations are general and valid for all types of substances in lakes.

The dimensionless moderator for surface water temperatures and evaporation (SWT in°C) is defined in the following manner:

$$\text{If SWT} < 9°C \text{ then } Y_{SWT} = 1 \text{ else } Y_{SWT} = (1 - 0.4 \cdot ((SWT/9) - 1)) \qquad (4.42)$$

Where 9 is the norm temperature related to the duration of the growing season (Håkanson and Boulion, 2002) and 0.4 is the amplitude value. This means that $Y_{SWT} = 1$ if the actual value of the mean monthly surface water temperature (SWT) is 9°C; Y_{SWT} will approach zero when the actual mean monthly surface water temperature approaches 33°C. Then the outflow of water (Q_{out}) also approaches zero. The dimensionless moderator for precipitation (Y_{prec}) is defined by:

$$\text{If Prec} < 650 \text{ mm/yr then } Y_{prec} = (1 + 1.8 \cdot (\text{Prec}/650 - 1)) \text{ else}$$
$$Y_{prec} = (1 + 0.5 \cdot (\text{Prec}/650 - 1)) \tag{4.43}$$

Where 650 is the norm for the mean annual precipitation (Prec); 1.8 is the amplitude value if Prec is smaller than 650; 0.5 the amplitude value if Prec is larger than 650 mm/yr. This means that $Y_{prec} = 1$ if Prec = 650; Y_{prec} will approach zero when the actual mean annual precipitation approaches 300 mm/yr, and then the outflow of water (Q_{out}) also approaches zero; Y_{prec} will increase when the actual mean annual precipitation increases. The outflow of water (Q_{out}) is twice as large when Prec is 2000 mm/yr compared to when Prec = 650 mm/yr.

Table 4.7 gives a compilation of the main equations making up the lake model.

Table 4.7. Compilation of equations making up the dynamic lake model for SPM. Abbreviations: M = mass; F = flow; R = rate; C = concentration; DC = distribution coefficient; T = time or age; Y = dimensionless moderator; D = depth; ET = ET areas; A = A areas; SW = surface water; DW = deep water; flux from SW to DW = F_{SWDW}, etc. Mass in SW = M_{SW}, etc.

Compartment, SPM in surface water (M_{SW}):

$M_{SW}(t) = M_{SW}(t - dt) + (F_{in} + F_{prod} + F_{ETSW} + F_{DWSWx} - F_{out} - F_{SWET} - F_{SWDW} - F_{SWDWx} - F_{minSW}) \cdot dt$

INFLOWS:

$F_{in} = Y_Q \cdot Q_{in} \cdot C_{in} \cdot 60 \cdot 60 \cdot 24 \cdot 30 \ [C_{in} = C_{TPin}/0.0203]$

$F_{prod} = (30.6 \cdot Chl^{0.927}) \cdot 0.45 \cdot 30 \cdot Area \cdot Sec \cdot 0.001 \cdot ((SWT+0.1)/9) \cdot (BM_{PL}/BM_{PH})$

$F_{ETSW} = M_{ET} \cdot (1-Vd/3) \cdot 1/T_{ET} \ [T_{ET} = 12 \text{ months}]$

$F_{DWSWx} = M_{DW} \cdot R_{mix} \cdot (V_{SW}/V_{DW}) \ [\text{for } V_{SW}/V_{DW} < 500]$

OUTFLOWS:

$F_{out} = Q_{in} \cdot C_{SW} \cdot Y_{SWT} \cdot Y_{prec}$

$F_{SWET} = M_{SW} \cdot ET \cdot (v_{def}/D_{SW}) \cdot Y_{SPMSW} \cdot Y_{salSW} \cdot Y_{DR} \cdot ((1-DC_{resSW})) + Y_{res} \cdot DC_{resSW}))$
$\qquad\qquad\qquad\qquad\qquad\qquad\qquad\qquad [v_{def} = 72/12 = 6 \text{ m/month}]$

$F_{SWDW} = M_{SW} \cdot (1-ET) \cdot (v_{def}/D_{SW}) \cdot Y_{SPMSW} \cdot Y_{salSW} \cdot Y_{DR} \cdot ((1-DC_{resSW})) + Y_{res} \cdot DC_{resSW}))$

$F_{SWDWx} = M_{SW} \cdot R_{mix} \cdot (V_{SW}/V_{SW})$

$F_{minSW} = M_{SW} \cdot R_{min} \cdot Y_{ET} \cdot (SWT/9)^{1.2} \ [R_{min} = 0.125]$

Compartment, SPM in deep water (M_{DW}):

$M_{DW}(t) = M_{DW}(t - dt) + (F_{SWDW} + F_{SWDWx} + F_{ETDW} - F_{DWA} - F_{DWSWx} - F_{minDW}) \cdot dt$

INFLOWS:

$F_{DWA} = M_{DW} \cdot (1-ET) \cdot (v_{def}/D_{DW}) \cdot Y_{SPMDW} \cdot Y_{salDW} \cdot Y_{DR} \cdot Y_{DW} \cdot ((1-DC_{resDW})) + Y_{res} \cdot DC_{resDW})$

$F_{SWDWx} = $ see above

$F_{ETDW} = M_{ET} \cdot (Vd/3) \cdot 1/T_{ET}$

OUTFLOWS:

$F_{DWA} = M_{DW} \cdot R_{DW}$; see below

$F_{DWSWx} = $ see above

$F_{minDW} = M_{DW} \cdot R_{min} \cdot Y_{ET} \cdot (DWT/9)^{1.2} \ [R_{min} = 0.125]$

Compartment, SPM on ET areas (M_{ET}):

$M_{ET}(t) = M_{ET}(t - dt) + (F_{SWET} - F_{ETSW} - F_{ETDW} - F_{minET}) \cdot dt$

INFLOWS:

$F_{SWET} = $ see above

OUTFLOWS:

$F_{ETSW} = $ see above

$F_{ETDW} = $ see above

$F_{minET} = M_{SET} \cdot R_{min} \cdot Y_{ET} \cdot (SWT/9)^{1.2} \ [R_{min} = 0.125]$

4.4. Model performance

The aim of this section is to demonstrate how the dynamic model predicts SPM concentrations in lakes using data from six systems. These lakes are briefly described in appendix 9.6 and table 4.8 gives a compilation of data used in the following simulations. Several comparisons between empirical SPM data and modelled values will be given and the presuppositions for these simulations will be discussed in the next section.

Table 4.8. Characteristics of the lakes used to test the dynamic SPM model.

Lake	Latitude °N	Altitude m.a.s.l.	Continen-tality km	Annual prec. mm/yr	Catch-ment km^2	Lake area km^2	Mean depth m	Max depth m	Mean TP µg/l	Sal-inity %
Batorino, Belarus	55	150	500	650	93	6.3	3	5.5	44	<1
Miastro, Belarus	55	150	500	650	130	13	5.4	11	34	<1
Naroch, Belarus	55	150	500	650	280	80	9	25	16	<1
Kinneret, Israel	32	-210	200	408	2560	168	24	42	27	1
Erken, Sweden	59	11	500	660	141	24	9	21	27	<1
Balaton, Hungary	47	106	1000	600	5280	596	3.2	11	63	<1
Balaton1, Hungary	47	106	1000	600	2750	38	2.3	4.5	87	<1

	Catchment ratio (ADA/A)	Dynamic ratio (DR)	Mean Secchi m	SPM inflow tons/yr	Mean SPM mg/l
Batorino, Belarus	14.7	0.84	1.5	160	9.9
Miastro, Belarus	10.2	0.67	2.9	140	4.1
Naroch, Belarus	3.5	0.99	6.8	170	1.6
Kinneret, Israel	15.2	0.50	2.9	40500	3.4
Erken, Sweden	5.8	0.54	4.5	130	4.2
Balaton, Hungary	8.9	7.62	0.55	15600	18
Balaton1, Hungary	72.4	2.58	0.36	8140	27

It should be stressed that the dynamic SPM model has not been validated (= blind tested). Most of the lake data have been used for calibrating and testing the model. However, there has been no "tuning" of the model to achieve the following results. All model variables have been kept as model constants and only the obligatory lake-specific driving variables (latitude, altitude, continentality, annual precipitation, catchment area, lake area, mean depth, maximum depth, and for most lakes also characteristic monthly chlorophyll data, and characteristic monthly TP concentrations in tributary water) have been changed for each lake.

4.4.1. Studied lakes and presuppositions

The six lakes represent a wide limnological domain: The areas vary from 6 to 600 km^2, the mean depths from 2 to 24 m, the mean TP concentrations from 16 to 87 µg/l (i.e., the trophic levels vary from low productive to very productive), and the latitudes from 32 to 59°N. There are also major differences in SPM values (table 4.8 and fig. 4.18. The highest SPM values appear in Lake Balaton, basin 1 (Balaton 1), Hungary (mean SPM = 27 mg/l), the lowest in Lake Naroch, Belarus (1.6 mg/l). So, there are major differences among these lakes in mean SPM values and there are also very marked seasonal variations (fig. 4.18A). This dynamic SPM model is constructed to predict these variations among and within lakes and the aim of this section is to demonstrate how that objective is achieved.

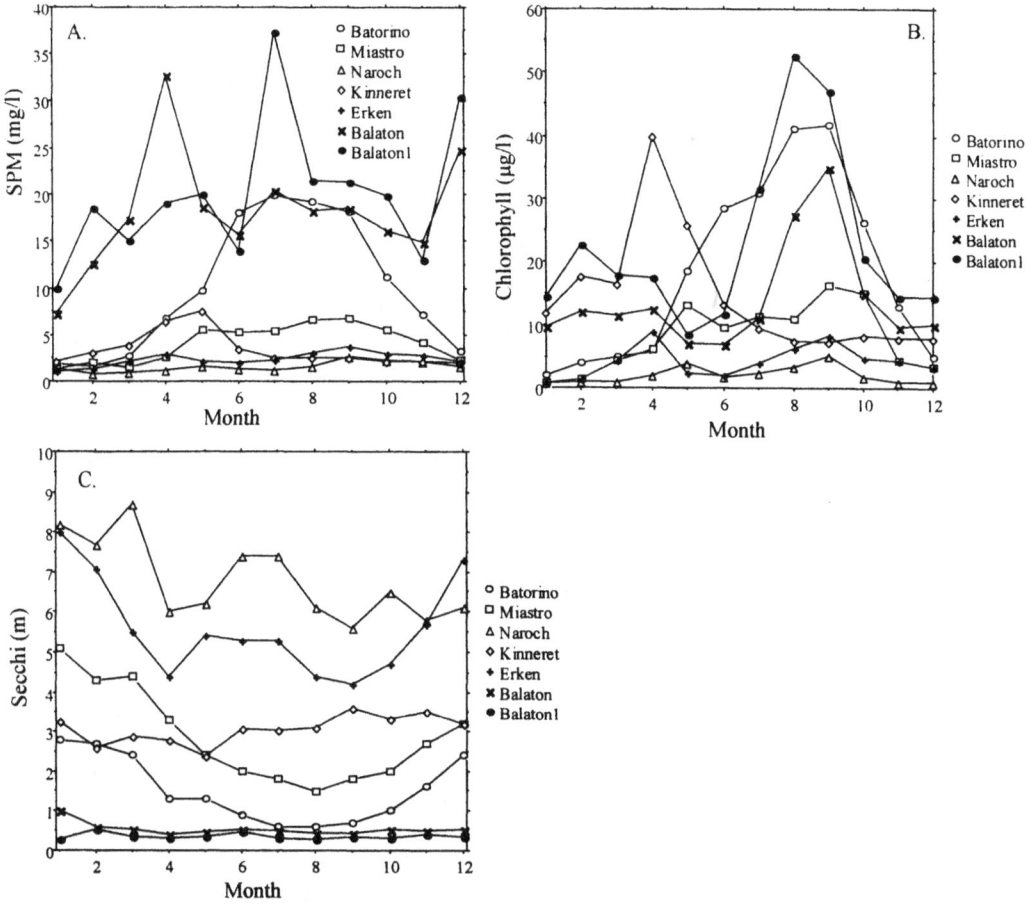

Fig. 4.18. Illustration of how median monthly empirical data for (A) SPM, (B) chlorophyll-a and (C) Secchi depth vary in the studied lakes.

These data emanate from the following sources:

- The Belarussian lakes Batorino, Miastro, and Naroch; basic data from several sources compiled by Dr. Alexander Ostapenia, Belarus State University, Minsk; the data have been presented and used by Håkanson and Boulion (2002). The data on chlorophyll, SPM in water, and Secchi depth represent median whole lake conditions from the 1990s.

- The data from Lake Kinneret, Israel, have been compiled by Dr. Arkadi Parparov, Kinneret Limnological Laboratory, Tiberias, and also these data have been used by Håkanson et al. (2000). The data used here represent median entire lake conditions for the period 1992-1996.

- The data from Lake Erken, Sweden, originate from the Erken Laboratory, Uppsala University (and Dr. Kurt Pettersson). They represent prevalent conditions during the 1990s.

- The data from Lake Balaton, Hungary, have been made available to the author from compilations done by Dr. Vera Istvanovics, Budapest University of Technology, and they have also been presented by Malmaeus and Håkanson (2003). They represent characteristics of the lake between 1989 and 1999.

The model calculates lake production of suspended particles using chlorophyll as a driving variable. These empirical chlorophyll data are shown in fig. 4.18B. The submodel for autochthonous SPM production has been used for all lakes in the following calculations and the Secchi depths needed in those calculations have not been the empirical Secchi depths shown in fig. 4.18C, but modelled values of Secchi depth since the idea is to illustrate how the model works and to create a feedback between modelled SPM values and the depth of the photic zone (using modelled values of SPM and the regression between those values and Secchi depth, fig. 2.5B). From fig. 4.18C, it is evident that large and shallow Lake Balaton has very low Secchi depths (often less than 0.5 m) and that Lake Kinneret has a relatively clear water and a Secchi depth of about 3 m. It is clear from fig. 4.18, that there are major variations among and within these lakes in SPM, chlorophyll, and Secchi depths.

Fig. 4.19A shows the modelled monthly values for lake production of SPM (in tons/month). These values represent the autochthonous part of SPM, as calculated using the submodel for lake production, which is driven by empirical chlorophyll data. The production model also accounts for surface water temperatures and production of SPM not just from phytoplankton but from all possible sources (zooplankton, bacterioplankton, feces, etc.). The calculation requires data on lake area and the depth of the photic zone (= Secchi depth). An interesting question in limnology concerns the role of the production to the total amount of SPM particles in lakes. This is illustrated in fig. 4.19C using empirical data (from fig. 4.18A, and not modelled SPM values). This fraction is generally high for all these lakes, especially during the growing season, and low during the winter period. Occasionally, as in Lake Kinneret, higher values than two appear. So, for all these lakes, autochthonous production is important for the SPM concentrations.

Similar data are given in figures 4.19B and D for SPM inflow, which is high in Lake Balaton (basin 1) during the winter when the fraction inflow to total amount in lake water is about 1. For most lakes, however, the inflow fraction is relatively low, especially during summer. It should be noted that the inflow has been quantified differently for different lakes. For the three Belarussian lakes, the inflow has been calculated from characteristic tributary TP concentrations using the LakeWeb model, as given by Håkanson and Boulion (2002; corresponding to tributary concentrations of 110, 65 and 40 µg/l respectively, for Lakes Batorino, Miastro and Naroch). These are the TP concentrations necessary to maintain empirical TP concentrations in water and sediments using the LakeWeb model (see appendix 9.2). For Lakes Erken and Balaton, the SPM inflows have been calculated in the same manner from empirical data on river TP concentrations (40 µg TP/l for Lake Erken and 200 µg TP/l for the inflow to Balaton 1 and for the entire Lake Balaton; see Malmaeus and Håkanson, 2003). For Lake Kinneret, empirical data on the SPM concentrations in River Jordan have been used; this inflow represents about 80% of the total inflow (see Håkanson et al., 2000).

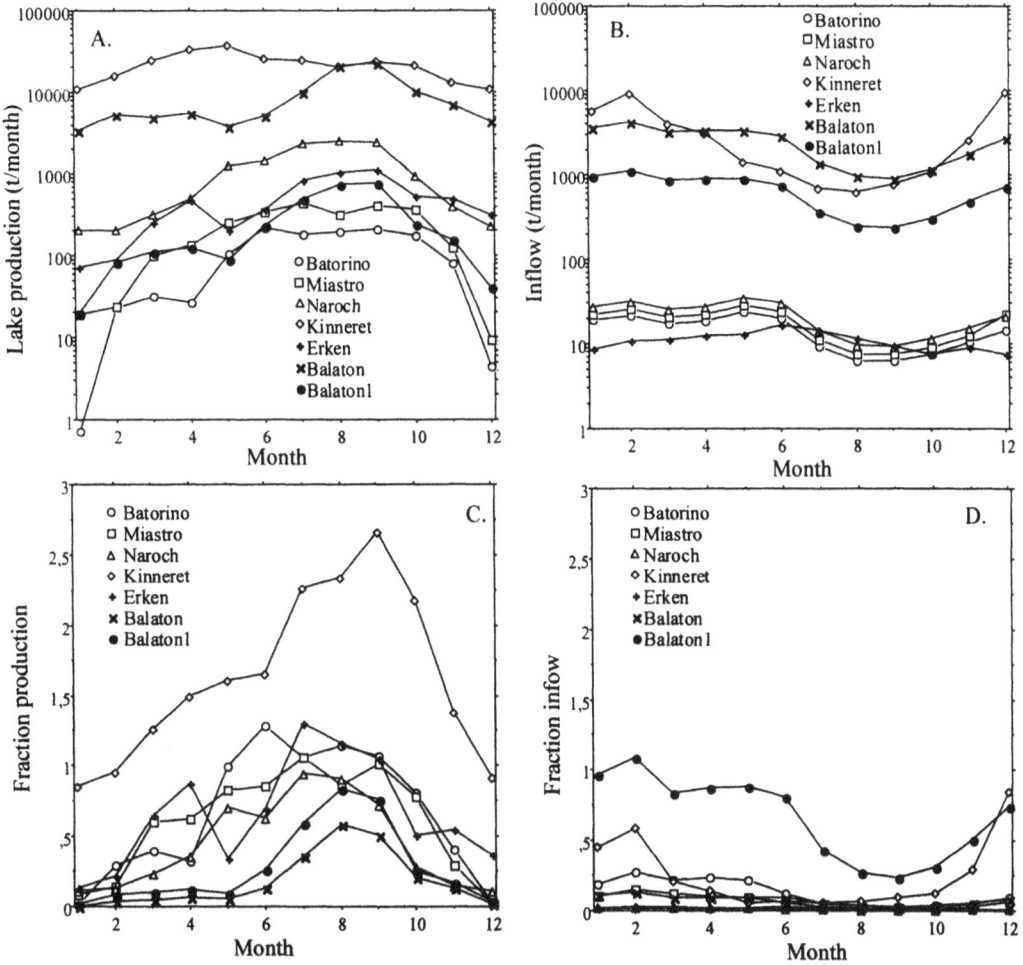

Fig. 4.19. Illustration of how production and inflow of SPM vary (figures A and B) and how the fraction of production and inflow vary (figures C and D) in the studied lakes (the fraction is calculated as the ratio between monthly SPM inflow in tons/month to total monthly SPM amount in lake water (tons).

4.4.2. Objectives

From these presuppositions one can ask: (1) How well will the model predict the empirical SPM data using the default set-up? (2) Will the model systematically predict too low or too high values in certain lakes or for certain periods? (3) How well can one expect the model to predict accounting for the uncertainties in the y variable (a general CV of 0.65 for individual SPM data in lakes and the available information on the empirical CVs for these studied lakes) and the uncertainties related to the driving variables and the given assumptions in the model structure?

The last question will be addressed first. Since the characteristics CV for individual SPM data is 0.65, and since there are generally between 1 and 8 empirical data per month in the different lakes, one can expect that the CV for the mean would be about $0.65/\sqrt{4} = 0.325$ if 4 samples would be at hand for each mean value. This is generally not the case, rather it is n = 3, which means a CV for the mean of $0.65/\sqrt{3} = 0.38$, corresponding to a highest reference r_r^2 of $(1-0.66\cdot0.38^2) = 0.90$. So, one should not expect the model to predict better than that when modelled values are compared to empirical data. As a reference, the empirical SPM model gives an r^2 value of 0.89.

In the following, the results will first be presented for each lake and the idea is to illustrate and exemplify how the model works by making sensitivity tests, uncertainty tests, and other types of tests.

4.4.3. Results

4.4.3.1. Belarussian lakes

A. Lake Naroch, the largest of the three Belarussian lakes.

Fig. 4.20A gives empirical SPM data and modelled-predicted values for Lake Naroch. One can see that the model predicts SPM values close to the empirical data. In fact, the model cannot be expected to predict better given all the uncertainties in the empirical SPM data and the data used to run the model, mainly chlorophyll and inflow. The uncertainties in the predictions in this lake – and in all lakes – are mainly related to the large fluxes. Fig. 4.21 gives a ranking of all SPM fluxes to, from, and within Lake Naroch. Lake production of SPM is the dominant flux, and also mixing from surface water to deep water, sedimentation from deep water to accumulation areas, upward mixing and mineralization in surface and deep water areas. The smallest flux in this lake is SPM outflow.

The results given in fig. 4.21 can also be seen as a result from a sensitivity analysis using uniform uncertainties, since this would give the same ranking among the fluxes. However, it is not realistic to use the same uniform uncertainties for all fluxes, since some of the fluxes have lower inherent uncertainties. Fig. 4.22 gives result from an uncertainty analysis using characteristic uncertainties for mineralization (a CV of 0.5), empirical chlorophyll values (a CV of 0.14 in Lake Naroch for mean monthly values), the calculation of ET (the fraction of ET areas in the lake, a CV of 0.2), the age of the material on ET areas (a CV of 0.5), sedimentation (CV = 0.5), inflow (CV = 0.5), mixing (CV = 0.5), outflow (CV = 0.25), and calculated surface water temperatures using the model given in appendix 9.4 (CV = 0.2). Utilizing these CV values, the total CV value for y (SPM in lake water) is 0.28 using Monte Carlo techniques and 100 runs (see section 3.2).

Fig. 4.22A provides all the curves related to the uncertainty in y when all uncertainties in the given model variables are accounted for and fig. 4.22B gives the corresponding box-and-whisker plot for the data from month 8, the month with the highest variation. Fig. 4.22C gives a ranking of the factors contributing to this uncertainty and mineralization contributes with the highest uncertainty. If the uncertainty for the mineralization rate is omitted, the uncertainty in the y variable is reduced from 0.28 to 0.10. The second most important uncertainty concerns the value for chlorophyll used to calculate lake production, which is the most important flux for SPM in this lake. All other uncertainties are of less importance in Lake Naroch in predicting SPM in the lake water.

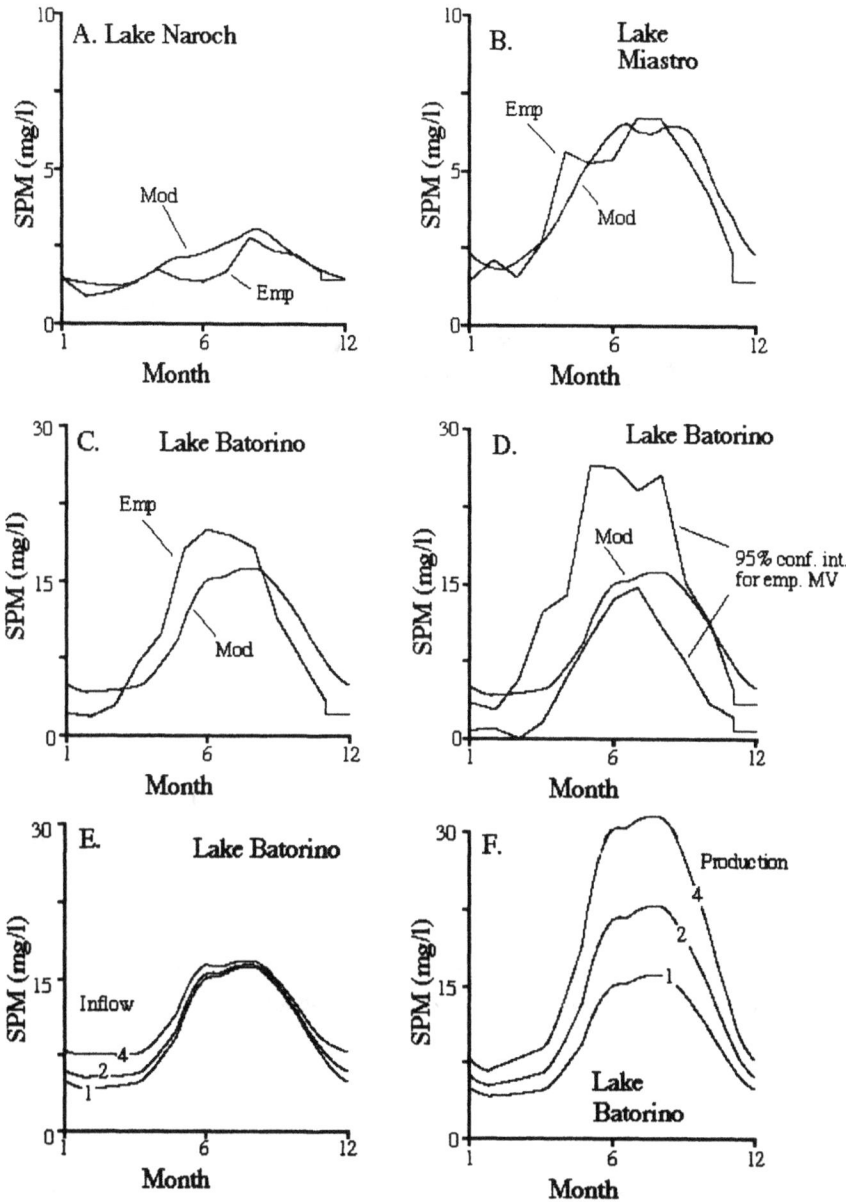

Fig. 4.20. Modelling results for the Belarussian lakes.
A. Empirical data versus modelled SPM values for Lake Naroch.
B. Similar results for Lake Miastro.
C. Similar results for Lake Batorino.
D. Modelled SPM values compared to the 95% uncertainty intervals for the empirical data (as given by the curve in fig. C) for Lake Batorino.
E. Sensitivity analyses showing modelled SPM values if the tributary inflow is increased by a factor of 2 and a factor of 4 and all else is kept constant in Lake Batorino.
F. Results from a similar sensitivity analysis when lake production is varied by a factor of 2 and a factor of 4 in Lake Batorino.

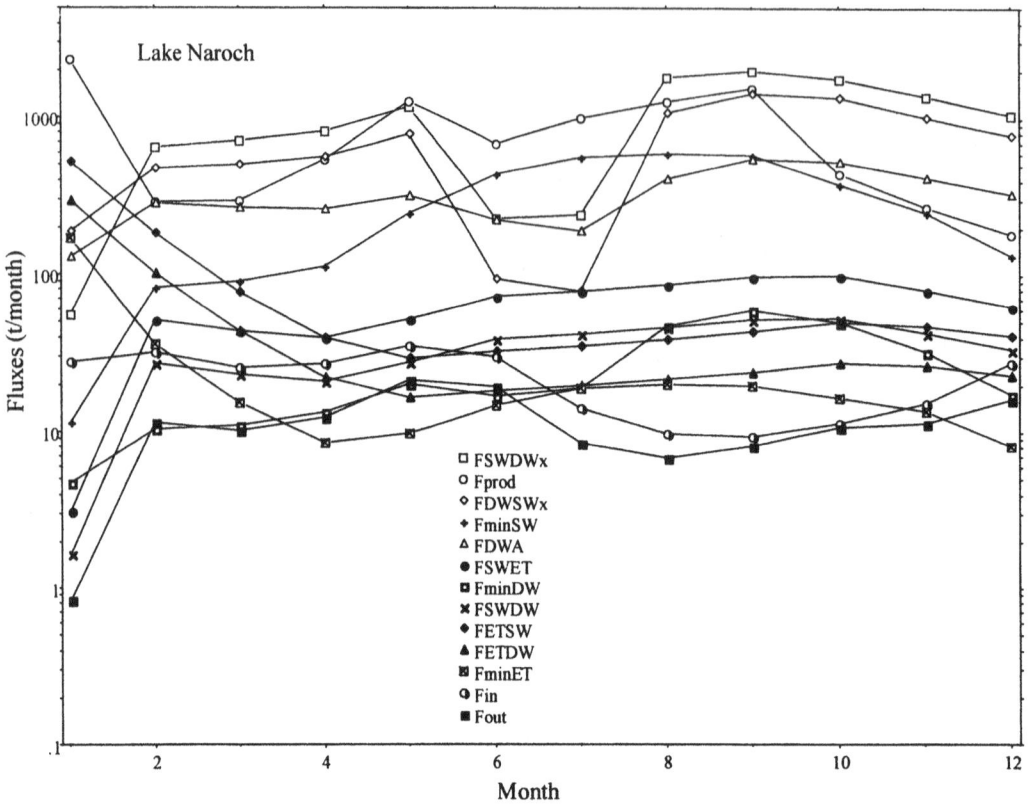

Fig. 4.21. Compilation of calculated monthly SPM fluxes in Lake Naroch.

B. Lake Miastro

Fig. 4.20B gives the results for Lake Miastro, the middle lake along the course of inter-connected Belarussian lakes. The model predictions are very good and well within the uncertainty bands of the empirical data.

C. Lake Batorino

The results for this lake are given in fig. 4.20C. This is one of the worst predictions. One can see that the model predicts too low values during the growing season and too high values during the winter, but that the average value agrees well with the empirical mean value. Fig. 4.20D shows that the model predictions, in fact, are not too bad since they are well within the 95% confidence interval for the uncertainties in the empirical data. Fig. 4.20E gives a simulation where the inflow has been increased by a factor of 2 and a factor of 4 to see if this would give a better correspondence between empirical data and modelled values for the summer period, but that is not the case because the inflow is not a dominating SPM flux in this lake, which is dominated by lake production. If, on the other hand, lake production is changed by a factor of 2 and a factor of 4, too high SPM values would be obtained (fig. 4.20F).

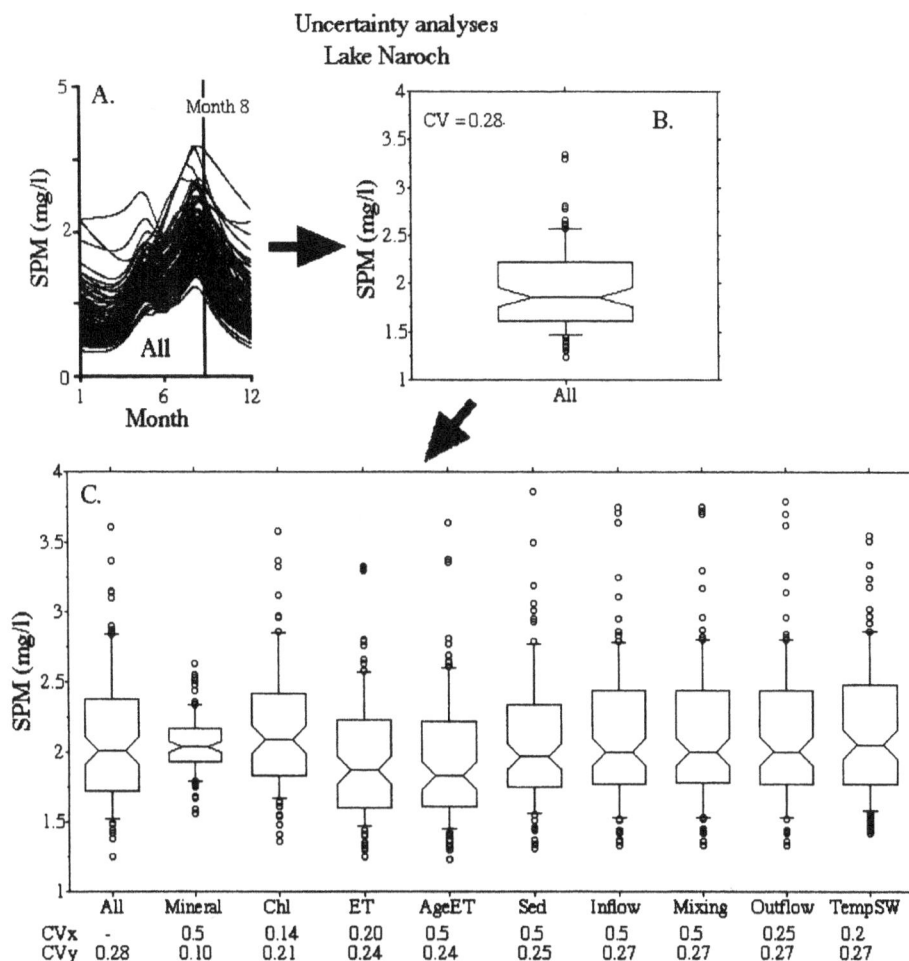

Uncertainty analyses
Lake Naroch

Fig. 4.22. Results of uncertainty analyses using characteristics CV values for selected model variables using Monte Carlo techniques and 100 runs for Lake Naroch.
A. Gives the total uncertainty for modelled SPM values in lake water;
B. Gives the corresponding box-and-whisker plot for the data from month 8 (the total CV for y is 0.28);
C. First gives the same results as in fig. B; then the ranking of the uncertainties contributing to the uncertainty in the y variable. If the uncertainties for mineralization are omitted, and all other uncertainties accounted for, CV for y is reduced the most, from 0.28 to 0.10. The next most important uncertainty in prediction SPM in this lake concern the empirical chlorophyll data used to predict lake production of SPM. If this uncertainty is omitted (and all else included), CV for y is reduced from 0.28 to 0.21.

To understand the role of SPM in aquatic systems and the factors influencing SPM, fig. 4.23 gives biomasses (using the LakeWeb model, see appendix, 9.2) of phytoplankton, bacterioplankton, herbivorous zooplankton, and total SPM amounts in lake water for (A) Lake Naroch, which has a low catchment area to lake area ratio (3.5), and for the upstream lake, Lake Miastro, which has a catchment area to lake area ratio of 10.2. The idea with this test is to stress the following important points:

1. Phytoplankton, bacterioplankton, and herbivorous zooplankton are all living, biotic components of SPM, but taken together, they form only a small fraction of the total SPM amount in lake water (note the scale in fig. 4.23).

2. In Lake Naroch, phytoplankton dominates over bacterioplankton, and in Lake Miastro, one finds the opposite, since the latter lake has a relatively much larger catchment area contributing allochthonous materials utilized for bacterioplankton production.

3. Herbivorous zooplankton feed on both phytoplankton and bacterioplankton. This predation is clear from fig. 4.23B where the spring peak in herbivorous zooplankton creates a reduction in phytoplankton biomass.

4. The main role that bacterioplankton play in aquatic systems is to decompose the organic fraction of SPM, and the very high biomass of bacterioplankton especially in Lake Miastro would not have been found if the SPM concentration would not have been so high (4.1 mg/l in this lake and 1.6 mg/l in Lake Naroch, on average). The very high biomasses of bacterioplankton indicates the important role of the mineralization process, which will be discussed in greater detail for some of the following lakes.

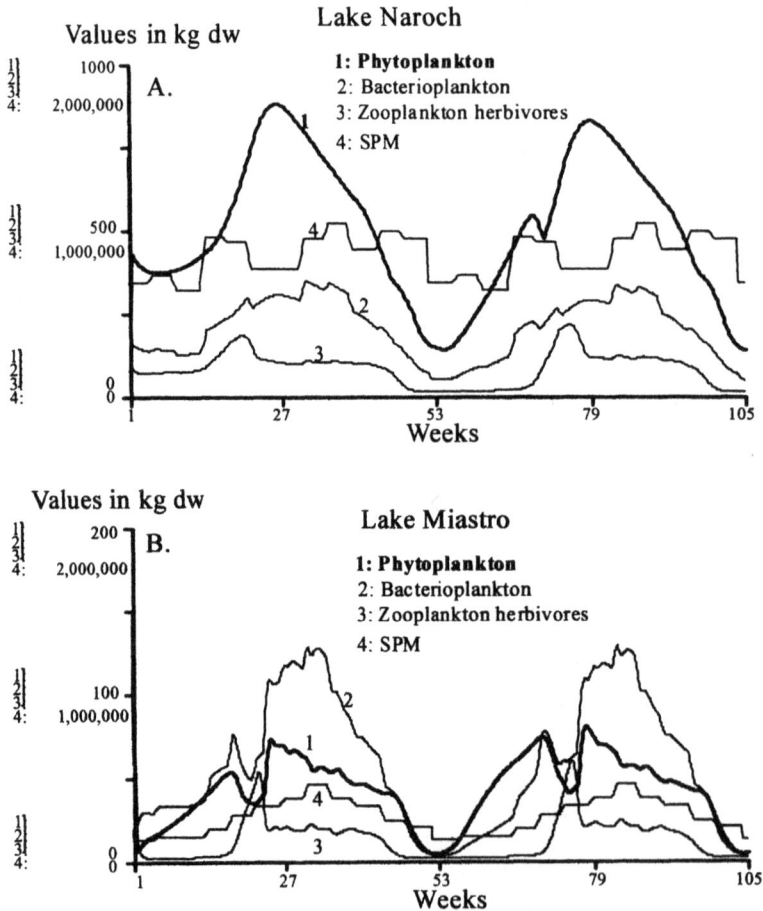

Fig. 4.23. Calculated biomasses of phytoplankton (curves 1), bacterioplankton (curves 2), herbivorous zooplankton (curves 3) and total SPM amount (curves 4, as calculated from the LakeWeb model using empirical data on Secchi depths) for (A) Lake Naroch and (B) Lake Miastro.

4.4.3.2. Lake Kinneret

The results for Lake Kinneret are shown in fig. 4.24. The first results (fig. 4.24A) have been obtained when empirical data on lake production (and not the lake production submodel) and on lake inflow were used (and not the inflow submodel; data from Malmaeus and Håkanson, 2003). One can first note the good correspondence between empirical and modelled values. Fig. 4.24B gives the results when the submodel for lake production has been used (using data on empirical chlorophyll values). Then, the mean SPM level is predicted well and also the seasonal pattern. These results reflect the seasonal pattern expected from the empirical chlorophyll values and the empirical inflow values and assuming all else would correspond to the conditions in a "normal" lake. It should be stressed that a poor covariation between modelled values and empirical data can be related either to deficiencies in the model or to uncertainties in the empirical data. The latter point is evident given the inherent uncertainty for empirical SPM values for lakes (a CV of 0.65). This point will be discussed in more detail in connection with the results for Lake Balaton.

It should be noted that the temperature submodel was not used to get the results in figures 4.24A and 4.24B. Instead those predictions utilized empirical temperature data for both the surface water and the deep water areas (from Håkanson et al., 2000). Lake Kinneret lies outside the domain of the temperature submodel, which applies to lakes situated between latitude 37°N and 65°N; and Lake Kinneret is at latitude 32°N. Also, the temperature submodel has not been tested for lakes situated lower than the sea level; and Lake Kinneret is at –210 m below the sea level. Anyhow, fig. 4.24C gives the results when the submodel for temperatures has been used, and one can note that the model then gives too high SPM values because the predicted temperatures are too low (compared to the empirically determined temperatures). The lake water temperature influences lake production, mixing, and mineralization, and the net result is shown in fig. 4.24C.

There are also uncertainties concerning the outflow of water and everything dissolved and suspended in the water, including SPM, since there have been major water level fluctuations in Lake Kinneret (see Håkanson et al., 2000). Fig. 4.24D gives results when lake outflow has been set to zero, to two times the default value, as given by the outflow submodel, and to five times the default value. This does not influence the result in any significant manner because lake outflow is not a dominating SPM flux in this lake (fig. 4.33).

The salinization of the water in Lake Kinneret is a major concern (see appendix 9.6.2) and fig. 4.24E gives a sensitivity analysis when the salinity has been changed (set to 1, 2 and 4%; 1% is the default value). A higher salinity will increase flocculation and sedimentation and reduce SPM from lake water, as shown in this figure.

A very important process in the warm water of Lake Kinneret is mineralization. This is demonstrated in the sensitivity analysis shown in fig. 4.24F, where the default value for the mineralization rate has been first reduced and then increased by a factor of 2. If the mineralization rate is low, higher SPM concentrations will appear in the lake water.

From these results for Lake Kinneret, one can note:

- The SPM predictions are quite good if empirical SPM data are used for lake production and inflow.

- The lake mean value is predicted quite well by the model using the lake production submodel, but the seasonal pattern does not predict the empirical SPM values well.

- There are, however, major uncertainties in the empirical mean monthly SPM data – the CV for the monthly mean values in this lake is 0.33, which means that a mean value of 3.4 has a 95% confidence interval between 1.2 and 5.6. The model predicts well within this uncertainty.

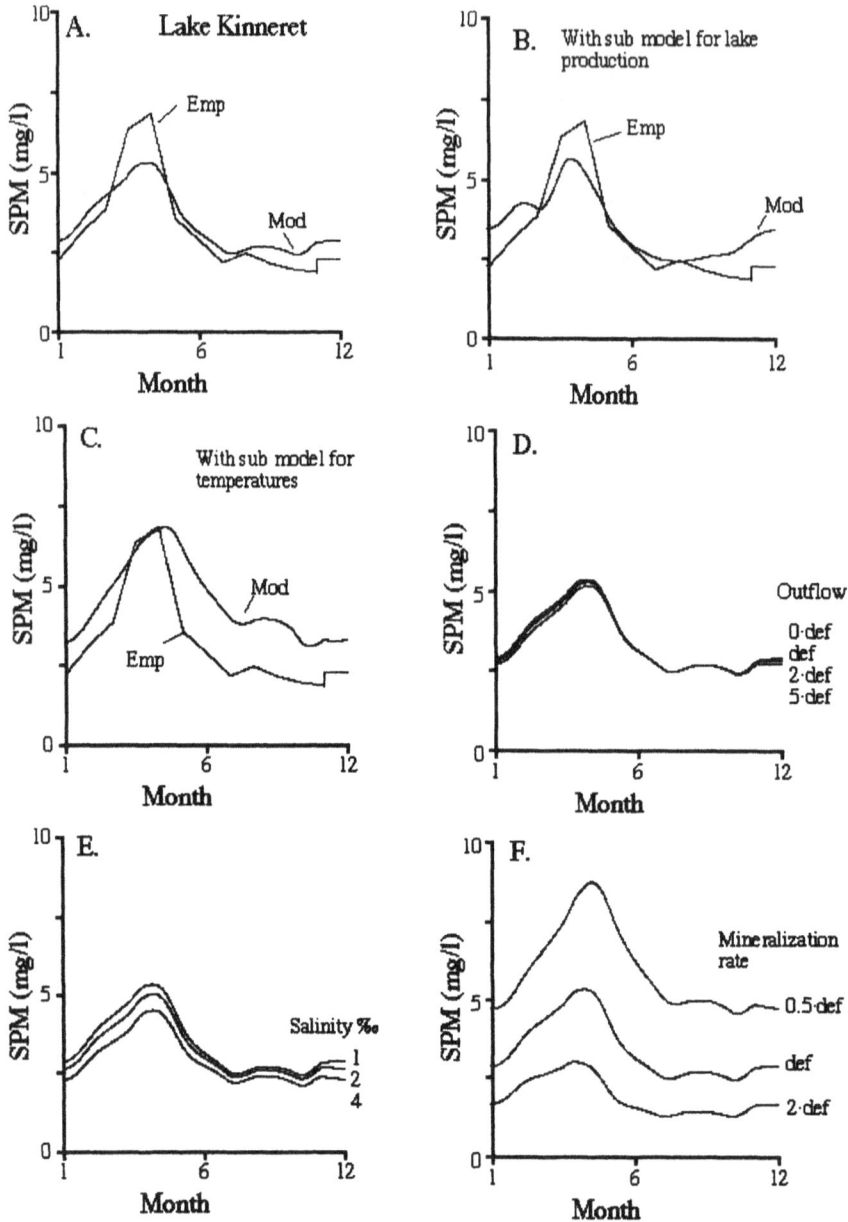

Fig. 4.24. Modelling results for Lake Kinneret.

A. Gives a comparison between modelled and empirical data.

B. Gives a comparison between modelled SPM data using not the empirical data for lake production (as in fig. A) but instead the submodel for lake production based on empirical chlorophyll data.

C. A comparison between modelled and empirical data as in fig. A but using modelled temperature data instead of empirical temperature data.

D. Sensitivity analyses when the outflow from the lake has been altered in steps (set to 0, 2, and 5 times the default values calculated by the outflow submodel) and all else kept constant as in fig. A.

E. Results from sensitivity analyses when the salinity has been altered in steps (set to 1, 2, and 4%; 1% is the default value for Lake Kinneret) and all else kept constant as in fig. A.

F. Results from sensitivity analyses when the default value of the mineralization rate has been increased in two steps (with a factor of 0.5 and a factor of 2) and all else kept constant as in fig. A.

4.4.3.3. Lake Erken

The results for Lake Erken are shown in fig. 4.25A. The SPM inflow has been calculated from the submodel based on empirical TP concentrations in the tributary. Lake production has been calculated using the submodel based on empirical chlorophyll data. To calculate the SPM inflow in this manner involves uncertainties, but the sensitivity analyses given in fig. 4.25B demonstrate that in this lake the SPM inflow is a relatively small flux (SPM in the lake is totally dominated by autochthonous SPM production; fig. 4.33) so the uncertainty in the inflow is not important. The default inflow has been changed in steps (1, 2, 4, and 8 times the default values) in fig. 4.25B.

There are also uncertainties in the equation for mixing. Fig. 4.25C gives results from a sensitivity analysis where the default mixing rate has been changed (by 0.2, 0.5, 2, and 5). This is not so important because mixing is not a very important transport process for SPM in this lake.

Mineralization is more important, which is shown by a similar sensitivity analysis in fig. 4.25D.

The results show that:

- The SPM predictions agree very well with the empirical SPM data.

- Lake Erken has a low catchment area to lake area ratio (5.8) and the SPM concentrations in the lake water are mostly influenced by lake production.

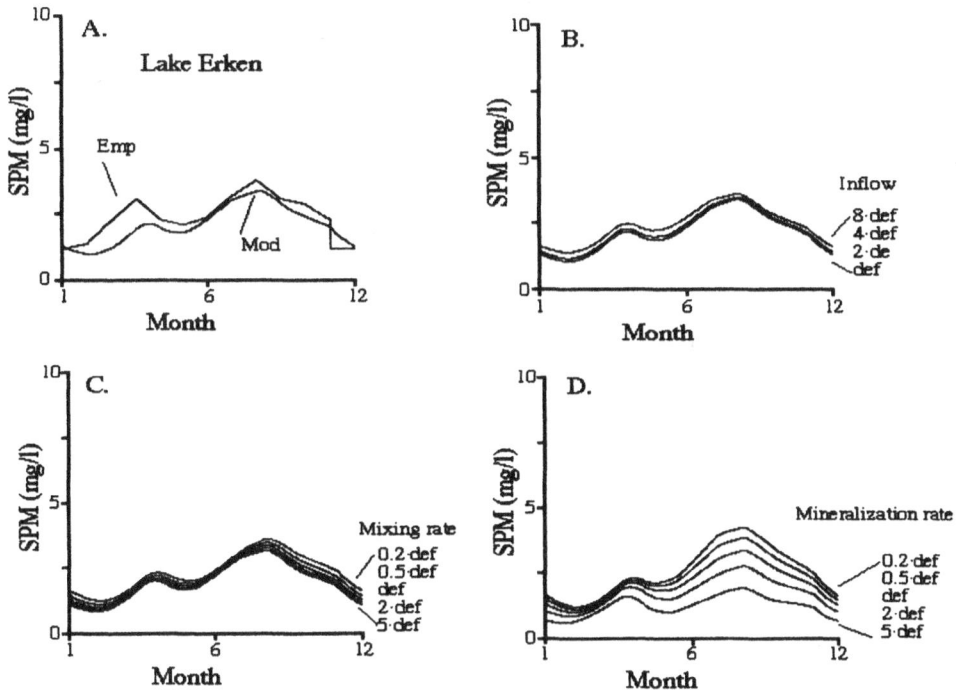

Fig. 4.25. Results for Lake Erken.
A. A comparison between modelled and empirical data.
B. Results from sensitivity analyses where the inflow to the lake has been altered in steps (set to the default values calculated by the outflow submodel and to 2, 4, and 8 times the default values) and all else kept constant as in fig. A.
C. Sensitivity analyses where the mixing rate has been altered in steps (set to 0.2, 0.5, 2, and 5 times the default values) and all else kept constant as in fig. A.
D. Sensitivity analyses where the mineralization rate has been altered (0.2, 0.5, 2, and 5 times the default values) and all else kept constant as in fig. A.

4.4.3.4. Lake Balaton

Lake Balaton is a large and shallow lake with four interconnected subbasins, Balaton 1, 2, 3, and 4 making up 4.3, 21.8, 31.6, and 42.2 % of the total lake area, respectively. Fig. 4.26 gives modelling results for the entire lake and for basin 1. The model does not capture the seasonal change given by the empirical data very well. The possible reasons for this will first be discussed. One can test if this depends on:

- Resuspension - from the fact that the lake has a dynamic ratio of 7.6 and that the ET areas totally dominate many, and maybe most, lake processes.

- Reduced SPM production from algal shading (see Håkanson and Boulion, 2002) given the fact that the Secchi depth is so very low (often < 0.5 m).

- Uncertainties in SPM outflow, since the evaporation may be large and the algorithm for evaporation and outflow may not work well for this lake.

- Major uncertainties in inflow since this is calculated from empirical tributary TP concentrations to Balaton 1 and from an assumption that the SPM transport from the rest of the catchment area is proportional to the transport from the catchment area of Balaton 1.

All these uncertainties could be expected, they have been tested, (fig. 4.26C to F) and the poor fit between empirical data and modelled values are not related to any of these processes influencing the model predictions. The reason for the poor fit instead lies in the empirical data, as the following results will demonstrate.

Fig. 4.27 first illustrates this by showing empirical CV values for Balaton 1 calculated from all existing 326 individual SPM determinations from this subbasin, a CV of 0.99. This is very, very high and much higher than the characteristic CV for SPM in lakes, which is also very high (0.65). It means that to determine the mean value with an error smaller than 20% (with a 95% certainty), 95 samples are needed. The CV value for the mean monthly SPM values from 1989 to 1999 is 0.93 based on data not from 120 months but from 109 months. It means that a given mean monthly value, e.g., 30 mg/l, with a 95% probability lies between zero and 86! The CV based on mean annual data is 0.41. This means that one year, the value can be 30 mg/l and the next year, the mean annual value is probably (95% probability) between 5 and 75 mg/l.

The results given in fig. 4.28 stress this point even more. It gives regressions between mean monthly SPM values in the four interconnected basins in Lake Balaton using the largest subbasin (Balaton 4) as a main reference. Had the empirical SPM data been reliable and representative for the given system rather than for the conditions at the given sampling site when the samples were taken, there would be a general systematic covariation in SPM values among these lakes – but there is none!

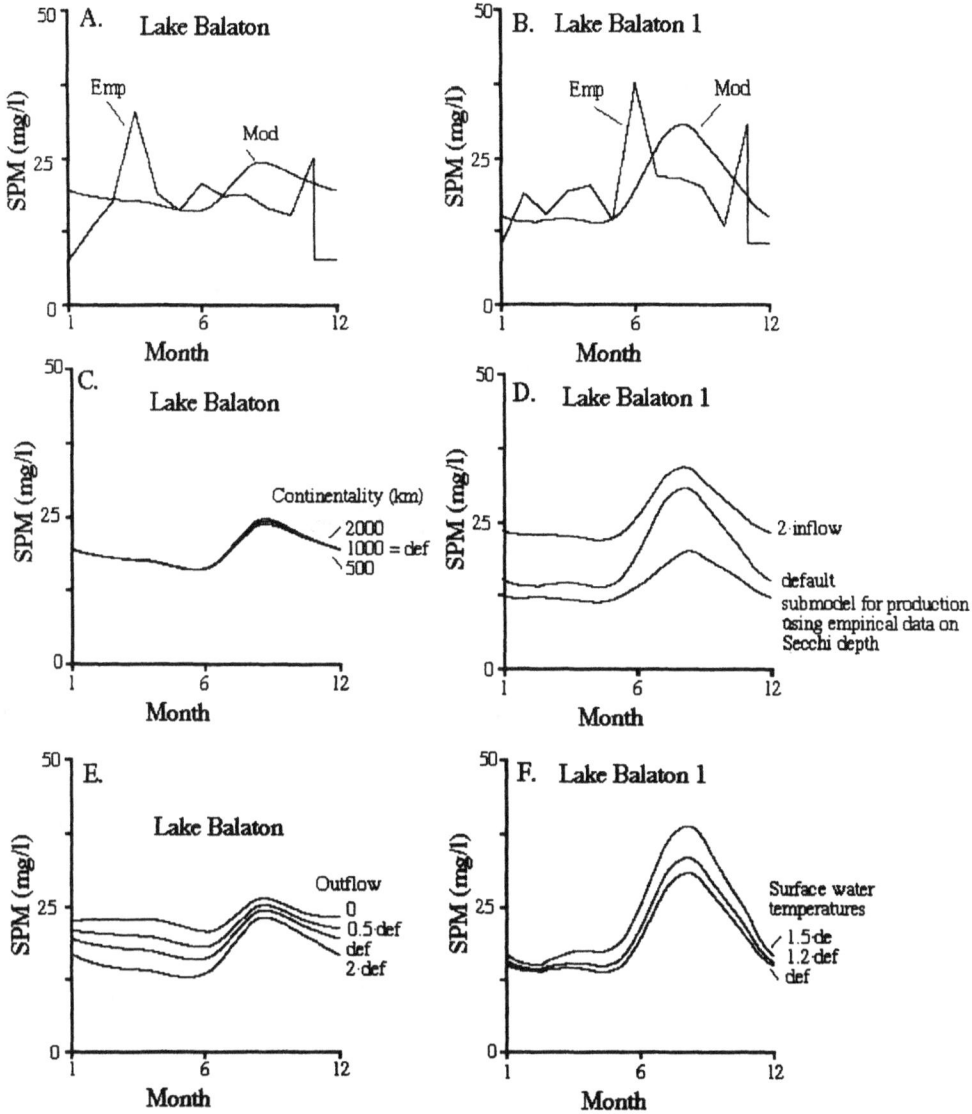

Fig. 4.26. Results for Lake Balaton and Lake Balaton 1.

A. Comparison between modelled and empirical data for Lake Balaton.

B. The same for Lake Balaton 1.

C. Results from sensitivity analyses where the continentality has been altered in steps (500, 1000, and 2000 m.a.s.l) and all else kept constant as in fig. A for Lake Balton.

D. Results when the inflow has been increased by a factor of 2 and when the submodel for production has been used with empirical data on Secchi depths rather than modelled values of Secchi depths and all else kept constant as in fig. B for Lake Balaton 1.

E. Sensitivity analyses where the outflow from Lake Balaton has been altered in steps (set to 0, 0.5, and 2 times the default values calculated by the outflow submodel) and all else kept constant as in fig. A.

F. Sensitivity analyses where the modelled surface water temperatures have been increased by a factor of 1.2 and a factor of 1.5 (compared to the default values calculated by the temperature submodel) and all else kept constant as in fig. B for Lake Balaton 1.

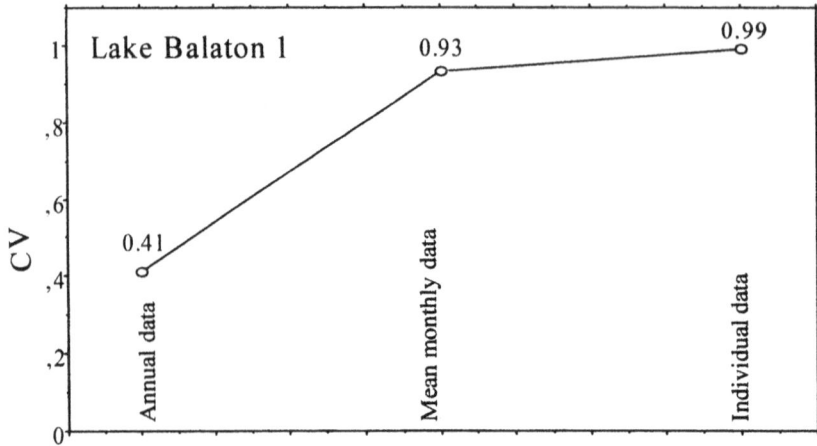

Fig. 4.27. CV values for SPM in Lake Balton 1 calculated from all individual SPM data, from mean monthly SPM data and from yearly mean values.

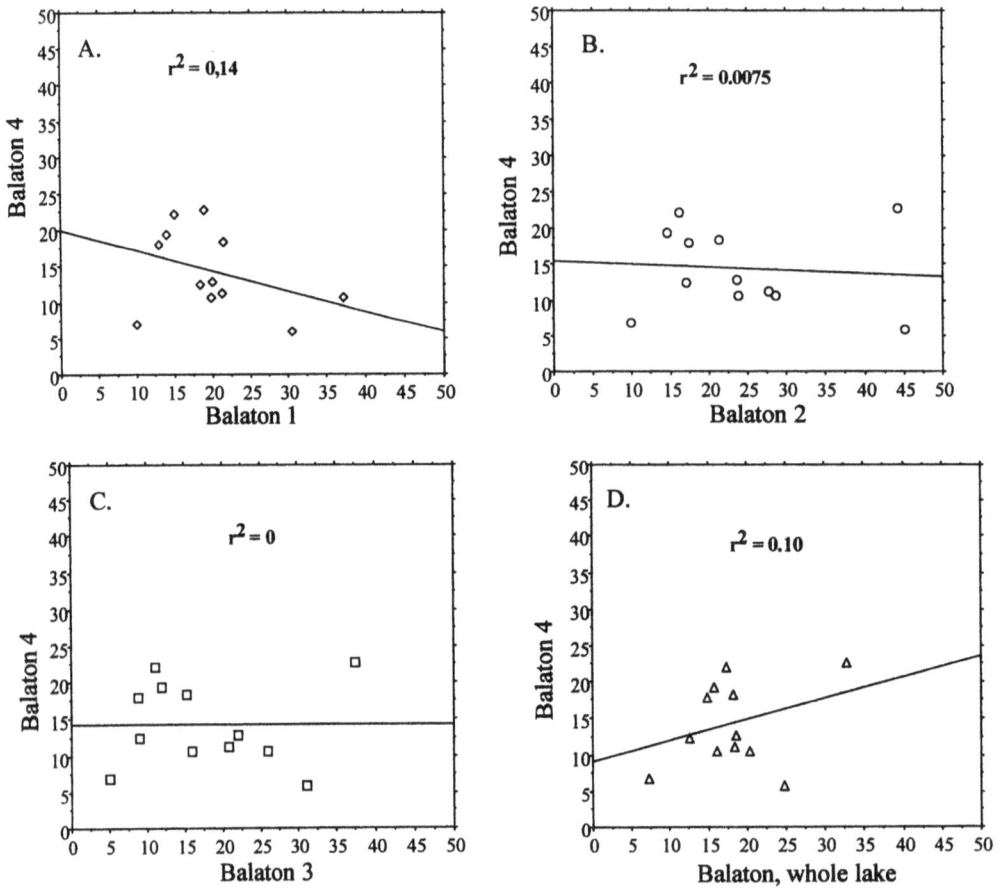

Fig. 4.28. Regressions between mean monthly SPM values from (A) Balaton 4 versus Balaton 1, (B) Balaton 4 versus Balaton 2, (C) Balaton 4 against Balaton 3, and (D) Balaton 4 against the whole lake.

One can stress this point from at least one more angle. Fig. 4.29 gives the data on Secchi depth and SPM using the same database as in fig. 2.4 but in this case only data with lower Secchi depths than 3 m have been used. Then, it is evident that the very marked relationship between Secchi depth and SPM is not at hand, and especially not for lakes or sites with Secchi depths below 1.6 m. Lake Balaton has a Secchi depth of about 0.2-0.5 m, and for such sites, the relationship between Secchi depth and SPM becomes very weak and uncertain basically because the spatial and temporal SPM variability becomes very large and each mean SPM value and Secchi depth registration carries very limited information.

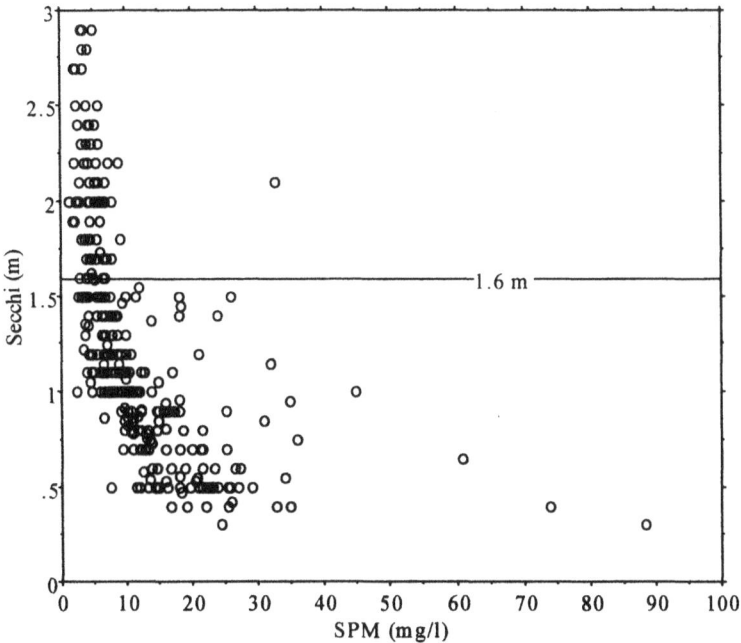

Fig. 4.29. SPM values versus Secchi depths for lakes/sites with Secchi depth smaller than 3 m.

A conclusion from this is that it is more appropriate to control how the model predicts the mean long-time average SPM conditions in Lake Balaton using a mean or median value based on many samples. Fig. 4.30 shows how the model predicts the mean annual SPM concentration in Lake Balaton. This figure gives a different view of the model behavior. The model predicts the mean long-time SPM value very well.

4.5. Conclusions and comments

To illustrate how the dynamic model works in predicting SPM concentrations in these lakes, fig. 4.31 gives a regression between the actual mean monthly empirical data (and not the yearly average for Lake Balaton) and the corresponding modelled values. The main reason why the more appropriate mean annual values are not used for Lake Balaton and Lake Balaton 1 is that it is rather pointless to regress a modelled time series of data against a constant for each lake.

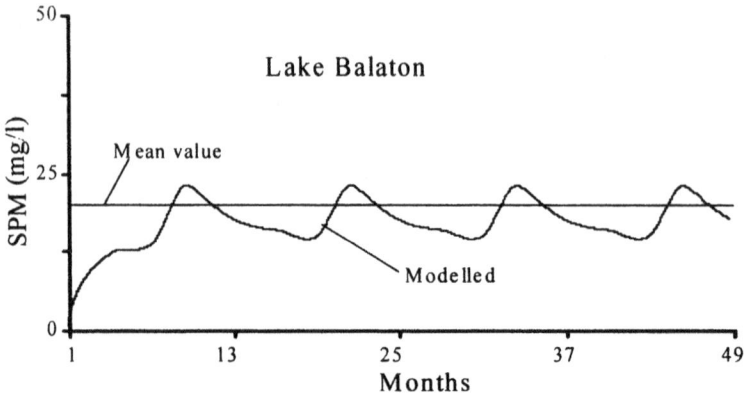

Fig. 4.30. Modelled SPM values versus the mean SPM value for Lake Balaton calculated from all individual data from 1989 to 1999.

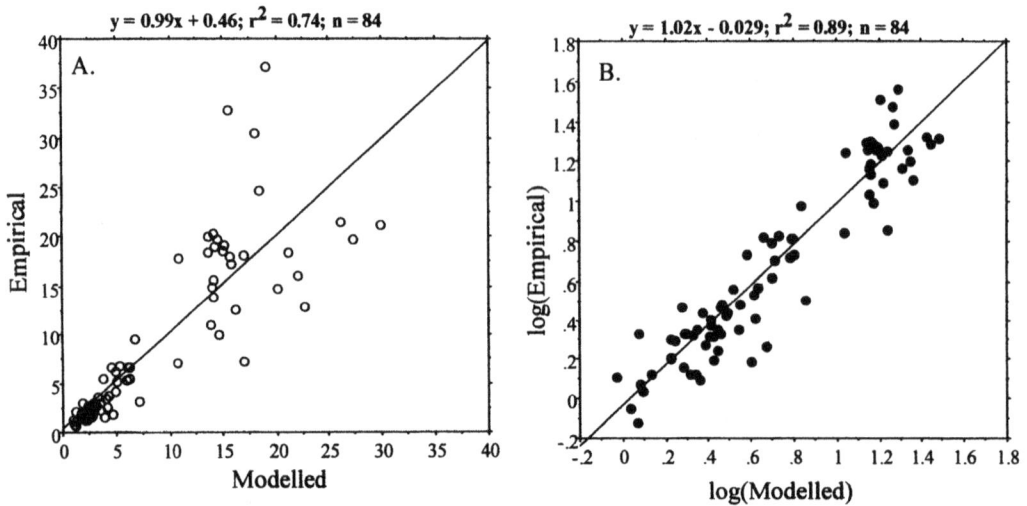

Fig. 4.31. Comparison between mean monthly empirical data and modelled monthly values for all 7 studied lakes using (A) actual data and (B) log data. The figure also gives the regression lines, r^2 and n.

From fig. 4.31A, it is clear that neither the x nor the y variable have a normal frequency distribution, so the regression equation and the r^2 value based on actual data have limited information. However, the slope is close to 1 (it is 0.99) and the r^2 value is 0.74. Fig. 4.31B gives the regression using logarithmic data. Then, the r^2 value is 0.89 and the slope 1.02 and the spread around the regression line is even. The highest reference r^2, r_r^2, is about 0.9 (as stressed before), so this is a very good result in spite of the fact that the empirical data for Lake Balaton (and Balaton 1) are very uncertain with a CV of 0.99. Fig. 4.32 gives the error function for the actual data (not the log values) and the mean error is close to zero (0.05), the standard deviation for the error is 0.48, and the error shows a normal distribution around the mean error. This means that the model is well balanced.

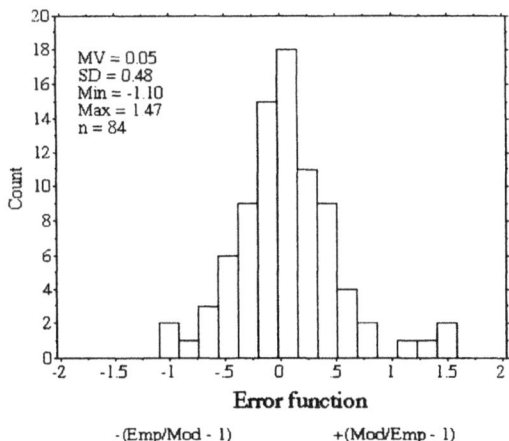

Fig. 4.32. The error function and basic statistics as calculated from the comparison between actual empirical and modelled mean monthly SPM values.

Fig. 4.33 gives a compilation of all annual fluxes for the studied lakes. One can note that:

- Upward and downward mixing are important transport fluxes in many lakes.

- Different fluxes are important in different lakes, so there is no meaning in generalizations such as "inflow is always important" or "mineralization is always important." Each lake is an "individual" but the individual characteristics may be expressed by the general transport fluxes given in fig. 4.33 and accounted for by this dynamic SPM model. These processes apply to all substances in all lakes, rivers, and coastal areas.

The following two chapters are meant to scrutinize if the last statement is valid.

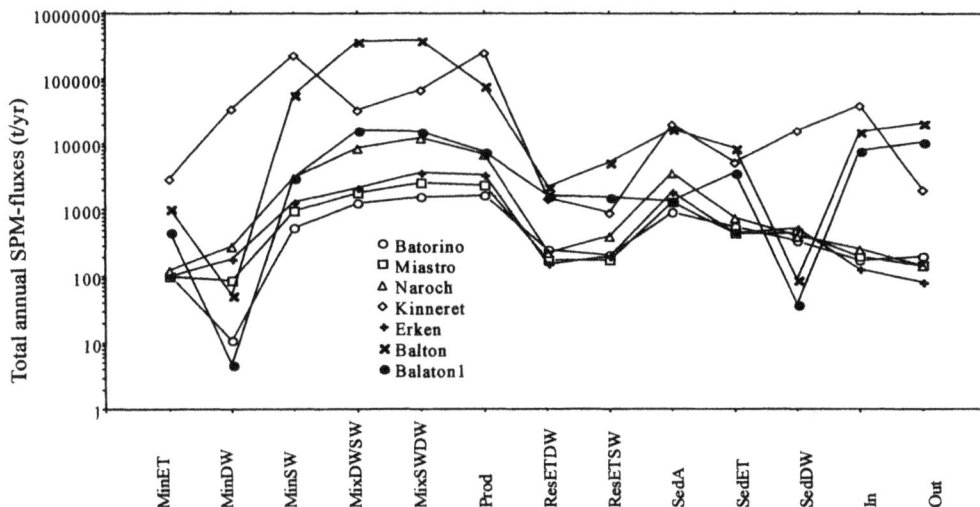

Fig. 4.33. Compilation of all calculated transport processes on an annual basis from the seven studied lakes. Min = mineralization, Mix = mixing, Res = resuspension, Sed = sedimentation, In = inflow, Out = outflow, ET = ET areas, DW = deep water areas, SW surface water areas, A = accumulation areas.

5. SPM in Rivers

This chapter will first give a short background on catchment areas and important concepts related to the transport of substances from land to water. The following sections discuss an empirical model and then a dynamic model for SPM in rivers. This dynamic model draws much from the dynamic lake model, and the idea here is not to repeat similarities but to stress the specific features of the dynamic river model.

5.1. Background on catchment areas and the transport from land to water

Fig. 5.1 gives a schematic description of a catchment area showing the lake, its tributaries, and the areal distributions of different features (upstream lakes, mires, open cultivated land, hard rock, morainic soils, and forests). The area of such features may be determined by planimetry, computerized digitizing, GIS-techniques (Geographical Information Systems), counting squares and dots, etc. (Gaudie, 1981, 1990). The chemical and biological characteristics of the river or lake water partly depend on these features and their distribution in the catchment area, as indicated in fig. 5.1.

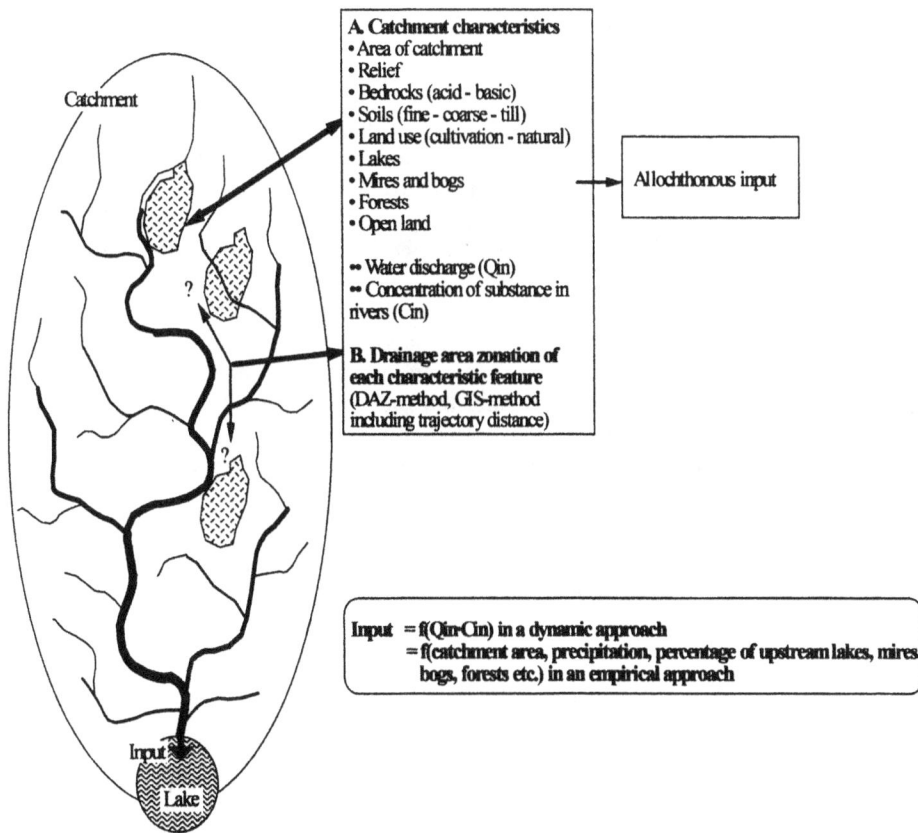

Fig. 5.1. Schematical illustration of a catchment area, fundamental features of catchment areas regulating the transport of substances from land to water, zonation of catchment areas, and modelling approaches.

The importance of bedrock geology for river and lake water chemistry has been stressed in many studies. For example, Newton et al. (1987) and Nilsson (1992) have shown that aquatic systems in areas with easily weathered, typically calcareous rocks generally have high pH, alkalinity, conductivity, and hardness. The reverse holds for lakes whose catchments are dominated by acid rocks that are highly resistant to chemical weathering (Lamm and Bobba, 1985). High concentrations of calcium and magnesium, high buffering capacity (i.e., high alkalinity), high pH and high conductivity characterize systems with catchments dominated by basic rocks [e.g., calcite, $CaCO_3$ and dolomite, $CaMg(C\,O_3)_2$]. The weathering of silicates is a slow process, but it still increases the buffering capacity in the soils. The abundance of soil water is considered (Ugolini, 1986) the most important factor controlling the rate of chemical weathering of silicates. This reaction also increases with temperature, so weathering is much faster in the tropics, and with soil water acidity, since H^+ consumes HCO_3- ions released during silicate weathering. Another important determinant of the ion exchange capacity of soils (Brady, 1984), and of lake water chemistry (Nilsson, 1992), is the weathering of K feldspars which leads to the formation of kaolinite, a clay mineral. Over 75% of the weathering in one of the most common soil type in Europe, podzolic soils, may be attributed to the weathering of minerals like epidote, hornblende, and feldspars (Sverdrup and Warfvinge, 1991). Fine grained soils have a higher specific surface area and a higher weathering potential than coarse grained soils (Birkeland, 1974). This has significant consequences for the ionic composition of the soil water and the lake water. The lowest pH usually occurs in coarse deposits, and the highest pH (alkalinity, conductivity, etc.) in the finest deposits (Jonasson et al., 1985). The depth of the soil also influences water chemistry. For example, pH (alkalinity, etc.) is usually higher in thicker deposits (Newton et al., 1987). Fine soils, dominated by different types of clay minerals, can influence the ion exchange processes in several ways, and this would also influence the water chemistry (Turner, 1984). The distribution of wetlands could also affect water quality. For example, water drained from wetlands generally has low pH (and low alkalinity, etc.) and low concentrations of metals, due to the capacity of acidic, humic materials to exchange ions and bind metals.

The literature on the influence of land use on water quality in extensive (e.g., van der Weijden et al., 1984; Harper and Stewart, 1987; Knoechel and Campbell, 1988; Schofield and Ruprecht, 1989; Nilsson, 1992). Agricultural soils are often fertilized and limed. This increases chemical concentrations in the drainage water and in the rivers and lakes. Fine-grained soils are generally used for agriculture. Modern industrial forestry, ditching operations to improve the drainage conditions of the catchment, selective or extensive tree cutting, other human activities in the catchment, and differences in precipitation also influence the quantity and quality of water leaving catchment areas (Andersson and Gustafsson, 1978; Foster, 1978; Edwards et al., 1984; Fritz et al., 1984). Several water quantity and quality models are available (Grip, 1982; Sverdrup and Warfvinge, 1990).

From a hydrological point of view, catchment areas are often divided into two parts (fig. 5.2):

1. Outflow areas (OA ≈ wetlands) dominated by a relatively fast turnover of substances and horizontal (land overflow) transport processes (see Eriksson, 1974; Nyström, 1985; Rodhe, 1987).

2. Inflow areas (IA ≈ dry land) dominated by vertical transport processes, first through the soil horizons, then ground water transport, and, finally, tributary transport to the lake.

Many textbooks treat these topics thoroughly (Kirkby, 1978; Hillel, 1982; Bear, 1979; Dunne and Leopold, 1978; Eriksson, 1985; Grip and Rodhe, 1985). Evidently, many factors and processes can influence the variability in substance concentrations in soils and river water within and among aquatic systems. Empirical/statistical methods provide a ranking of different x variables influencing a target y variable. Very complex mechanistic models quantifying fluxes to, within and from different catchment area compartments have been presented (see, e.g., Calmon,

2002). Models of this kind generally provide poor predictions but they can be important tools in science.

Outflow areas = Wetland areas = OA = Upstream lakes, rivers, bogs and mires
Inflow areas = Dry land areas = IA

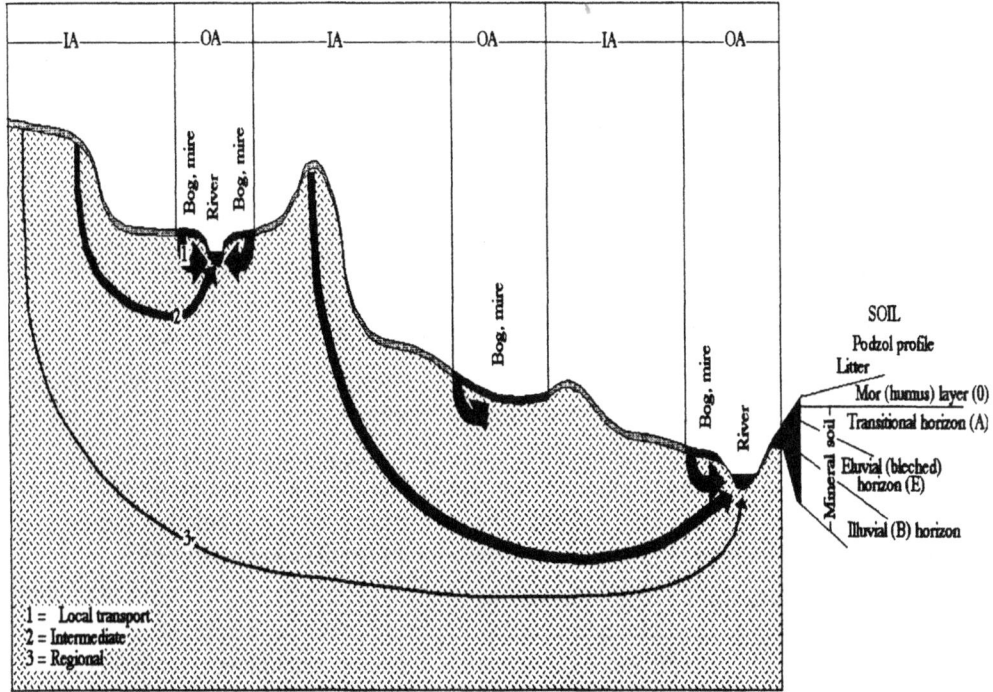

Fig. 5.2. Illustration of fundamental factors regulating the transit time of substances in catchment areas, surface and ground water fluxes and outflow and inflow areas.

5.1.1. Catchment area influences and simplifications

From this general background, the question is: Will the predictive power of Secchi depth (as an y variable closely related to SPM) increase if one increases the resolution of the catchment area description by increasing the number of x variables describing the catchment area and by accounting for catchment "zonation" in a more detailed manner using, for example, GIS approaches? Is there an optimal degree of resolution? Is there an optimal simplification?

Five degrees of resolution have been tested. All data used in the following test emanate from 95 catchment areas and lakes that are described in appendix 9.1.

1. The simplest way of describing the catchment area is when only the percentages of forests, upstream lakes, mires, open land, soil types, bedrock types, etc. are used and no consideration or correction for zonation (i.e., the distance between the lake and the given feature in the catchment) is used.

2. The second simplest catchment area description in this test accounts for the zonation in a simple manner. The DAZ method (drainage area zonation) has been used and it gives a correction factor (Cf) accounting for the mean distance (x in km) between the lake and the catchment feature, as given by:

    $$Cf = (3/x)^{2/3} \tag{5.1}$$

3. Using the DAZ method, one can also define a "near area" to the lake (see Håkanson and Peters, 1995). One can then test if the features close to the lake, in the near area, will influence the lake water quality more than features far away from the lake.

4. One could also test if various combinations of noncorrected, DAZ-corrected and near area-corrected features can increase the predictive power.

5. Finally, one can test if very detailed descriptions of the catchment areas based in digitized maps (GIS maps) of the features in the catchment, and by accounting for trajectory distance (the ways the water actually runs), will increase the predictive power of the models. GIS data for the same 95 catchments comes from Thierfelder (1999).

The results are given in fig. 5.3. One can note that:

- Secchi depth can generally be predicted rather well from catchment area parameters ($r^2 \approx$ 0.7).

- There are no significant improvements in the predictions when the very extensive corrections are being made. This means that one obtains almost as good predictions for Secchi depth (or SPM in lakes) without any correction, and if corrections are to be made, it is sufficient to use simple corrections.

Note that the latter statement is valid for lake Secchi depths based on mean annual values. It may not be valid for short-term predictions of daily runoff from very small catchments.

From the results shown in fig. 5.3, a structure of the catchment area should be sought that is simple, mechanistically correct, relevant for the target y variable, and useful in water management. Most catchment areas larger than, say 100 km^2 are very complex – the soil characteristics and soil cover vary spatially and vertically, and so does the geological characteristics. Generally, there is also a significant geographical variation in the vegetation cover, land-use and topography within larger catchments. This means that even very detailed catchment area descriptions are gross simplifications, and this fact explains the results in fig. 5.3. This knowledge is used in building the dynamic river model for SPM presented later in this chapter.

5.2. Empirical river modelling

Fig. 5.4 gives a schematic compilation of important channel characteristics. Channels generally increase in width and depth along the path from spring to the outlet in the sea. Algorithms for this will be presented in this section. There are also typical changes in water discharge, water velocity, suspended particulate matter concentration and gain size characteristics of the sediments on the river-bed. Also the biota changes along the river stretch, as indicated in fig. 5.4. Mountain streams are often dominated by salmonides, whereas, e.g., cyprinids are often caught in lowland rivers. The literature of river ecosystems is extensive (Vannote et al., 1980; Ward, 1989, 1998; Ward and Stanford, 1995; Bloesch, 1997).

This section will first present the empirical SPM model, the next section will discuss the dynamic model. The main results discussed in this section come from Håkanson et al. (2004b).

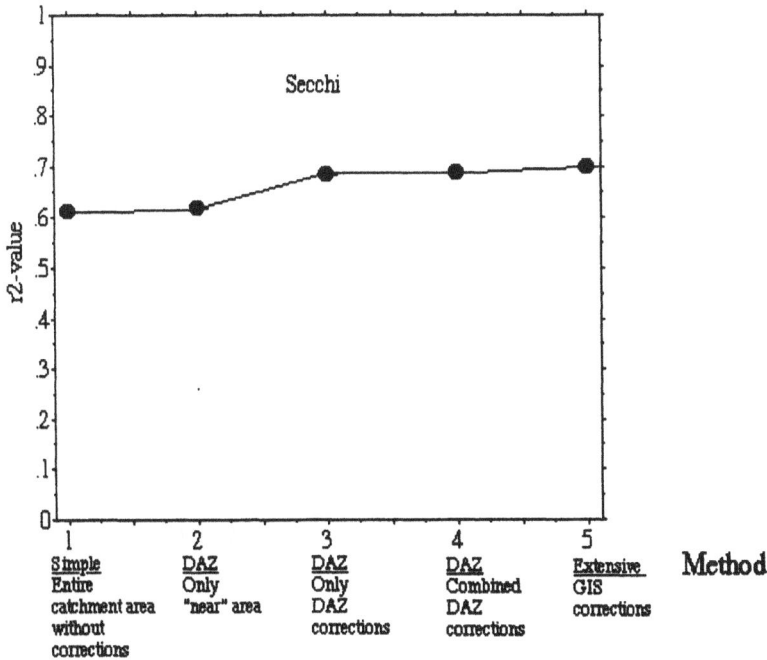

Fig. 5.3. A comparison between results from five increasingly detailed approaches to describe catchment areas and the obtained r^2 value in predicting mean characteristics lake Secchi depths based on multivariate regression analyses and data from 88 catchments and lakes. Data from Håkanson and Peters (1995).

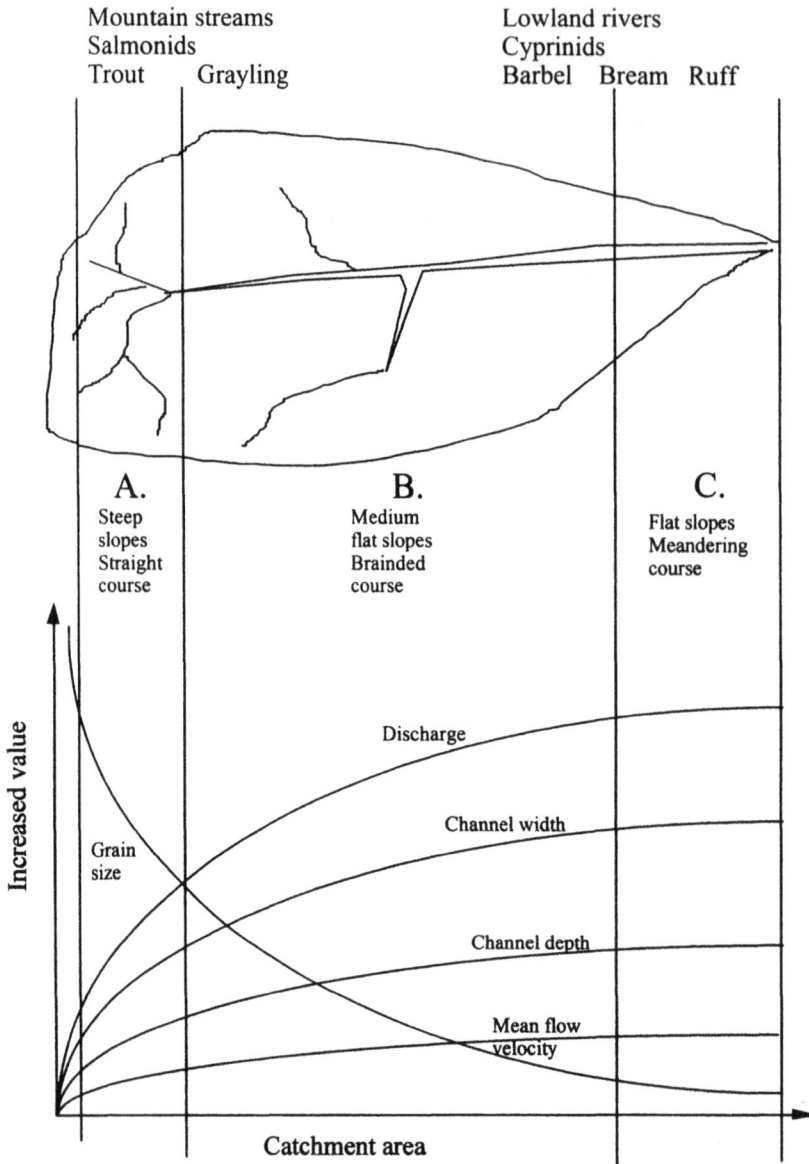

Fig. 5.4. Illustration of typical characteristics of running waters from the spring to the outflow in the sea (see Huet, 1949; Illies, 1961; Calow and Petts, 1994).

5.2.1. Databases

Three databases (see appendix 9.1.2.) will be used to discuss and present the empirical model for SPM in rivers. The model is mainly meant to be used as a submodel in, e.g., ecosystem modelling or in the modelling of ecosystem effects of water pollutants.

1. The European database includes data from 89 stations in Belgium, Finland, France, Germany, Hungary, the Netherlands, Poland, Portugal, Russia, Switzerland, and UK on: Mean water discharge (Q in m^3/s), SPM (mg/l), latitude (Lat,°N), longitude (Long,°E), continentality (Cont, i.e., distance from the ocean in km), and altitude (Alt, m.a.s.l.).

2. The UK database provides a range of data for 79 monitoring sites in the 2210 km^2 catchment of the Warwickshire River Avon, upstream of the town of Evesham. 3375 samples were taken over a period of several years.

3. The Swedish database for 95 lakes and catchments is used to address questions related to how catchment factors influence the variability among sites in Secchi depth (and hence also SPM). The variability in Secchi depth and SPM is generally much higher in rivers than in lakes (table 3.1), and lakes may be regarded as integrators of river transport. The idea here is to identify the most important features of the catchment areas regulating variability among lakes in Secchi depth and rank the relative role of these features. The Swedish database is extensive and includes data on mean water discharge (Q, m^3/s), mean monthly Secchi depth, comprehensive water chemistry, and comprehensive catchment area descriptions, including drainage area zonation (DAZ) of many features (soil types, bedrocks, vegetation, etc.), and data of catchment characteristics close to the lake ("near area", as defined by the DAZ method).

The European database has been used for a statistical treatment to find out if there exist any interesting correlations between SPM (as y variable) and Q, Lat, Long, Cont and Alt (as x variables). Such a statistical treatment might also give insights regarding the factors influencing the variability in SPM among river sites. The UK database will be used to address the fundamental problem of temporal variability within river sites (weekly data are available for measurement series from 79 sites covering several years), and hence also the representativeness of the samples, and the predictive power of river models for SPM.

5.2.2. Working hypotheses

Many factors influence the variability among and within rivers in SPM (fig. 2.3), for example, sampling phase in relation to high or low water discharge or to high/low bioproduction period, and if there are high/low point source emissions close to sampling sites.

For lakes, chapter 4 demonstrated that characteristic SPM values logically increase with lake phosphorus concentrations (i.e., with increasing autochthonous production), with increasing resuspension, and with lake pH (as a more complex measure of both autochthonous and allochthonous sources). In this section, the aim is to discuss an analogous empirical model for SPM in rivers.

Many papers and books have discussed the relationship between water discharge and SPM in rivers (see literature compilation by Jansson, 1982; Walling and Amos, 1999; Walling, 2000), but few models to predict characteristic SPM values in rivers meet the criteria for general applicability, easy access of the driving variables, high predictive power, and suitability as a submodel in an overall model for water pollutants discussed in chapter 1 (fig. 1.4).

Working hypothesis for this SPM model are:

- The SPM concentration in a river should, generally, increase with water discharge (Q).

- The relationship should be different in different rivers depending on catchment area soil type, climatological conditions, the existence of upstream lakes that would act as sediment traps, land use, vegetation, etc. (fig. 5.5)

SPM (mg/l)

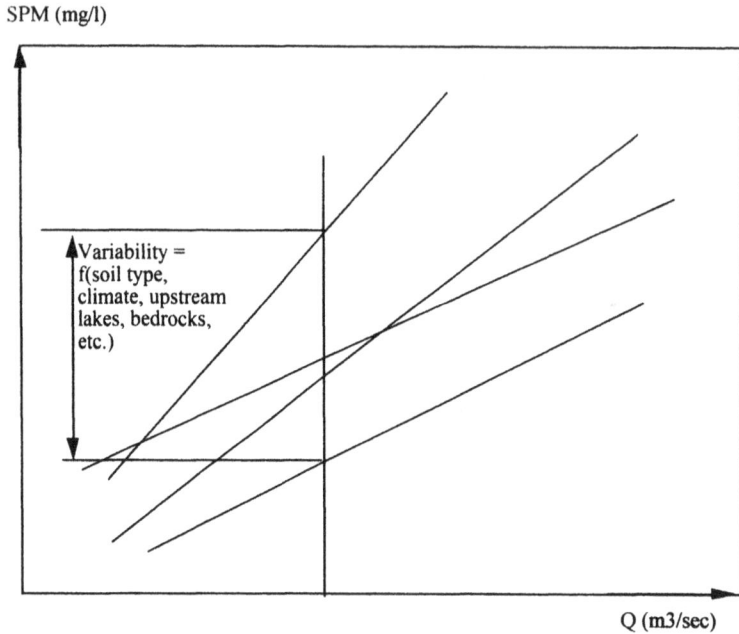

Fig. 5.5. Illustration of working hypotheses that SPM concentrations in rivers would depend on water discharge (Q), but differently in different rivers depending on variations in soil type, climate, upstream lakes that may act as sediment traps, bedrocks, etc.

The next section will focus on variations with time at sample sites in rivers. Those results will identify the limits to the predictive power of the empirical SPM model. If several samples are collected at a sampling site during a day, week, month, or a year, the CV value will increase significantly - data collected from a sampling site in River Danube (for chlorophyll and several types of phytoplankton) gave CVs of 0.3, 0.35, 0.55, and 0.95, respectively (Håkanson et al., 2003; fig. 1.4). This trend should also be true also for SPM even if the actual figures may not be the same. In this context, it is important to stress that the variability in mean, median, or characteristic SPM values among sites in different rivers is primarily related to differences in catchment area characteristics (soil type, land use, vegetation, latitude, altitude, etc.), whereas the temporal variability in individual data at a given site in a river primarily depends on climatological factors.

5.2.3. Uncertainties of SPM data in rivers

Fig. 5.6 illustrates SPM variations at two rivers sites in UK, Avon (Evesham) and Avon (Stare Bridge), for a period of almost 3 years. Fig. 5.6A gives the actual data. The coefficient of variation (CV) is 2.6 and 1.5, respectively; these are extremely high CV values. A characteristic CV for SPM in lakes is about 0.65 and the CV for SPM in Lake Balaton 1 is 0.99 (see chapter 4). The very high CV for individual data (CV_{ind}) in frequency distributions based on randomly distributed data may be reduced if one creates mean values based on a certain number of data (n) and then calculates the CV for the mean value (CV_{MV}) since $CV_{MV} \approx CV_{ind}/\sqrt{n}$ (see eq. 3.3). From fig. 5.6B, one can see that, for the two given sites, creating mean values would reduce the CV to 1.1 and 0.84, respectively, if n is 9. This indicates that these time series of data are not randomly distributed. The CV values for the mean are still very high, which implies that many samples are required in order to determine a mean value with a high certainty (see sampling formula in section 3.2.3.). If, for example, the CV is 0.84, about 70 samples are required to establish a site-typical mean value for the given variable provided that one accepts an error of 20% in the mean value. It one accepts a larger error, fewer samples would be required.

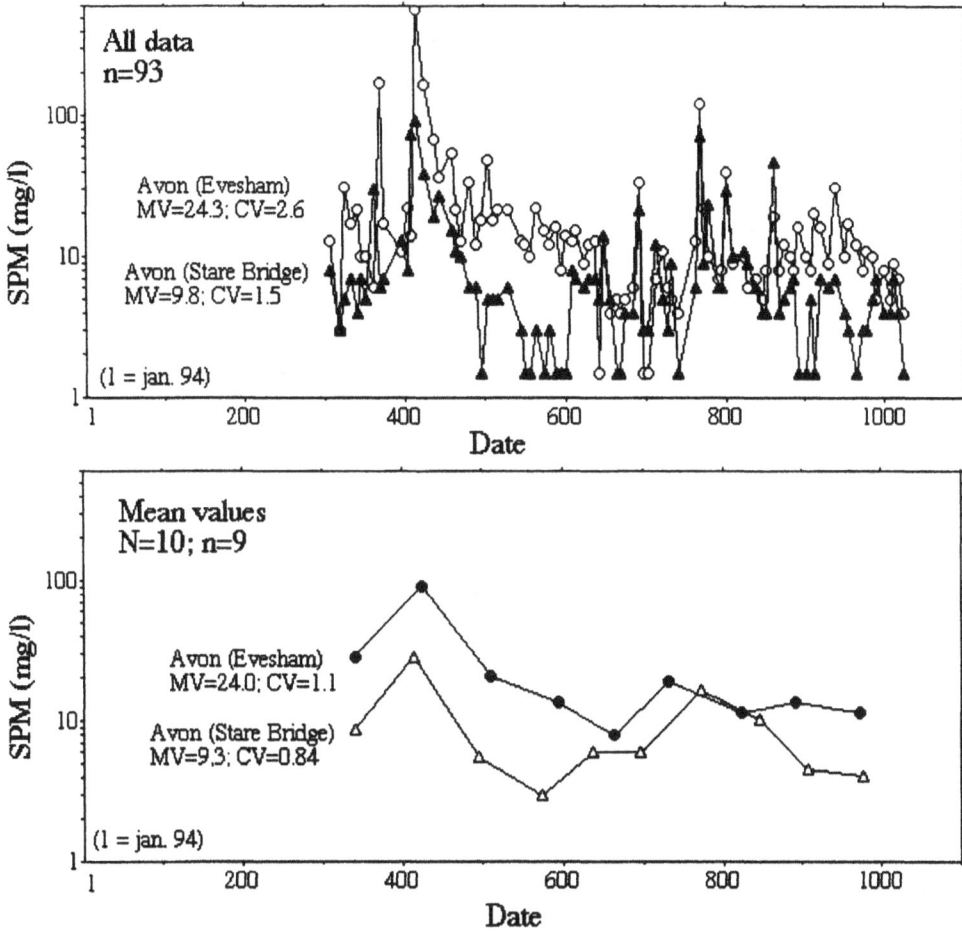

Fig. 5.6. Variations in SPM at two river sites (UK database) for a period of about 3 years.

Table 5.1 gives a compilation of CV values for several variables from the river sites in the UK database and the corresponding r_r^2 values. From this table, both Q and SPM are highly variable. These results are based on data from many sites in UK rivers but the general message about variability is probably valid for all rivers, although the CV values may not necessarily be the same in rivers from other parts of the world. McMahon et al. (1987) and Dedkhov and Mozzherin (1996), for example, suggest that mid-latitude rivers, such as those of the UK probably have lower variability in variables such as SPM and Q than those of Mediterranean, semiarid and arid environments. The high CV values will strongly influence the expectations one would have on model predictions of SPM: If it is difficult to measure a given y variable so that representative mean or median values are hard to determine, then it will also be very problematic to develop predictive model yielding a high predictive power.

Table 5.1. Coefficients of variation (CV) and highest reference r^2 values (r_r^2) calculated for the variables for the sites in the UK database. Q = water discharge, SPM = suspended particulate matter, DO = dissolved oxygen, Temp = water temperature, Eh = redox potential, BOD = biological oxygen demand, COD = chemical oxygen demand, TP = total phosphorus, Chl = chlorophyll-a concentration, PP = particulate phosphorus (mg/l), PartP = particulate phosphorus (mg/g), DP = dissolved phosphorus and PF = particulate fraction of phosphorus. Note that by definition (see section 3.2.2.) r_r^2 can attain values smaller than 0, but r^2 cannot.

	CV (median)	r_r^2
Q, m³/s	1.42	-0.33
SPM, mg/l	1.71	-0.93
DO, mg/l	0.25	0.96
Temp,°C	0.41	0.89
pH	0.03	1.00
Cond, mS/m	0.44	0.87
Eh, mV	0.18	0.98
BOD, mg/l	0.54	0.81
COD, mg/l	0.65	0.72
TP, mg/l	0.51	0.83
Chl, mg/m³	1.15	0.13
PP, mg/l	1.59	-0.67
PartP, mg/g	1.78	-1.09
DP, mg/l	0.77	0.61
PF	0.31	0.94

5.2.4. Catchment area influences

The aim of the following statistical treatment is to identify and rank the features in the catchment areas influencing the variability in characteristic Secchi depths among lakes (as integrators of river influences). The 3-year mean lake Secchi depth (Sec in m) is the target y variable. Data for Sec exist for 88 of the 95 catchments/lakes in the Swedish database (appendix 9.1.1).

Table 5.2 gives the results for stepwise regressions of log(Sec) vs catchment area parameters using the methods described by Håkanson and Peters (1995) and already applied for the empirical lake model for SPM (chapter 4). Table 5.2 shows that:

- As the cover of bare rocks increases (Rock%), Secchi depth increases. Rocky catchments are infertile and export little colored substances, suspended materials, and nutrients.

- The ratio area/A_{DA} (A_{DA} = the area of the catchment area in km²) has a positive correlation with Secchi depth. This is expected since large lakes with small catchments generally have clear waters.

- Tributary water discharge (Q) – the higher the water discharge, the higher (note higher, not lower) the Secchi depth. This relationship to water discharge will be discussed in greater detail in the next section.

- The relief of the catchment (RDA = dH/√A_{DA}; dH = the highest height difference in the catchment area in m) – the higher the relief, to lower the Secchi depth, and the more suspended particulate matter.

- The upstream lakes also have a positive effect on Secchi depth, which is expected since the larger the percentage of upstream lakes in the catchment, the greater the possibilities for entrapment for all types of materials, the lower the load of suspended matter to the target lake, and the greater the Secchi depth.

Table 5.2. An r rank table for Secchi depth (Sec; 3-year mean values) versus different catchment area parameters (see appendix 9.1 for definitions). r = the correlation coefficient, p = the statistical certainty; and n = 88 catchments.

n=88	r	p
Sec	1.00	
Area/A_{DA}	0.44	0.001
ForestN	0.35	0.001
RockN	0.28	0.0082
RockZ	0.27	0.0099
Rock%	0.27	0.0125
ForestZ	0.26	0.0139
Forest%	0.24	0.0243
Lake%	0.09	
Acid bedrockZ	0.09	
LakeZ	0.08	
Acid bedrock%	0.08	
Acid bedrockN	0.08	
ReliefN	0.06	
Q	0.03	
TillN	0.03	
Coarse sedimentZ	0.02	
Coarse sediment%	0.02	
Fine sediment%	0.02	
Relief (RDA)	0.01	
Fine sedimentZ	0.00	
Coarse sedimentN	-0.01	
A_{DA}	-0.03	
Intermediate rock%	-0.05	
Intermediate rockN	-0.06	
LakeN	-0.06	
Intermediate rockZ	-0.06	
Open land%	-0.08	
Open landZ	-0.08	
Fine sedimentN	-0.08	
TillZ	-0.11	
Till%	-0.11	
Open landN	-0.12	
Basic rockN	-0.14	
Basic rock%	-0.14	
Basic rockZ	-0.15	
Mire%	-0.24	0.0239
MireZ	-0.25	0.017
MireN	-0.28	0.0092

The factors affecting a low transport of SPM will cause significant correlations rather than the factors that one would assume would create a high transport of SPM from land to water, like the percentage of fine sediments, the percentage of open, cultivated land, or the percentage of basic rocks.

The characteristic coefficient of variation (CV) for lake Secchi depth is 0.15 (table 3.1). This means that the highest reference r^2 (r_r^2) is 0.985. Regression models for lake Secchi depth based on catchment parameters (table 5.3) can statistically explain about 60% of the variability among these sites. This means that the unexplained residual term includes all other factors that could potentially influence the long-term mean value of Secchi depth. The results given in table 5.3 will be used to define a system of criteria to categorize catchment areas into classes which are likely to exert different influences on SPM in rivers in a general way.

Table 5.3. Results of the stepwise multiple regression for Secchi depth (Sec; mean value for 3 years) using the Swedish database and catchment area variables; $n = 88$ catchments; $F > 4$; y variable = log(Sec).

Step	r^2	Model variable	Model
1	0.15	$x_1 = \log(1 + \text{Rock\%})$	$y = 0.12 \cdot x_1 + 0.30$
2	0.39	$x_2 = (\text{area}/A_{DA})^{0.5}$	$y = 0.16 \cdot x_1 + 0.73 \cdot x_2 + 0.094$
3	0.47	$x_3 = \log(Q)$	$y = 0.17 \cdot x_1 + 0.92 \cdot x_2 + 0.10 \cdot x_3 + 0.15$
4	0.54	$x_4 = 1/RDA$	$y = 0.15 \cdot x_1 + 1.0 \cdot x_2 + 0.12 \cdot x_3 - 1.75 \cdot x_4 + 0.23$
5	0.58	$x_5 = \text{Lake\%}$	$y = 0.16 \cdot x_1 + 1.11 \cdot x_2 + 0.12 \cdot x_3 - 1.61 \cdot x_4 + 0.013 \cdot x_5 + 0.16$
6	0.61	$x_6 = \text{Forest\%}$	$y = 0.142 \cdot x_1 + 1.048 \cdot x_2 + 0.125 \cdot x_3 - 1.606 \cdot x_4 + 0.0139 \cdot x_5 +$
			$0.0024 \cdot x_6 + 0.009$

5.2.5. Water discharge and SPM

This section addresses the working hypothesis concerning a relationship between water discharge (Q) and SPM in rivers. Fig. 5.7B-I gives examples from several of the UK measurement stations (these are examples of within-site relationships between individual data on Q and SPM in one river); fig. 5.7A gives the corresponding results using data from the 60 stations in the European database (this is an example of an among-site relationship between averaged values of Q and SPM using data from many rivers); and fig. 5.7I gives the results for 3303 data from the entire UK database. Note that these data cover several years and a relatively wide range in Q values.

The main conclusion is evident: There is no general, strong, positive correlation between Q and SPM. Note, however, that this does not exclude that one might find such correlations, as assumed in the working hypothesis, for specific measurement stations or for data from a wider geographical domain. But generally, these results are rather conclusive: Q is a poor predictor of SPM both for different sites in one river and among sites in different rivers. This will be discussed also in chapter 6. An important reason for the poor relationship between Q and SPM in rivers is that SPM will increase with Q in an initial high-water phase after periods of low water transport when relatively much SPM may be been deposited on the river bed. In contrast to this, there will be little material to be resuspended in the second phase with declining Q. So, depending on whether the water discharge increases or decreases, the SPM concentrations may be high or low for one and the same Q value.

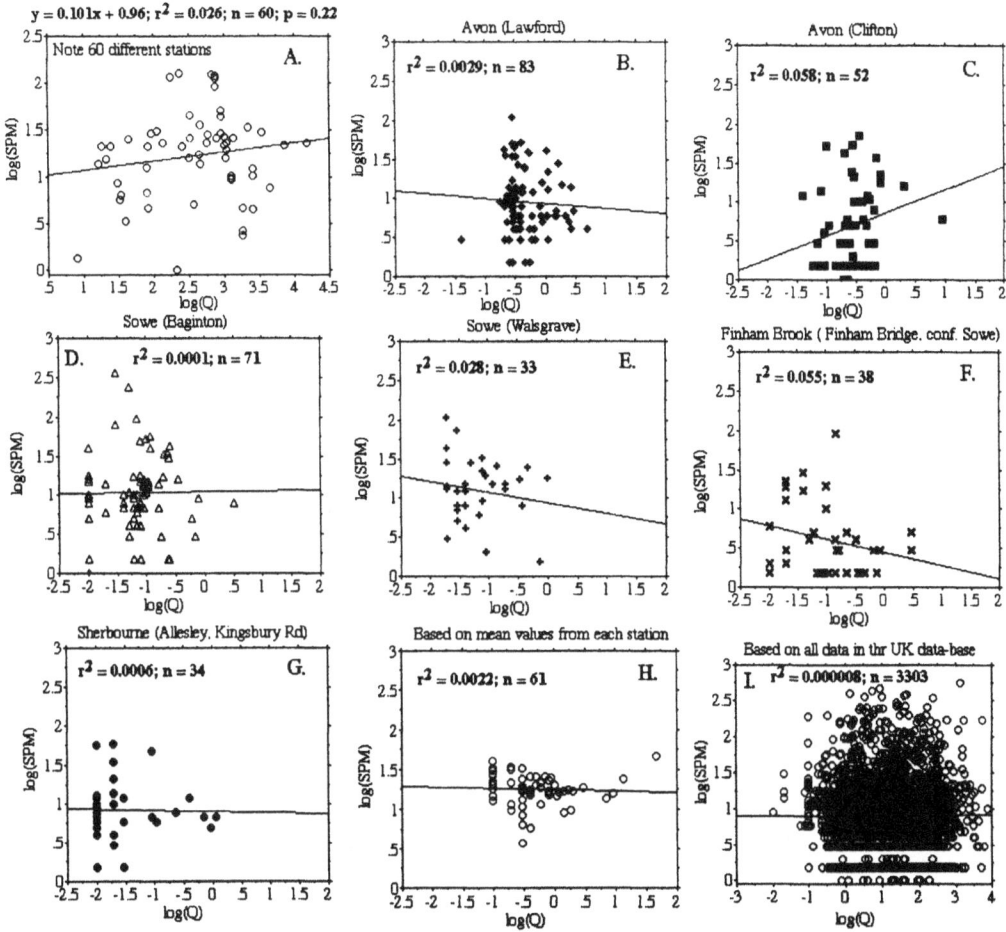

Fig. 5.7. Selected relationship between SPM [log(SPM) on the y axis] and water discharge [log(Q)] from different stations (B to H) from the UK database, for data from the European database (A) and from the entire UK database (I). Note that the data from the UK database cover a wide range in Q values from time series covering several years.

5.2.6. The empirical model

The European database has been used for the derivation of the model for characteristic SPM values in rivers. Table 5.4 gives a correlation matrix (r, p, and n) for transformed parameters and the highest r values and lowest p values are obtained for SPM versus latitude (r = 0.42; see also fig. 5.8A). Using this database, which gives a very wide range in rivers from many parts of Europe, there is a weak positive relationship between SPM and Q (see also fig. 5.8D). There are also relatively strong internal correlations between several of the potential x variables, e.g., between latitude and altitude (r = 0.51) and between Q and longitude (Q = 0.64), and longitude and latitude (r = 0.61). This is important information in the derivation of the SPM model, which should be based on as unrelated (orthogonal) variables as possible.

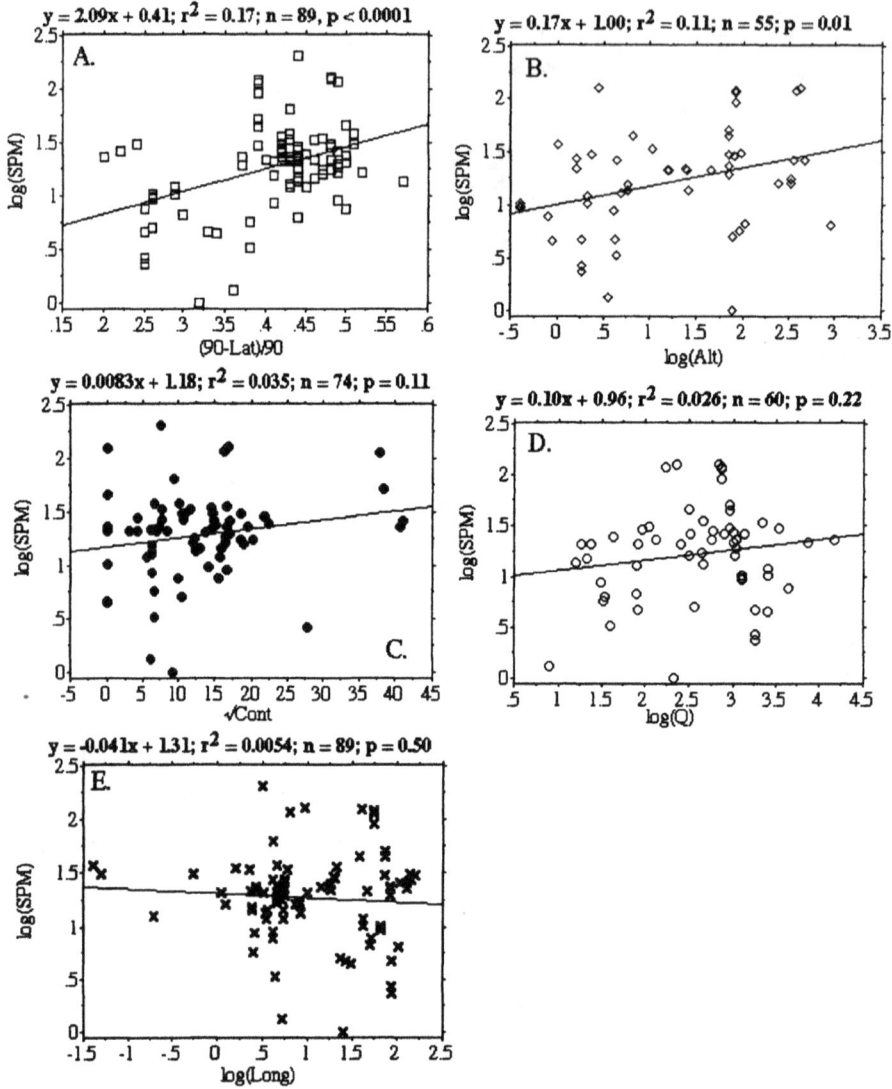

Fig. 5.8. Regressions between SPM [log(SPM) on the y axis] and all x variables (latitude, altitude, continentality, water discharge, and longitude) from the European database.

Table 5.4. Correlation matrix showing linear correlation coefficients (r), number of data pairs (n), and statistical certainties (upper part of the matrix) for the data in the European database.

r/n and p	log(Q_{MV})	(90-Lat)/90	log(Alt)	√Long	√Cont	log(SPM)
log(Q_{MV})	1	0.0008	0.063	0.0001	0.19	0.22
(90-Lat)/90	-0.42/60	1	0.0001	0.0001	0.39	0.0001
log(Alt)	-0.26/53	0.51/55	1	0.68	0.063	0.012
√Long	0.64/60	-0.61/89	-0.06/55	1	0.017	0.49
√Cont	0.20/45	0.1/74	0.20/45	0.28/74	1	0.11
log(SPM)	0.16/60	0.42/89	0.34/55	-0.07/89	0.19/74	1

Fig. 5.8 gives pair wise scatter plots (and statistics) for all relationships from the European database. The results of the stepwise multiple regression analyses are given in table 5.5. At step 1, 22% of the variations in SPM values, log(SPM), among these rivers sites can be statistically explained by variations in latitude (transformation (90-Lat)/90). At higher latitudes, where the mean temperatures are lower, the bioproduction of autochthonous particles should be lower, and the chemical weathering of bedrocks and soils, and hence also the transport of particles from land to water, should be smaller. However, the physical weathering related to freeze/thaw cycles would be higher. The net result, as given by this statistical/empirical analysis, is shown in table 5.5. At the second step, mean water discharge (Q_{MV}) appears - the higher the discharge, the higher SPM. This is in agreement with the basic working hypothesis and in disagreement with the main results from the UK database. However, this result does not falsify the conclusions from the UK database, but it clarifies the presuppositions when it is possible to find a relationship between SPM and Q. The European database covers a wider range of rivers and in this wider domain, there is a weak relationship between mean SPM and mean Q.

Table 5.5. Results from stepwise multiple regression analyses using the European database to rank the factors influencing the variability in SPM concentration among the 37 stations in Europe.

Step	r^2	x variable	Model
1	0.22	(90-Lat)/90	$y_1 = 2.76 \cdot x_1 + 0.11$
2	0.38	$\log(Q_{MV})$	$y_2 = 3.55 \cdot x_1 + 0.28 \cdot x_2 - 0.89$
3	0.40	$\sqrt{Cont}$	$y_3 = 3.44 \cdot x_1 + 0.24 \cdot x_2 + 0.0066 \cdot x_3 - 0.83$

The third and last factor is continentality - the further away from the ocean the river is, the more continental, the greater the annual temperature variations and the greater the weathering and the greater the potential SPM concentration in the rivers (if all else is constant). This is also logical. The regression model given in table 5.5 can statistically explain 40% of the variability in the mean SPM values.

This information will now be combined with the results from the statistical treatment of the Swedish database summarized in table 5.3. 61% of the variations in Secchi depth/SPM in lakes that could be statistically explained by variations in the six given catchment area parameters. However, water discharge has already been accounted for (in table 5.3) and the ratio between lake area and size of catchment (area /A_{DA}) should not be accounted for since for the new river model, one does not have any lake area and A_{DA} is mechanistically related to water discharge (fig. 5.10). About 50% of the variability explained by the six factors given in table 5.3 can be related to the remaining four factors. Subsequently, a simple rule system will be introduced to account for how variations in bare rocks (Rocks%), relief (RDA), upstream lakes (Lake%) and catchment area cover of forests (Forest%) would influence SPM values in rivers. That system is summarized in table 5.6. There are three classes: 1 = small influence, 2 = average influence and 3 = relatively large potential influence on SPM in rivers. The class limits are derived from information given by Håkanson and Peters (1995). This means that if less than 5% of the catchment area is dominated by bare rocks (this is a relatively small value), the transport of suspended matter from land to water should be relatively high (class 3), and the Secchi depth relatively low. If more than 20% of the catchment is bare rocks, this is categorized as class 1. If the RDA is high, the Secchi depth should be high (table 2) and SPM low. The class limits are 1 if RDA > 60, 2 if RDA is between 20 and 60, and 3 if RDA < 20 (i.e., a relatively flat terrain). If the percentage of upstream lakes is higher than 5%, SPM should be low, and vice versa. If the catchment is covered by a large area of forests, SPM should be low. Class 1 is given by a Forest% > 80%, class 3 by Forest% < 50%.

Table 5.6. Algorithms and rule systems to estimate how variations in catchment area features (Rock%; relief = RDA = $dH/\sqrt{A_{DA}}$; dH = the maximum height difference in the drainage area in m; A_{DA} = the area of the drainage area in km^2; Lake%; and Forest%) are likely to influence SPM concentrations in rivers. The table gives the weighting factors for the correction and the dimensionless moderator (Y) expressing the combined effects of variations in these catchment area characteristics on river SPM values.

SPM Class/Weight	Rock% (% of A_{DA}) 1	RDA 0.75	Lake% (% of A_{DA}) 0.5	Forest% (% of A_{DA}) 0.25	Correction factor (CF)	
Low 1	>20	>60	> 5	> 80	$1=(1\cdot1+1\cdot0.75+1\cdot0.5+1\cdot0.25)/2.5$	=Min
2	5-20	20-60	1-5	50-80	$2=(2\cdot1+2\cdot0.75+2\cdot0.5+2\cdot0.25)/2.5$	=Norm
High 3	< 5	< 20	< 1	< 50	$3=(3\cdot1+3\cdot0.75+3\cdot0.5+3\cdot0.25)/2.5$	=Max

1 = Small value	**Dimensionless moderator (Y)**
2 = Mean, normal value	$Y = (1 + 0.55\cdot(CF/2 - 1))$
3 = Large value	$Y_{max} = 1.275$; $Y_{min} = 0.725$; $Y_{max}/Y_{min} = 1.76$
	Empirical variation = $(0.61/2)\cdot6.3/1.1 = 1.75$

A correction factor (CF) is also defined using a weighting factor accounting for the ranking order given in table 5.3. This means that Rock% gets full weight (1.0), relief has weight 0.75, Lake% weight 0.5, and Forest% weight 0.25. The total empirical variations in Secchi depths among the 88 lakes in the Swedish database are given by a factor of 6.3/1.1 = 5.7. The regression model in table 5.3 explained 61% of this variability, and the remaining four factors about 50% of the 61%. So, these four factors should cause a variability of about 1.75 in SPM values. This is achieved by the dimensionless moderator defined in table 5.6 as:

$$Y = (1 + 0.55\cdot(CF/2 - 1)) \qquad (5.2)$$

Where CF is the correction factor accounting for the different weights. If all four catchment area parameters are classified as 1, CF is 1, and Y = 0.725. If all four factors are normal (= 2), Y = 1, and if all of them are 3, Y = 1.275. The latter case gives the highest SPM values, 1.76 higher than the minimum situation.

The model is shown in fig. 5.9. To calculate mean monthly water discharge, one needs data on annual precipitation (Prec in mm/yr), catchment area (A_{DA} in km^2), and the specific runoff rate, which may be set to 0.01 $m^3/km^2\cdot s$ (fig. 5.10). Multiplication with the seasonal moderator for Q (Y_Q) provides a seasonal (= monthly) pattern. The norms to do this are also defined in fig. 5.9. Data on latitude and continentality, which are easily accessed for any river site, are used to estimate SPM (from the regression model given in table 5.5). Finally, the model uses the rule system given in table 5.6 to estimate how the four catchment area characteristics (Rock%, RDA, Lake%, and Forest%) are likely to influence SPM. If reliable empirical data on mean monthly Q are available, such data are evidently preferable to the modelled values given by:

$$Q = A_{DA}\cdot0.01\cdot(Prec/650)\cdot Y_Q \qquad (5.3)$$

Fig. 5.11 illustrates how latitude, water discharge, and continentality influence mean monthly values of SPM in rivers, as this is given by the new model. In fig. 5.11A, latitudes have been altered from 35°N to 75°N, while all else is constant. So, this is a sensitivity analysis of the model. The major differences for predicted SPM values occur for changes at lower latitudes. Fig. 5.11B gives changes for SPM if mean monthly water discharges are altered five orders of magnitude. This would create marked changes in predicted SPM values. Fig. 5.11C gives similar results when the continentality has been changed four orders of magnitude. The greatest changes occur at high continentalities, i.e., for rivers with marked seasonal variations in temperature.

Fig. 5.12 gives similar results if the factors influencing mean monthly water discharge are altered. Fig. 5.12A shows the predicted changes in SPM values if the size of the catchment area is changed four orders of magnitude. This will cause relatively large changes in SPM values. Fig. 5.12B shows the predictions if mean annual precipitation is changed in four steps (3·650 mm/yr, 2·650, 650 and 650/2). This would not influence the predicted SPM values very much (if all else is constant). Changes in altitude (fig. 5.12C) influence SPM even less.

Fig. 5.13 illustrates how changes in Rock%, relief, Lake%, and Forest% are predicted to affect SPM values. Fig. 5.13A gives SPM for the three classes of Rock% (1, 2, and 3; for which the weight is 1). Such changes will affect SPM values relatively little. Changes in Forest% (figures 5.13B and 5.13C) would influence SPM even less. Fig. 5.13D illustrates the changes in SPM if all four factors are altered as much as possible.

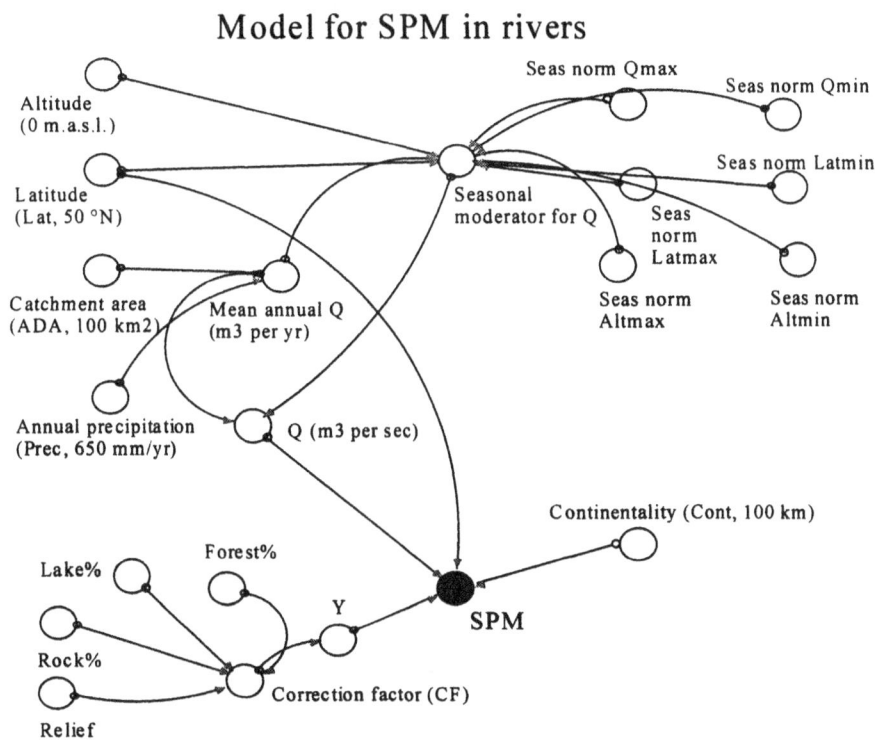

Model for SPM in rivers

Definition of norms						
Month	Qmax	Qmin	Latmax	Latmin	Altmax	Altmin
1	-0.71	0.58	-1.000	1.040	-0.97	0.47
2	-0.48	0.81	-1.000	1.370	-0.98	0.51
3	-0.17	0.84	-1.000	0.56	-0.58	0.22
4	-0.17	1.580	-1.000	0.38	-0.69	0.24
5	0.62	-0.1	2.170	-0.29	2.110	0.18
6	1.740	-1.000	2.510	-0.23	1.870	-0.32
7	0.52	-1.000	0.63	-0.62	0.51	-0.42
8	0.09	-1.000	0.24	-0.71	0.07	-0.49
9	-0.16	-0.82	0.05	-0.79	0.03	-0.38
10	-0.2	-0.56	-0.03	-0.74	-0.06	-0.2
11	-0.63	0.11	-0.66	-0.28	-0.62	0.07
12	-0.44	0.54	-0.92	0.32	-0.68	0.13

Fig. 5.9. An outline of the empirical river model for SPM. The submodel for water discharge (Q) is presented in more details in appendix 9.3.

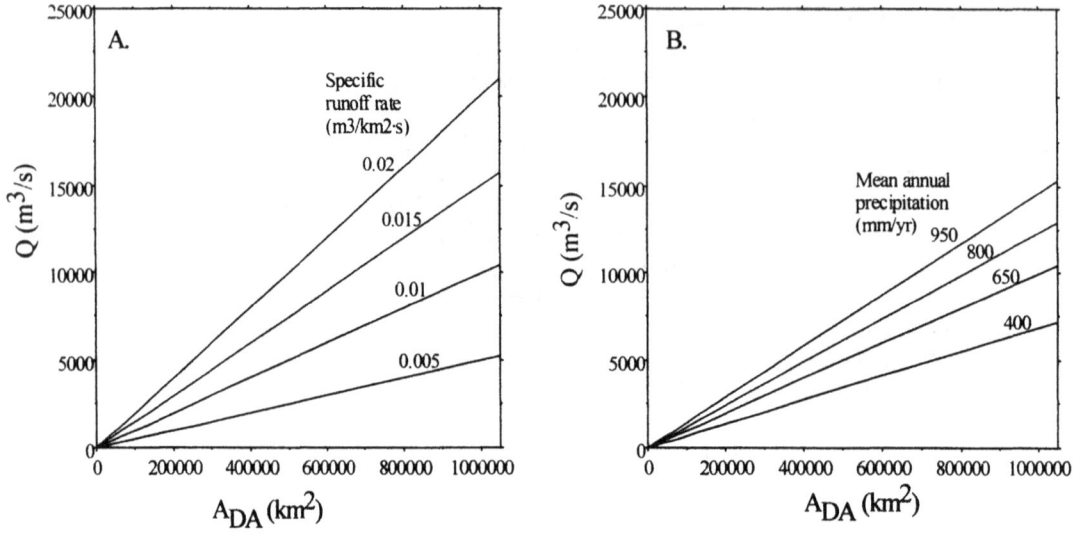

Fig. 5.10. A. Gives water discharge, Q, as a function of catchment area (A_{DA}) and four specific runoff rates (0.005, 0.01, 0.015, and 0.02; 0.01 is the default value) for a mean annual precipitation of 650 mm/yr.
B. Gives water discharge, Q, as a function of catchment area (A_{DA}) and four values for the mean annual precipitation (400, 650, 800, and 950 mm/yr; 650 is the default value) for a catchment area with a specific runoff rate of 0.01.

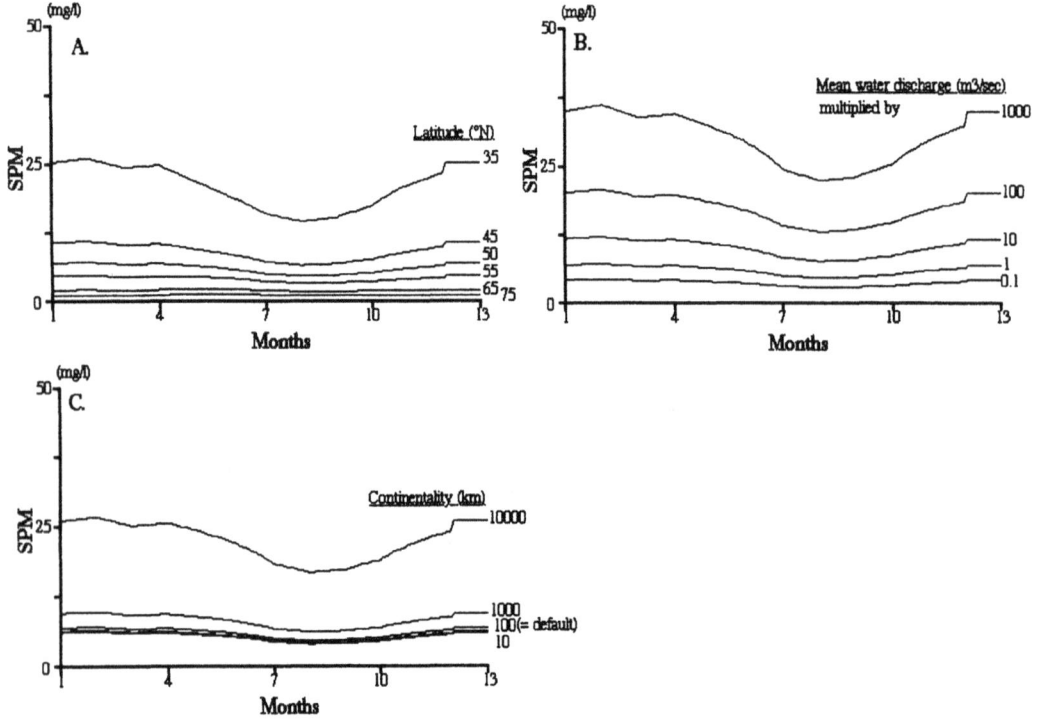

Fig. 5.11. Illustration of how variations in latitude (A), water discharge (B), and continentality (C) influence monthly variations in mean SPM concentrations as predicted by the model. These are the three main catchment area characteristics influencing SPM as given by the data from the European database.

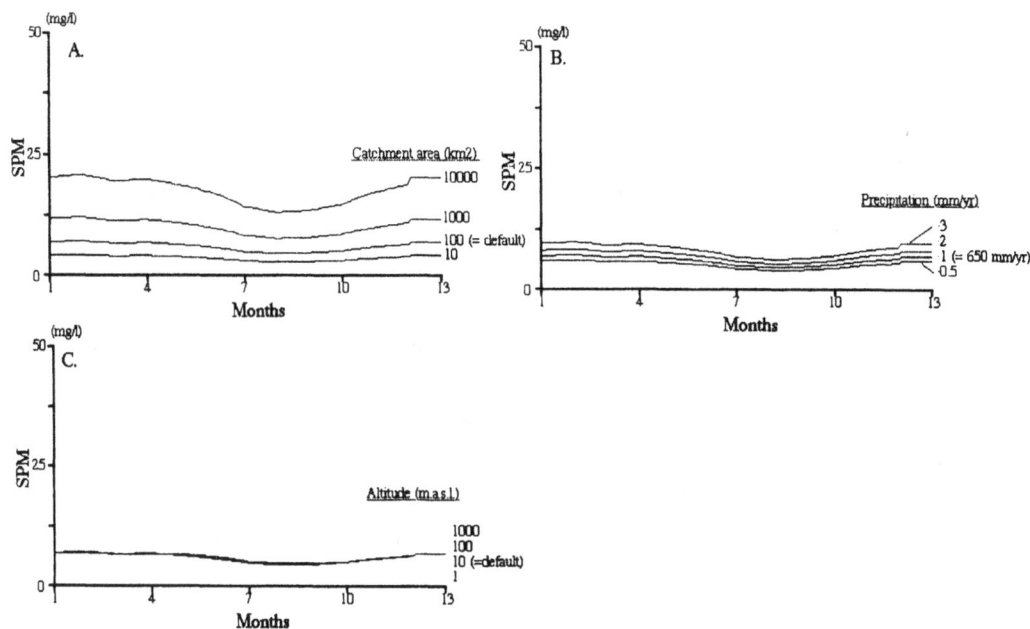

Fig. 5.12. Illustration of how the factors regulating water discharge also influence monthly variations in mean SPM concentrations as predicted by the new model: (A) variations in size of catchment area, (B) in mean annual precipitation, and (C) in altitude.

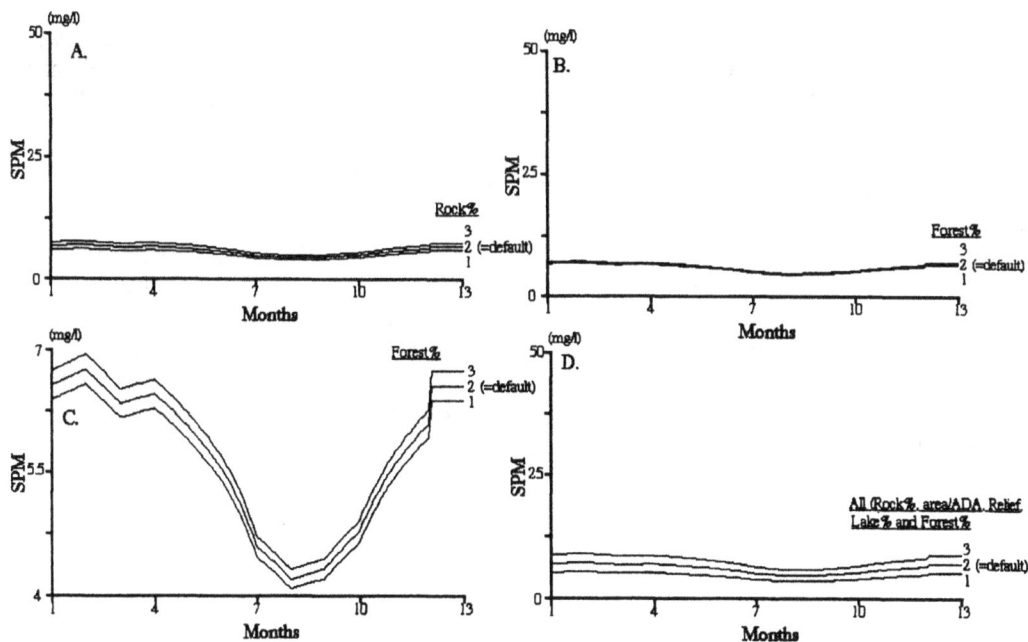

Fig. 5.13. Illustration of how variations in bare rocks in the catchment area (A), forests (B and C; note that these two figures show the same results but there is a difference in the scale on the y axis) and "All" catchment area features (D) influence monthly variations in mean SPM concentrations as predicted by the new model. These are the three main catchment area characteristics influencing SPM as given by the data from the Swedish database. The variations in bare rocks influence SPM variations most (weighing factor 1) and the variations in forest least (weighting factor 0.25).

5.2.7. Comments

The characteristic CV for within-site variations in rivers is 1.71, a very high value. This will restrict the predictive power of any model targeting SPM in rivers, and any model that uses SPM in rivers as an x variable to predict a given y variable.

The Swedish database has been used to identify and rank the catchment area features influencing the variability among sites in Secchi depth/SPM. From this information, a simple rule system has been applied in the presented model for SPM in rivers. The new European database has been used to identify and rank the overall factors (latitude, mean water discharge, and continentality) influencing characteristic values of SPM in rivers.

A very important demand for the model for SPM in rivers is that all driving variables should be easily accessed from standard monitoring programs and/or maps.

The measured river characteristic SPM values can be predicted from the following variables, (1) latitude (as a measure of mean temperature, primary production, and weathering), (2) mean water discharge, (3) continentality (SPM increases with increased continentality), and from a simple system accounting for (4) the percentage of bare rocks, the relief of the catchment, the percentage of upstream lakes, and the forest cover of the catchment area. The latter factors are categorized into three classes (1, 2 and 3) having little, average or relatively large influences on the transport of suspended particulate matter from land to water. The presented model does not include any climatological variables or factors influencing daily to seasonal variations in SPM.

5.3. Dynamical modelling

The dynamic river model presented here is meant to predict monthly SPM concentrations in water based on processes and mechanistic principles. It should also be as small as possible and the obligatory driving variables should be readily accessed. The tests of the model will use empirical data on the tributary inflow to the lakes discussed in chapter 4 (the three Belarussian lakes, Lake Erken, Lake Kinneret, and Lake Balaton), and the calibrations will use the empirical SPM model discussed in the previous section. This river model is partly based on the lake model discussed in chapter 4 but also on the catchment and river models for toxic substances, which have been tested with good results for radiocesium and radiostrontium (see Håkanson, 2000a, b). The following section discusses the modifications as compared to the lake model and the previous river model for toxic substances. The second part gives results from calibrations and sensitivity tests of the new dynamic river model and the third part presents results from the validations. It should be noted that the model discussed here is meant to be a generic SPM model for defined river stretches (and not for river segments). It is intended for river stretches with catchment areas larger than 100 km^2.

5.3.1. Modifications of the lake model

Fig. 5.14 gives a schematic description of the modelling approach. One can see the target river site, the upstream river stretch, and the left and right side catchment areas. There are major differences between lakes and rivers, and one fundamental difference concerns the relationship between the catchment area, the water surface area of the river upstream at a given site and the water discharge at this site. The ratio between the catchment area (A_{DA}) and the lake area can vary from about 2 to over 2000, whereas the river water discharge (Q) is directly related to catchment area ($Q = SR \cdot A_{DA}$, where SR is the specific runoff rate in $m^3/m^2 \cdot time$; fig. 5.10). An evident change compared to the lake model is to define the upstream river stretch and its geometry.

1. The algorithm in the lake model to calculate the fraction of bottom areas where erosion and transport processes of fine sediments occur (the ET areas) has been omitted and replaced by a constant, ET = 0.95, for all river stretches.

2. The algorithm to calculate hypolimnetic (= deep water) temperatures has been omitted since the river water is not treated as being stratified. This means that there is only one water compartment, the river water compartment (RW). This also implies that upward and downward mixing are omitted.

3. The algorithm to calculate sedimentation in lakes has been modified since there is no longer two water compartments and since the river flow velocity regulates the turbulence of the river water and hence also sedimentation and resuspension of SPM in rivers. This will be quantified by a dimensionless moderator (Y_v), which also replaces the algorithm in the lake model that relates the settling velocity of SPM to the resuspended fraction and the age of the ET sediments.

4. The distribution coefficient (Vd/3), which distributes the resuspended matter from ET areas either to surface water areas or to deep water areas in the lake model, has also been omitted. In the river model, there is resuspension only from ET areas to river water (RW).

5. The geometry (mean depth, mean width and length) of the entire upstream river stretch is estimated by simple equations. Evidently, if empirical data characterizing the geometry of the upstream river stretch (URS) are available, such data can preferably be used. The algorithms to estimate river geometry have, however, been used in all the following calibrations and validations.

The next section will present the catchment area submodel. Fig. 5.15 gives an outline of the two compartments and the main fluxes in the dynamic river model. For hydrological reasons, the catchment area is (as discussed) divided into outflow areas and inflow areas, the upstream river stretch is divided into ET areas and river water (RW) compartments.

Outline of the dynamic river model

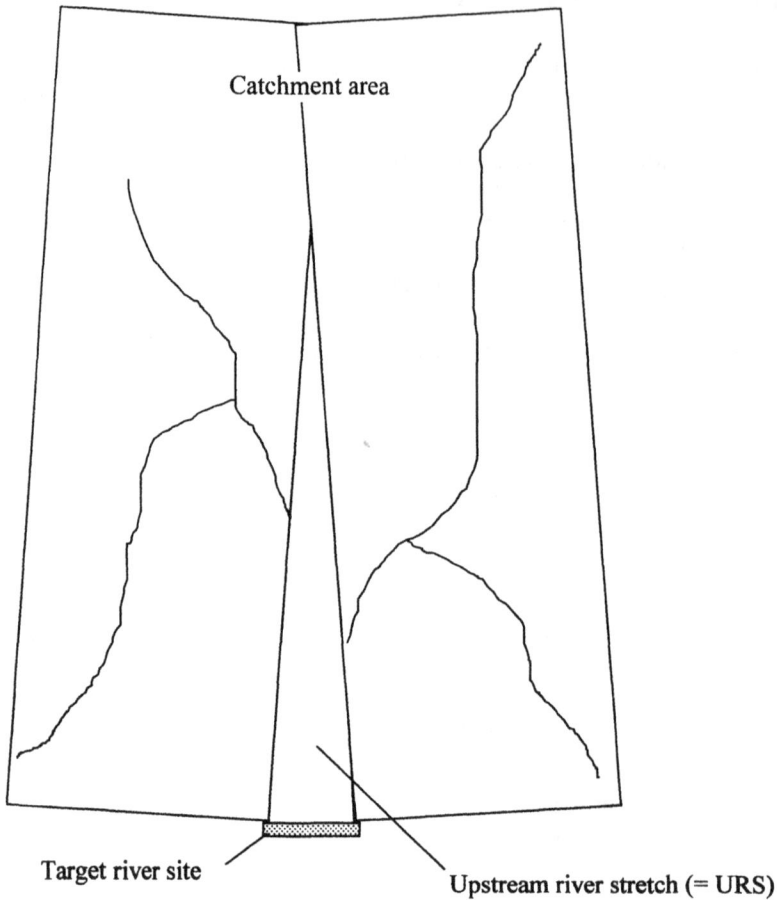

Obligatory driving variables
Catchment area (km2)
Mean annual precipitation (mm/yr)
Latitude (°N)
Altitude (m.a.s.l.)
Continentality (km)

The model includes submodels to estimate the following variables for the upstream river stretch
River length (m)
Mean depth of URS (m)
Mean width of URS (m)
Mean velocity (m/s)
Mean monthly water discharge (m3/month)
Mean monthly water temperature (° C)

Taget variables for the river site:
•Mean monthly SPM concentration in URS (mg/l)

Fig. 5.14. An outline of the river model. It is based on a submodel for the river stretch upstream a defined sampling site and a corresponding catchment area submodel. The figure also lists the obligatory driving variables, the target variables and the fact that the river model also includes submodels to predict water discharge and water temperature based on readily accessible data.

Outline of the submodel for catchment areas and rivers

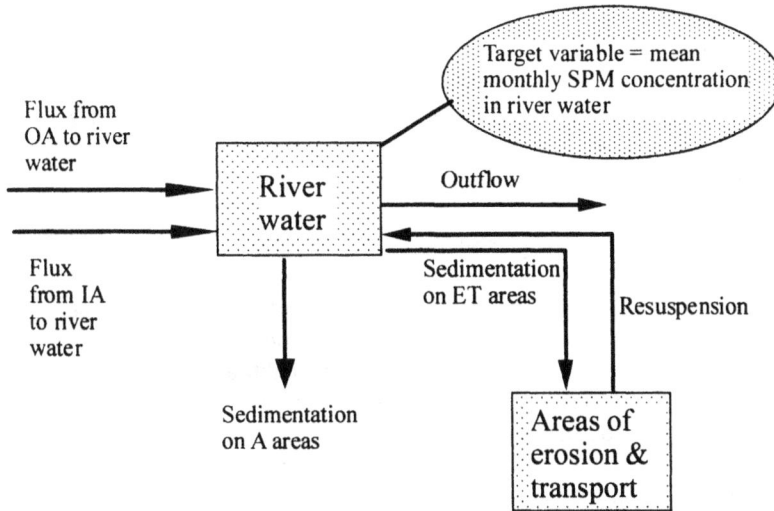

Fig. 5.15. Illustration of the river model with inflow and outflow areas (from the catchment) and the compartments for river water and ET areas characterizing the upstream river stretch (URS).

5.3.2. The geometry of the upstream river stretch

There are several simple algorithms to calculate the geometry of the upstream river stretch. These calculations are done in steps and one can start with the expression by which the total length of the upstream river stretch (L) is estimated from catchment area (A_{DA}). Just like the diameter of a circle is related to the square root of the area, the river length is estimated from the square root of the catchment area. But the river length should be shorter at high altitudes and high latitudes and longer for very continental rivers (see Strahler, 1963 and fig. 5.4). How differences in altitude, latitude, and continentality influence L will be expressed by three dimensionless moderators, which also have been used to regulate the specific runoff rate of suspended particulate matter from catchment areas (see later). Many alternative approaches to define the moderators have been tested during the calibrations and several of those tests will be given in the following section.

The length of the upstream river stretch (L in m) is first estimated from:

$$L = 10 \cdot Y_{ADA} \cdot Y_{lat} \cdot Y_{alt} \cdot Y_{cont} \qquad (5.4)$$

Where Y_{ADA} (in m) is simply given by the square root of the catchment area (A_{DA} in m^2) as $\sqrt{(A_{DA}/1)} = \sqrt{A_{DA}}$. Y_{lat} is the dimensionless moderator expressing how changes in latitude likely influence L, where:

$$Y_{lat} = (75/(Lat-35)) \qquad (5.5)$$

This moderator is meant to apply for rivers at latitudes between 75 and 40°N. Y_{lat} attains values between 15 and 75, which means that L (i.e., the total length of rivers in the entire upstream river stretch) can be expected to be about 5 times longer at a latitude of 40°N than at 75°N – if all else is constant. In the following calibrations, the default latitude was set to 50°N, which means that $Y_{lat} = 5$.

The moderator for altitude is defined by:

$$Y_{alt} = (1/(Alt+25))^{0.5} \qquad (5.6)$$

In the following calibrations, the default altitude is set to 0 m.a.s.l., which gives $Y_{alt} = 0.2$. If altitude is 1000 m.a.s.l., $Y_{alt} = 0.03$ and the river 6.7 times shorter, compared with altitude = 0.

Continentality will influence L in the following manner:

$$\text{If Cont} < 500 \text{ km then } Y_{cont} = 1 \text{ else } Y_{cont} = (Cont/500)^{0.5} \qquad (5.7)$$

This means that a continentality less than 500 km will not influence the L value; if Cont = 1000 km, $Y_{cont} = 1.4$; if $Y_{cont} = 5000$ km, $Y_{cont} = 3.2$, etc. So, continentality influences L less than altitude and latitude. The default continentality for the following calibrations was set to 100 km.

Note that these rules apply for the entire upstream river stretch (and not for river segments) and they have been used in the following calibrations and model tests. Fig. 5.16 illustrates how the algorithms work. Fig. 5.16A shows the relationship between the length of the upstream river stretch (L) and the area of the catchment area (A_{DA}) for river sites at different latitudes, fig. 5.16B shows the same thing for sites at different altitudes and fig. 5.16C the corresponding nomograms for different continentalities.

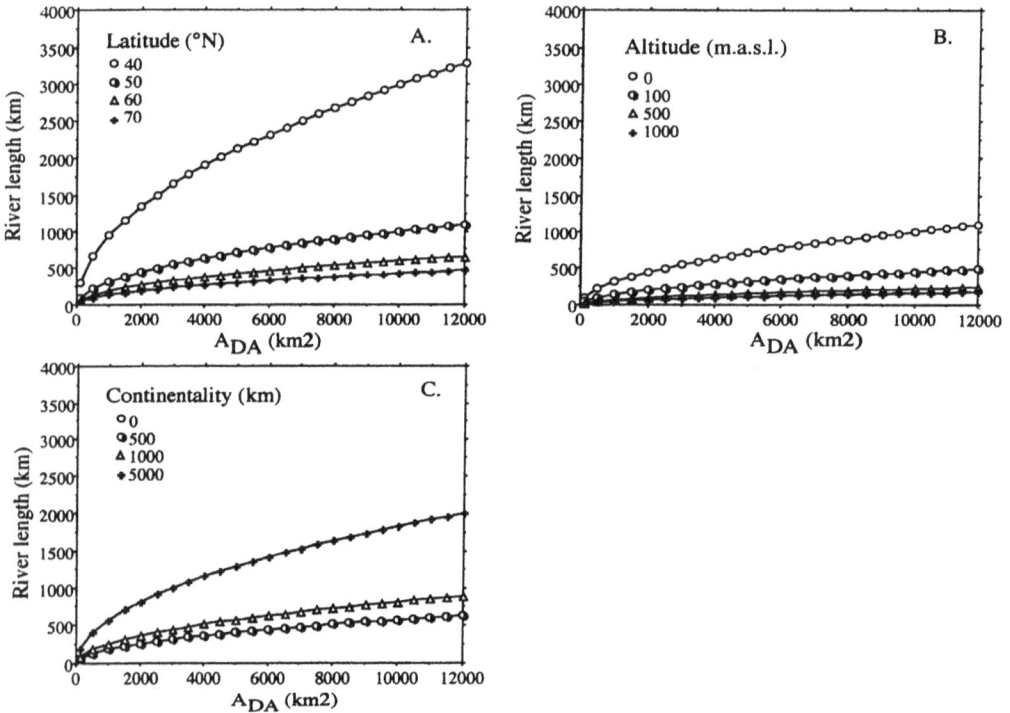

Fig. 5.16. Nomograms illustrating the (A) relationship between length of upstream river stretch (L) versus catchment area (A_{DA}) and latitude (for altitude = 0 m.a.s.l. and continentality = 500 km), (B) the relationship between L versus A_{DA} and altitude (for latitude = 50°N and continentality = 500 km) and (C) the relationship between L versus A_{DA} and continentality (for altitude = 0 m.a.s.l. and latitude = 50°N).

The mean depth of the upstream river stretch (D_m in m) and the mean flow velocity of the water in the upstream river stretch (v in m/s) are estimated from mean monthly water discharge (Q; fig. 5.17A and fig. 5.17C); the mean river width (B in m), in turn, is estimated from mean depth (D_m; fig. 5.17B) using algorithms presented and tested by Håkanson (2004c). The volume of the upstream river stretch is calculated from mean depth (D_m) and the area of the upstream river stretch (Area = B·L in m^2). The volume is needed to calculate the water concentration of SPM ($C_{RW} = M_{RW}/V$). The flow velocity (v) is needed both in the new algorithm for sedimentation and in the new algorithm for resuspension. There should be very little sedimentation if v is higher than 50 cm/s and better possibilities for sedimentation of suspended particles if v is smaller than 10 cm/s (fig. 5.18). The information given in fig. 5.18 is used to define the dimensionless moderator expressing how flow velocity (v) influences sedimentation and resuspension (as given by Y_v).

Fig. 5.19 gives a compilation of the submodel used to calculate the geometry of the upstream river stretch.

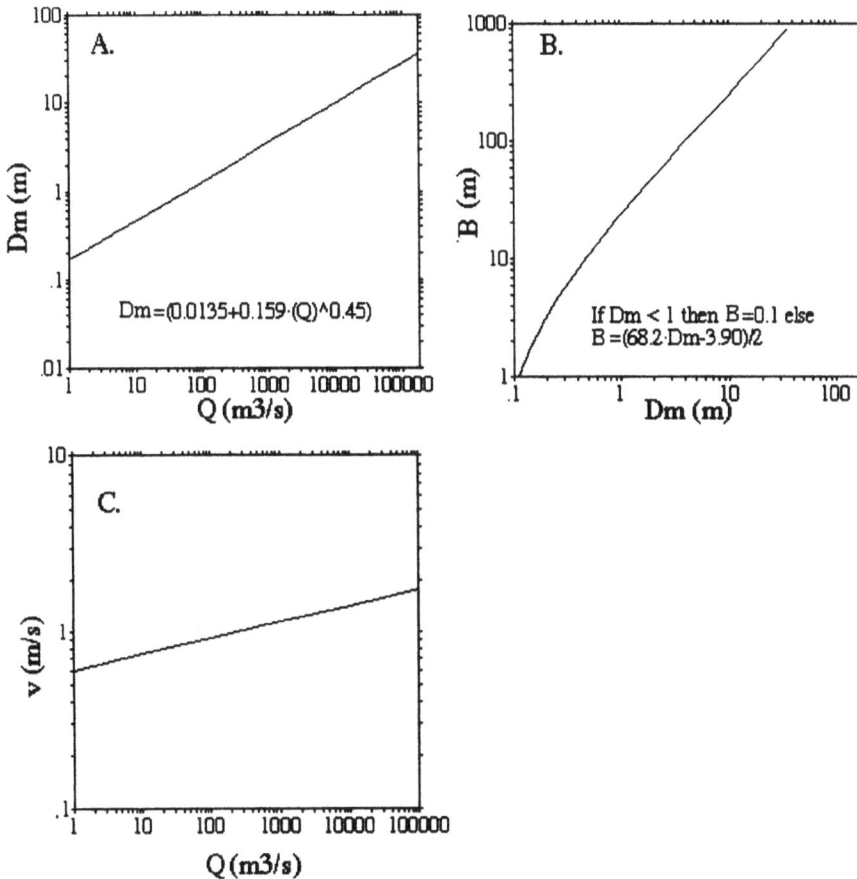

Fig. 5.17. Illustration of the relationship between (fig. A) mean monthly water discharge (Q in m^3/s) versus mean depth of upstream river stretch (D_m), (fig. B) mean depth of upstream river stretch (D_m) and mean width (B), and (fig. C) mean monthly water discharge (Q in m^3/s) and mean velocity (v in m/s) in upstream river stretch.

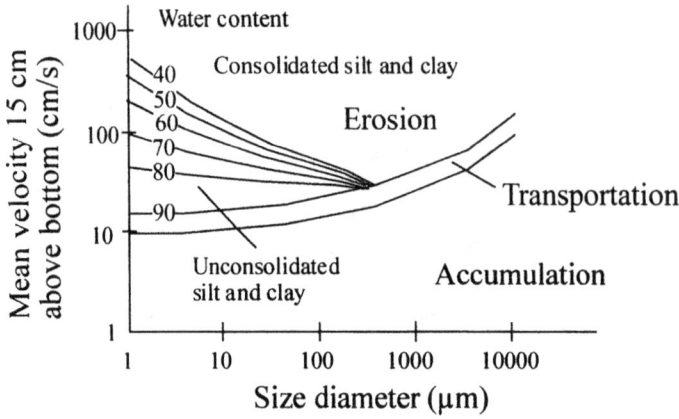

Fig. 5.18. Erosion, transportation, and deposition (accumulation) velocities for different grain sizes. Values for different stages of consolidation (water content) are also indicated (modified from Postma, 1967, 1982).

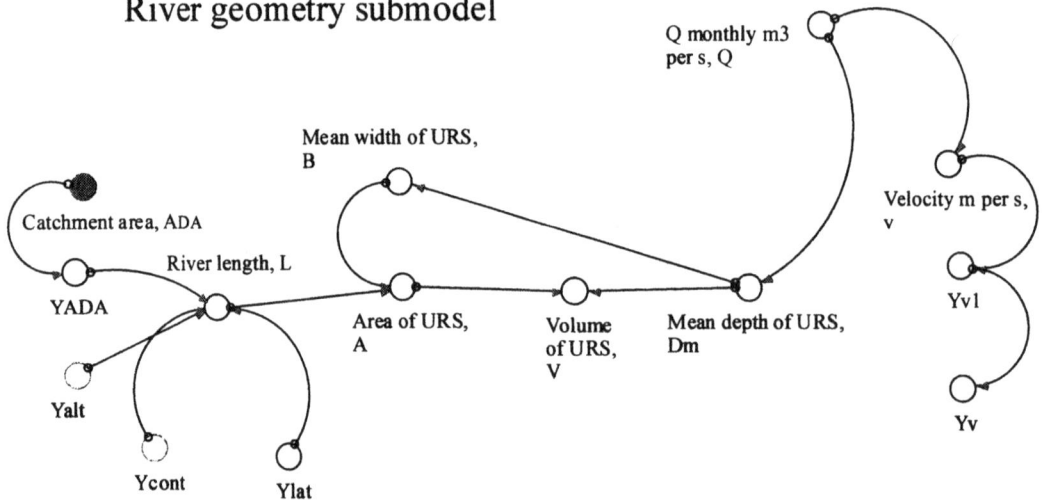

Equations:
YADA = ADA^0.5
Yalt = (1/(Alt+25))^0.5
if Cont < 500 km then Ycont = 1 else Ycont = (Cont/500)^0.5
Ylat = (75/(Lat-35))
L = 10·Ycont·Ylat·YaltRR·YADA
A = L·B
B = (68.2·Dm-3.90)/2
V = A·Dm
Dm = (0.0135+0.159·(Q)^0.45)
v = (0.532·(Q)^0.1 + 0.068)
Yv1 = if v < 0.1 m/s then Yv1 = 1 else Yv1 = (1-0.25·(Vv/0.1-1))
If Yv1 < 0.01 then Yv = 0.01 else Yv = Yv1

Fig. 5.19. Outline of the submodel defining the geometry of the upstream river stretch (length, mean depth, water flow velocity, water discharge, etc.) and how the geometry depends of latitude, area of catchment area, and altitude.

5.3.3. The catchment area submodel

As motivated in the first part of this chapter, the catchment area is divided into two parts:

1. The outflow areas (i.e., wetlands = upstream lakes, rivers, bogs and mires), where there may be a relatively quick horizontal transport of SPM from land to river water, and the remainder;

2. The inflow areas (= the dry land areas), where there is a relatively less transport of SPM and mainly from the ground water to the river water.

This submodel (fig. 5.20) quantifies SPM fluxes from land to water. Compared to the previous catchment area submodel for radionuclides (see Håkanson, 2004a, b), this model is modified in several ways.

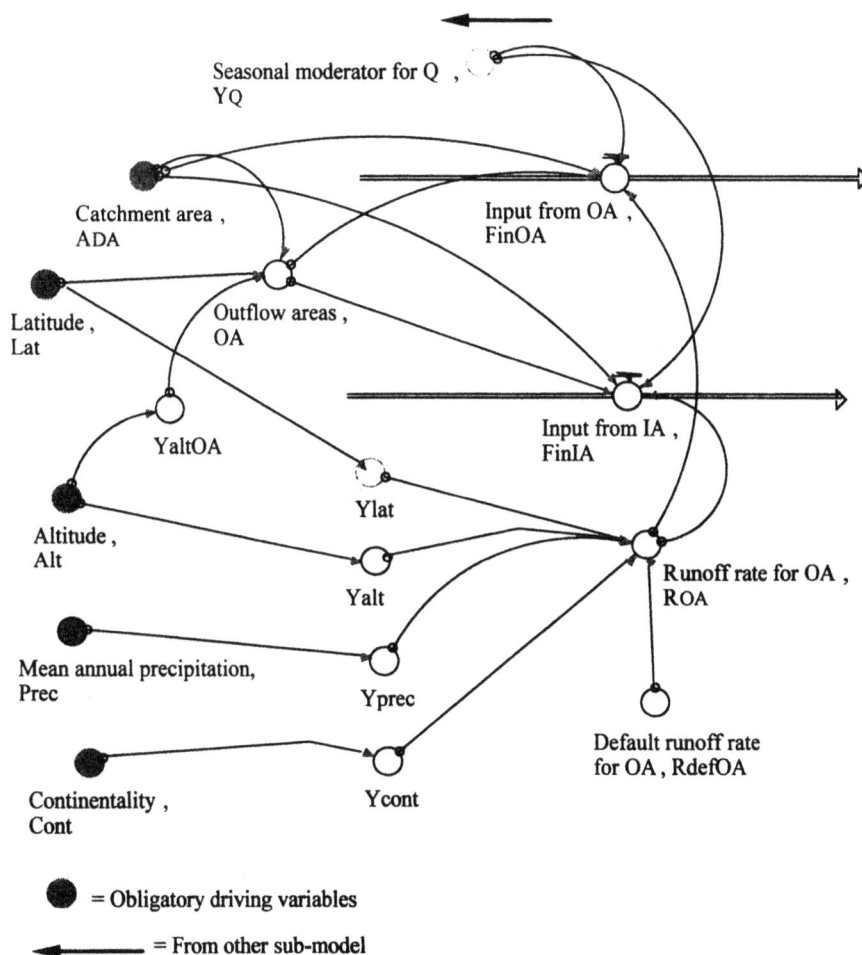

Fig. 5.20. Outline of the catchment area submodel. The calibrations of the dynamic river model have focused on the value for the default runoff rate (R_{defOA} in g/m^2·month) and the characteristics of the dimensionless moderators for how variations in latitude, altitude, mean annual precipitation, and continentality influence the runoff of SPM from land to river water.

The requirement to include the fraction of outflow areas (OA) as an obligatory driving variable has been omitted, and OA is now predicted from first catchment area (A_{DA}; fig. 5.21) and the value also depends on latitude and altitude. This approach to predict OA has been presented by Håkanson (2004b) and the main reason for using this algorithm is that it is sometimes difficult to obtain reliable data on OA, especially for large and topographically complex catchments. A_{DA}, latitude and altitude, on the other hand, can be determined easily and more accurately. Note that the equation given in fig. 5.21 to estimate OA from A_{DA} is not a regression. It is a deterministic relationship based on the boundary requirements that OA should be about 0.25 for very small catchments and approach 0.02 for very large catchments (Håkanson and Peters, 1995). The equation to predict OA from A_{DA} has been tested against empirical data and it describes the data quite well (fig. 5.21), so it has been used in this river model. Evidently, if reliable empirical data on OA are available, it is preferable to use such data rather than the estimate of OA given by the approach in fig. 5.21.

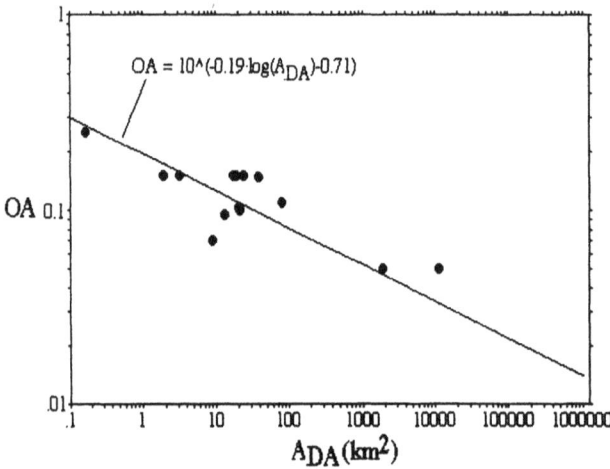

Fig. 5.21. The relationship between catchment area (A_{DA}) and outflow areas (OA = wetland areas) used in this dynamic river model. The figure also gives data of OA and A_{DA} from Håkanson (2004a, b).

From the relationship given in fig. 5.21, OA may be estimated by the following equation:

$$OA = (Lat/60) \cdot (1+0.0025 \cdot (Alt/1-1)) \cdot 10^{(-0.19 \cdot \log(ADA)-0.71)} \qquad (5.8)$$

Where (Lat/60) is a simple dimensionless moderator which shows that OA will increase with latitude (fig. 5.22B).
The moderator for altitude is given by:

$$Y_{altOA} = (1+0.0025 \cdot (Alt/1-1)) \qquad (5.9)$$

Which means that (fig. 5.22A) OA will increase at higher altitudes; if Alt = 0, $Y_{altOA} \approx 1$ and if Alt = 1000 m.a.s.l., $Y_{altOA} = 3.5$ and OA 3.5 times larger than at the sea level.

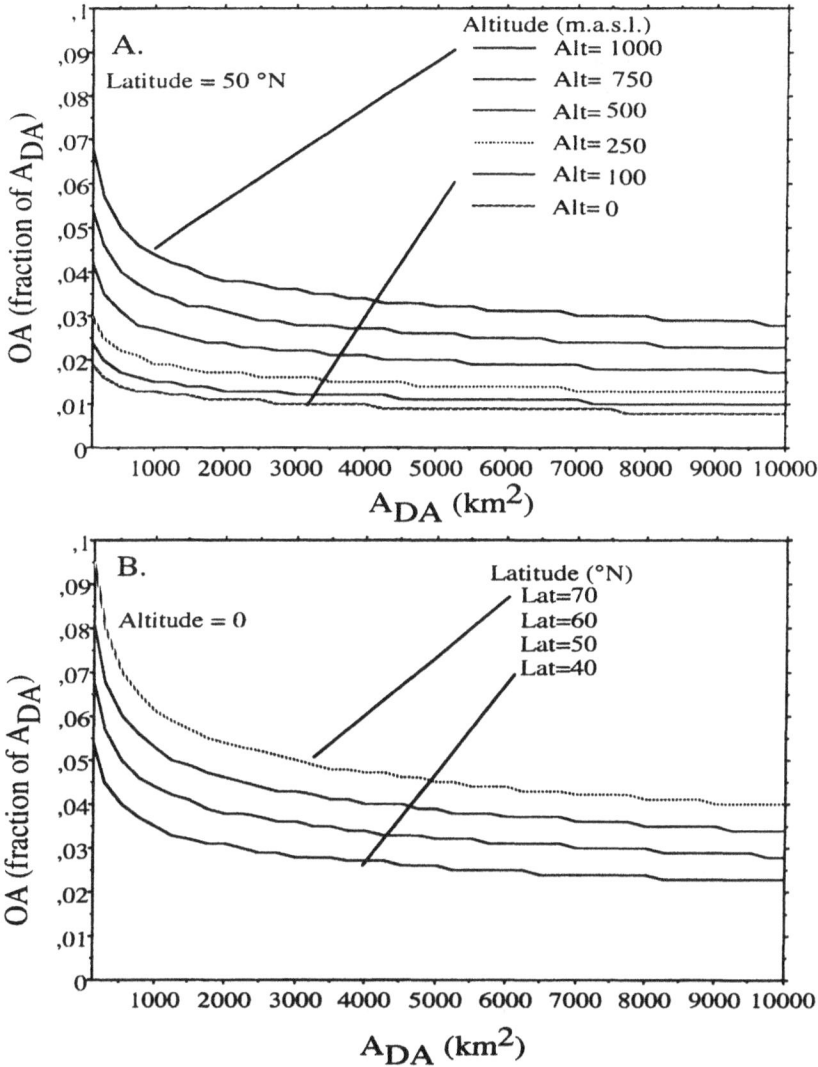

Fig. 5.22. The relationship between (A) the fraction of outflow areas (OA), the catchment area (A_{DA}) and altitude for rivers at a latitude of 50°N, and (B) the same for rivers at different latitudes (40, 50, 60, and 70°N) at an altitude of 0 m.a.s.l.

An important part of the catchment area submodel is the approach to predict river discharge (the Q model; see appendix 9.3 and fig. 5.10) and monthly variations in river discharge (as given by the dimensionless moderator for Q, Y_Q). Differences in hydrological regime evidently affect the transport of substances from catchments (Brittain et al., 1994). In spite of the fact that river discharge is a variable with great temporal and spatial variability, this submodel has yielded good predictions ($r^2 = 0.84$, n = 119) of mean monthly Q. The best results were achieved for rivers with a mean annual discharge in the range 1-500 m^3/s.

The monthly flow of SPM from inflow areas (IA) to the upstream river stretch (F_{IARW} in g/month) can then be calculated from:

$$F_{IARW} = (A_{DA} \cdot (1 - OA) \cdot Y_Q \cdot R_{OA})/12 \qquad\qquad (5.10)$$

where

A_{DA} = the catchment area (m^2);
OA = the fraction of outflow areas [IA = (1-OA), dimensionless];
Y_Q = the dimensionless moderator for water discharge giving mean monthly Q values (m^3/month) from mean annual values (division by 12);
R_{OA} = the specific runoff rate for SPM (g SPM/m^2·month). The specific runoff rate depends on latitude, altitude, continentality and precipitations. This is given by:

$$R_{OA} = R_{defOA} \cdot Y_{prec} \cdot Y_{lat} \cdot Y_{alt} \cdot Y_{cont} \qquad (5.11)$$

The calibrations have focused on the value for the default specific runoff rate (R_{defOA} in g SPM/m^2·month); the dimensionless moderators for latitude (Y_{lat}), altitude (Y_{alt}) and continentality (Y_{cont}) have already been defined. The specific runoff rate of SPM will be higher at lower latitudes and altitudes and higher if the continentality is higher than 500 km. The dimensionless moderator for precipitation (Y_{prec}) comes from the river model for radionuclides (Håkanson, 2004a, b) and gives:

$$Y_{prec} = (Prec/650)^2 \qquad (5.12)$$

There is an exponential and not a linear increase in SPM runoff (the exponent is 2) and the normal mean annual precipitation is set to 650 mm/yr in this moderator.

The monthly flow of SPM from outflow areas to the upstream river stretch (F_{IARW} in g/month) is calculated in a similar way. The main difference is that the specific runoff rate should be higher and in this modelling, it is set to be 10 times higher than the value defined for the inflow areas. This gives:

$$F_{OARW} = (10 \cdot A_{DA} \cdot OA \cdot Y_Q \cdot R_{OA})/12 \qquad (5.13)$$

Fig. 5.23 illustrates the calibration routine and also how the default value for the specific runoff rate ($R_{defOH} = 2$ g/m^2·month) has been obtained. The calibrations have been done along gradients; in this case along a catchment size gradient (A_{DA} set to 10, 100, 1000, and 10000 km^2 in four steps in fig. 5.23A; this is the driving variable in this test). Three different equations for the dimensionless moderator for how A_{DA} influences SPM concentrations in river water have been tested. Fig. 5.23B also gives the data along this gradient, as calculated using the empirical SPM model. The idea has been to see if there is a logical relationship between the SPM values given by the empirical model and the dynamic model. Note that in all these tests, there have been no consideration to variations among the rivers in terms of catchment area percentages covered by forests, lakes, or bare rocks, and there have been no changes in the relief. The moderator (Cf; table 5.6) accounting for such variations in the empirical SPM model has been held constant to the normal default value of 2.

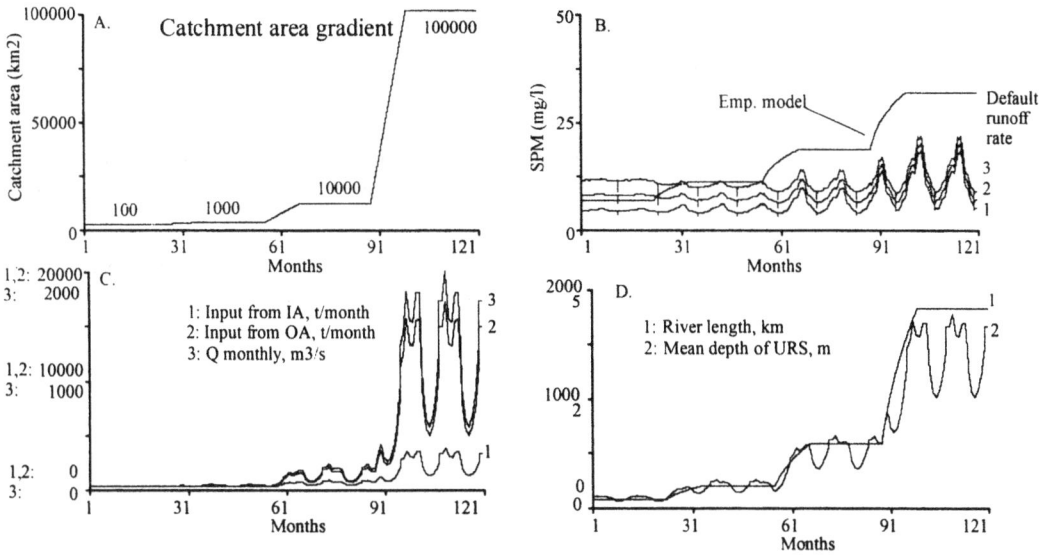

Fig. 5.23. Calibrations and sensitivity analyses of the river model along a catchment area size gradient (fig. A; values set to 10, 100, 1000, and 10000 km in four steps) for three different equations for the dimensionless moderator for how latitude influences SPM concentrations in river water. B. gives modelled SPM values and also data calculated using the empirical SPM model along this gradient. C. gives the corresponding modelled input of SPM from inflow areas (curve 1), from outflow areas (curve 2) and the calculated mean monthly water discharge (curve 3). D. gives the calculated values for river length (curve 1) and mean depth (curve 2) along this catchment area gradient.

From fig. 5.23B, one can note:

- The seemingly good correspondence between the SPM values predicted by the empirical model and the dynamic model for small catchment areas (when A_{DA} is set to 100 km^2), but the poorer correspondence for the larger A_{DA} values.

- The higher the value for the default runoff rate (R_{defOH}), the higher the SPM inflow from inflow and outflow areas (fig. 5.23C).

- Fig. 5.23C also gives the calculated changes in river discharge (here mean monthly Q in m^3/s) related to these changes in the size of the watershed (fig. 5.23A).

- To understand the difference between the predictions from the empirical model and the dynamic model in fig. 5.23B, one must also look at the changes in river geometry associated with the given changes in A_{DA}. Fig. 5.23D shows how river length (L) and mean depth of the upstream river stretch (D_m) increase with increasing catchment area.

- It should be noted that this is also a sensitivity test where then size of the catchment area has been varied while all else, including precipitation, have been kept constant.

From this, one can conclude that the empirical model predicts the SPM concentrations from changes in water discharge (Q) at a given river site, but water discharge is basically calculated from A_{DA} and precipitation (as $Q_{yr} = 0.01 \cdot A_{DA} \cdot Prec/650$, where Q_{yr} is the mean annual Q in m^3/s, 0.01 is the default value of the specific runoff rate for water in m^3/km^2·s and Prec is the mean annual precipitation in mm/yr; fig. 5.10). Fig. 5.24B shows a similar comparison between the empirical SPM model and the dynamic river model along a precipitation gradient (fig. 5.24A; mean annual precipitation set to 400, 600, 800, and 1000 mm/yr). In this test, the exponent in the dimensionless moderator for precipitation has also been varied in three steps (1, 2, and 3). One can note that:

- The default exponent of 2 gives a rather good correspondence with the empirical SPM model along the precipitation gradient, although the values from the dynamic model are initially low; and for higher precipitation, the predicted SPM values are higher than what the empirical model gives.

- Fig. 5.24C illustrates how the SPM transport from inflow and outflow areas depend on variations in precipitation (using the default values, e.g., the exponent in precipitation algorithm = 2).

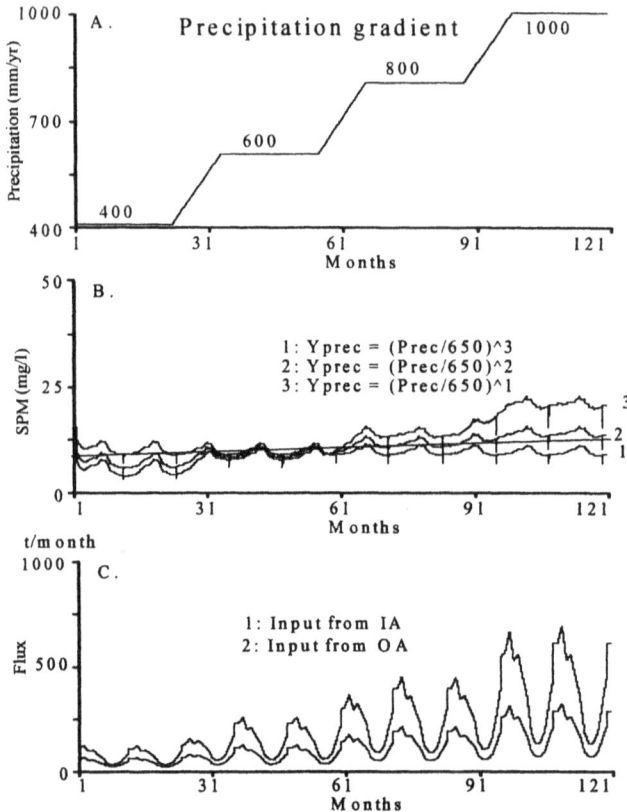

Fig. 5.24. Calibrations and sensitivity analyses of the river model along a gradient in precipitation (fig. A; values set to 400, 600, 800, and 1000 mm/yr in four steps) for three different equations for the dimensionless moderator for how precipitation influences SPM concentrations in river water. B. also gives the data calculated using the empirical SPM model along this gradient, and C. shows how this will influence the transport of SPM from inflow and outflow areas.

With the dynamic model, one can quantify the fluxes causing the given SPM concentrations. The basic idea is that the two models should give similar results when tested on real systems. These initial calibrations and sensitivity tests are meant to illustrate the difference between the two modelling approaches. In the dynamic model, the changes in catchment area influence the geometry of the upstream rivers stretch, and hence also the outflow from the upstream river stretch. The different roles that variations in catchment area and precipitation have may be studied using the dynamic model, not just for the target SPM concentrations in river water but also for the transport processes, i.e., the flow from inflow and outflow areas, outflow, sedimentation, resuspension, mineralization in river water and on ET areas and primary production in the upstream river stretch.

5.3.4. The submodel for the upstream river stretches

Basically, the upstream river stretch (URS) submodel handles (fig. 5.25, which shows the entire model and table 5.7, which gives a compilation of all key equations) input from the catchment area (one flux from outflow areas to the river water, F_{OARW}, another flux from inflow areas, F_{IARW}), internal fluxes (sedimentation on ET areas and A areas; i.e., areas of fine sediment erosion plus transportation and accumulation, respectively), primary production of SPM in the upstream river stretch (F_{prod}; calculated in the same manner as in the lake model; see chapter 4 and below), resuspension from ET areas back to river water (F_{ETRW}), mineralization, i.e., losses of SPM from bacterial decomposition in river water (F_{minRW}) and in from sediments on ET areas (F_{minET}) and outflow to downstream river stretches (F_{out}).

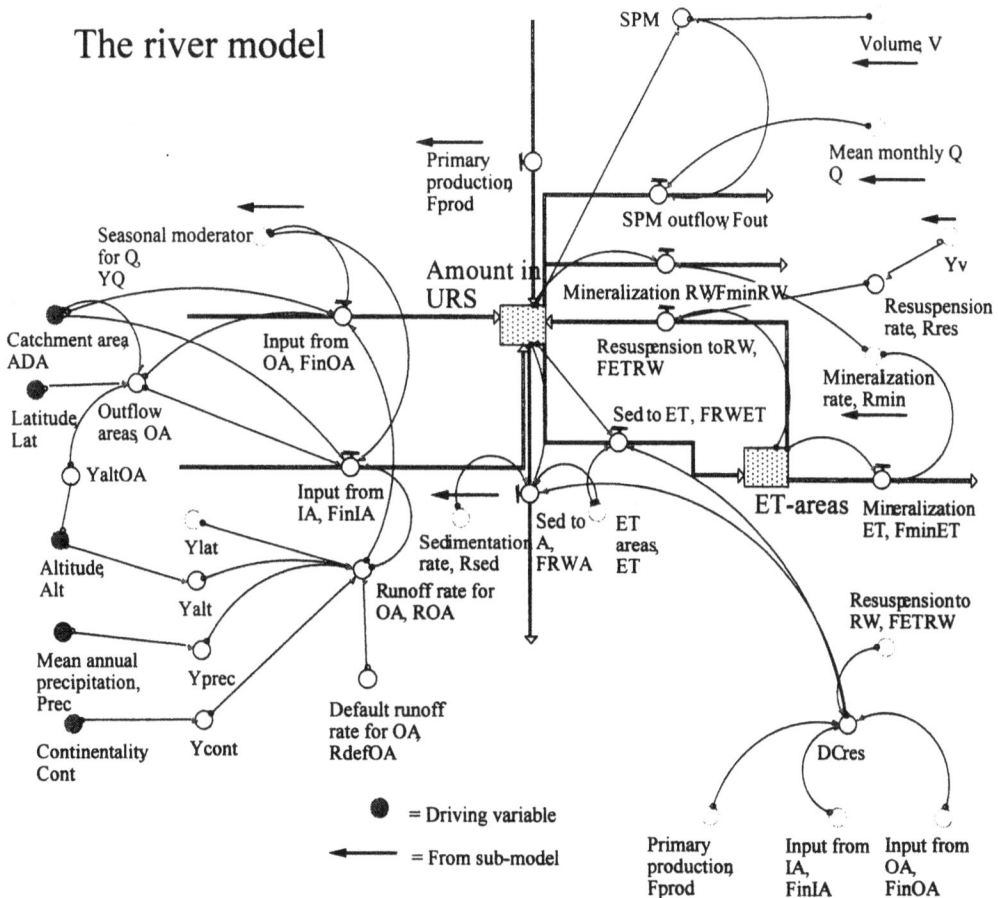

Fig. 5.25. Illustration of the dynamic river model.

Table 5.7. Basic equations making up the dynamic river model for SPM. M = mass; F = flow; R = rate; SWT = river water temperature; OA = outflow areas; Q = water discharge; ET = ET areas; A = A areas; RW = river water; flux from A to B = F_{AB}, etc. Mass in RW = M_{RW}, etc.

Compartment, SPM in river water (M_{RW}):

$M_{RW}(t) = M_{RW}(t - dt) + (F_{prod} + F_{inIA} + F_{inOA} + F_{ETRW} - F_{out} - F_{RWA} - F_{RWET} - F_{minRW}) \cdot dt$

INFLOWS:

F_{prod} $= (30.6 \cdot Chl_{def}^{0.927}) \cdot (A_{DA}/10^{10})^{0.7} \cdot 0.45 \cdot 30 \cdot Area \cdot 10^{-(0.7 \cdot log(SPM)+0.77)} \cdot 0.001 \cdot ((SWT+0.1)/9) \cdot (BM_{PL}/BM_{PH})$

F_{inIA} $= (A_{DA} \cdot (1-OA) \cdot Y_Q \cdot R_{defOA})/12$

F_{inOA} $= 10 \cdot A_{DA} \cdot OA \cdot Y_Q \cdot R_{defOA}/12$

F_{ETRW} $= M_{ET} \cdot R_{ET}$

OUTFLOWS:

F_{out} $= Q \cdot C_{RW}$

F_{RWA} $= (1-ET) \cdot M_{RW} \cdot R_{sed}$

F_{RWET} $= ET \cdot M_{RW} \cdot R_{sed}$

F_{minRW} $= M_{RW} \cdot R_{minRW}$

Compartment, SPM in ET areas (M_{ET}):

$M_{ET}(t) = M_{ET}(t - dt) + (F_{RWE} - F_{ETRW} - F_{minET}) \cdot dt$

INFLOWS:

F_{RWET} $=$ see above

F_{ETRW} $=$ see above

F_{minET} $= M_{RW} \cdot R_{minET}$

Some key rates, velocities and dimensionless moderators:

Resuspension rate = R_{res} = $(1-Y_v)$; if v < 0.1 m/s then Y_v = 1 else Y_v = $(1-0.25 \cdot (v/0.1-1))$; if Y_v < 0.01 then 0.01

Runoff rate for OA = R_{defOA} = $2 \cdot Y_{prec} \cdot Y_{lat} \cdot Y_{alt} \cdot Y_{cont}$

Sedimentation rate = R_{sed}=$(72/12) \cdot (1/D_m) \cdot Y_{SPM} \cdot Y_v$ [default settling velocity for SPM=72/12 m/month]

Mineralization rate for SPM in RW = R_{minRW} = $R_{minSPM} \cdot (SWT/9)^{1.2}$

Mineralization rate SPM = R_{minSPM} = $0.125 \cdot (0.99/ET)$

v $= (0.5319 \cdot Q^{0.1} + 0.0677)$ [Q in m^3/s; v in m/s]

Y_{ADA} $= \sqrt{A_{DA}}$

Y_{alt} $= (1/(Alt+25))^{0.5}$

Y_{altOA} $= 1/(1+0.0025 \cdot (Alt/1-1))$

Y_{cont} $=$ if Cont < 500 km then 1 else $\sqrt{(Cont/500)}$

Y_{lat} $= (75/(Lat-35))$

Y_{prec} $= (Prec/650)^2$

Y_{SPM} $= (1+0.75 \cdot (SPM/50-1))$

Compared to the lake model, there are some rather evident simplifications:

- The temperature submodel (see appendix 9.4) is used to predict only surface water temperatures (SWT in °C) since the deep water compartment has been omitted. SWT is, like in the lake model, predicted on a monthly basis from latitude, altitude, continentality, and water volume.

- As mentioned, the algorithm to calculate the fraction of bottom areas where erosion and transport processes of fine sediments occur (ET areas) has been omitted and replaced by a generic constant, ET = 0.95. In all rivers, there are topographically shelters parts, macrophyte beds, or deep holes (5% of the river area in this modelling) where SPM may be retained.

- Sedimentation is calculated in a modified way accounting for river water velocity and the fact that little fine sediments will settle if the velocity is higher than about 10 cm/s (fig. 5.18); if the velocity is higher than 50 cm/s, there is likely no net sedimentation but rather erosion and transportation of all types of fine particles. This means that the sedimentation rate (R_{sed} in 1/month) is given by the following approach:

$$R_{sed} = (72/12) \cdot (1/D_m) \cdot Y_{SPM} \cdot Y_v \tag{5.14}$$

Where the default settling velocity for SPM is the same as in the lake model (72/12 = 6 m/month). The main difference between the lake model and the river model concerns the dimensionless moderator for the flow velocity (Y_v), which influences both sedimentation (eq. 5.14) and resuspension.

If $v < 0.1$ m/s then $Y_v = 1$ else $Y_v = (1-0.25 \cdot (v/0.1-1))$; if $Y_v < 0.01$ then 0.01 (5.15)

The resuspension rate (R_{res}) is given by:

$$R_{res} = (1-Y_v) \tag{5.16}$$

The next section will give more calibrations of the dynamic river model and the aim is also to motivate the new algorithms and illustrate the model behavior.

- Primary production of SPM in the upstream river stretch is calculated using eq. 4.22. However, if no empirical river data on chlorophyll are available, the first term in eq. 4.22 may be estimated in the following way:

$$(30.6 \cdot Chl^{0.927}) \approx (30.6 \cdot Chl_{def}^{0.927}) \cdot (A_{DA}/10000)^{0.7} \tag{5.17}$$

Where Chl_{def} are the default (=reference) values for mean monthly chlorophyll (in μg/l; the following values from River Danube at a site near Regensburg have been used in all the following simulations; from Håkanson et al., 2003a).

Jan	Feb.	Mar.	April	May	June	July	Aug	Sep.	Oct.	Nov.	Dec.
2.6	9.6	9.1	26.7	33.6	21.9	21.8	19.6	6.7	6.6	3.8	3.7

The ratio between A_{DA} (the catchment area of the given river site in km^2) and the size of the reference site (area = 10000 km^2) illustrates that the requested chlorophyll values should be higher in rivers with larger catchments than in a rivers with small catchments. The exponent 0.7 indicates that there is not a linear increase in SPM production with A_{DA}.

This approach is evidently a simplification. It should be used if river data on chlorophyll are not available. It means that if A_{DA} = 1000 km^2, if Chl$_{ref}$ = 25 µg/l (as it is for the reference site for the growing season), then eq. 5.17 gives 600·0.2 (µg C per liter per day); if A_{DA} = 10000 km^2, the left side is equal to the right side = 600; if A_{DA} = 100000 km^2, the value is 600·5. The other factors in the production algorithm (0.45·30·Area·Sec·0.001·((SWT+0.1)/9)·(BM$_{PL}$/BM$_{PH}$)) are the same for the lake model and the river model.

5.3.5. Further calibrations of the dynamic river model

Fig. 5.26 shows calibration results along a latitude gradient (latitudes set to 40, 50, 60, and 70°N in four steps and fig. 5.26A gives the driving variable). Three different approaches for how differences in latitude influence SPM transport from catchment areas and river geometry (Y$_{lat}$) have been tested and compared to results using the empirical SPM model as a reference. Note that in all the tests in this section, there have been no considerations to variations among the rivers in the cover by forests, lakes, bare rocks or relief. The moderator (Cf; table 5.6) accounting for such variations in the empirical SPM model has been held constant to the normal default value of 2.

From fig. 5.26B, one can note:

- The good correspondence between the SPM values predicted by the empirical model and the dynamic model for the default moderator (curve 1), except that the dynamically modelled values are higher than the values given by the empirical model at low latitudes.

- The other two algorithms for Y$_{lat}$ give lower values as compared to the empirical SPM model for rivers at low latitudes.

The default latitude in these calibrations was set to 50°N.

The next figure (fig. 5.27) gives calibrations along an altitude gradient (altitudes set to 0, 100, 500, and 1000 m.a.s.l in four steps and fig. 5.27A gives the driving variable; the default value in these calibrations is set to Alt = 0). Four dimensionless moderators expressing how differences in altitude influence SPM transport from catchment areas and river geometry (Y$_{alt}$) have been tested and compared to the empirical SPM model. Fig. 5.27B shows that:

- The empirical model does not account for differences in altitude, so there is just one reference line.

- There is a good correspondence between the SPM value given by the empirical model and the different SPM values calculated by the dynamic model for the default moderator (curve 1).

- All moderators give higher SPM concentration for low-altitude rivers than for high altitude rivers, but the SPM values vary very much using the moderator expressed by curve 4.

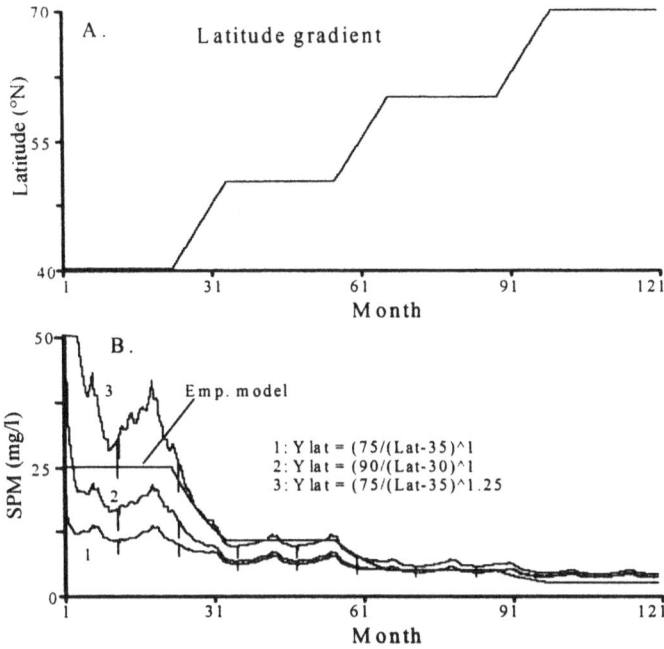

Fig. 5.26. Calibrations and sensitivity analyses of the dynamic river model along a latitude gradient (fig. A; latitudes set to 40, 50, 60, and 70°N in four steps) for three different equations for the dimensionless moderator for how latitude influences SPM concentrations in river water. As a reference, fig. B also gives the data calculated using the empirical SPM model along this latitude gradient.

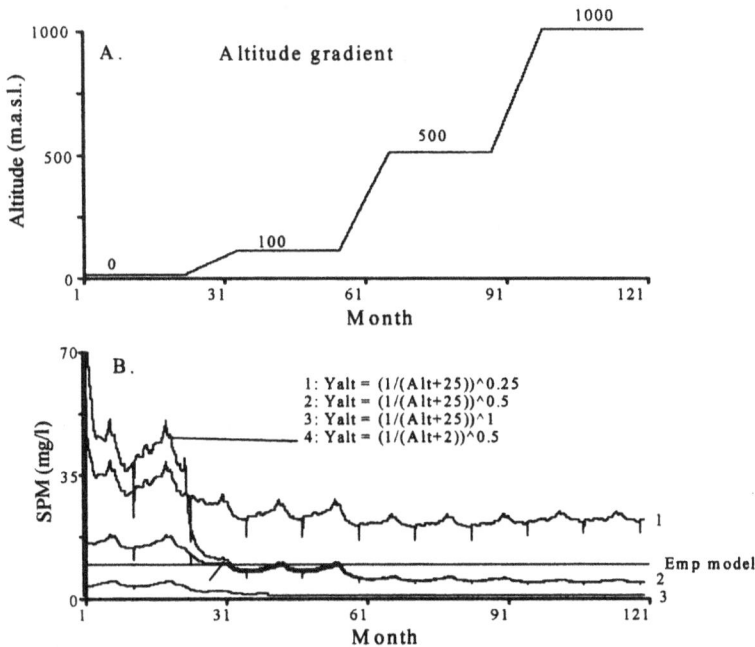

Fig. 5.27. Calibrations and sensitivity analyses of the river model along an altitude gradient (fig. A; altitudes set to 0, 100, 500, and 1000 m.a.s.l. in four steps) for four different equations for the dimensionless moderator for how differences in altitude may influence SPM concentrations in river water. Fig. B also gives the data calculated using the empirical SPM model along this altitude gradient and that the empirical model does not account for how variations in altitude influence SPM.

- The last calibration here is along a gradient in continentality (values set to 10, 100, 1000, and 10000 km from the ocean in four steps and fig. 5.28A gives the driving variable; the default value in these calibrations is set to Cont = 100 km). Three moderators expressing how differences in continentality would influence SPM transport from catchment areas and river geometry (Y_{cont}) have been compared to values given by the empirical SPM model. Fig. 5.28 shows that there is an excellent correspondence between the values predicted by the two models for the default set-up (curve 2).

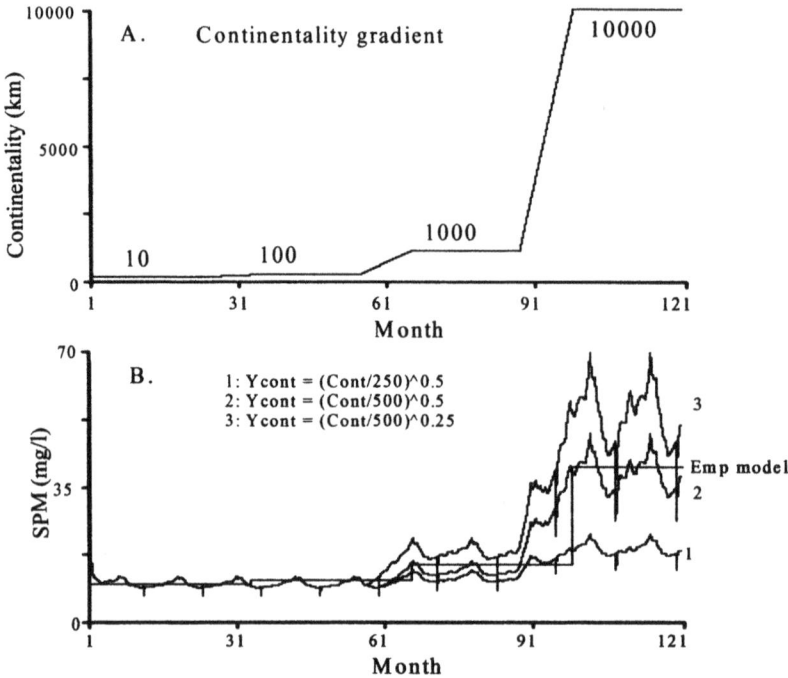

Fig. 5.28. Calibrations and sensitivity analyses of the river model along a continentality gradient (fig. A; values set to 10, 100, 1000, and 10000 km in four steps) for three different equations for the dimensionless moderator for how continentality influences SPM concentrations in river water. Fig. B also gives the data calculated using the empirical SPM model along this gradient.

5.3.6. Validations

From the results of the calibrations, one can ask: How will the dynamic model work when validated, i.e., if the model predictions are blind tested against independent data? The following validation will use the data on SPM inflow to the lakes discussed in chapter 4 (table 4.8; these data are also given in table 5.8 for direct comparative purposes). First, one should reiterate that these SPM inflows are based on empirical data (either on SPM inflow, as to Lake Kinneret, or on empirical data on TP inflow recalculated as SPM inflows to Lake Erken, or on well-tested mass-balance calculations based on TP inflow, as in the three Belarussian lakes and Lake Balaton and Lake Balaton 1). But these data on SPM inflow, although based on measurements and independent of the dynamic river model, are not "cut in stone," but uncertain, at least by a factor of 1.5.

Table 5.8 gives the results and fig. 5.29 gives four regressions based on the data given in table 5.8.

- First (fig. 5.29A), there is a direct comparison between the actual values from the dynamic river model (on the x axis) and the empirically based inflow values to the seven lakes (on the y axis). The correspondence is excellent ($r^2 = 0.9996$, slope = 1.03), but it is must be stressed that there are only 7 cases.

- Fig. 5.29B gives the same results but for log transformed data. The results are also excellent ($r^2 = 0.997$, slope = 1.01).

- The regression in fig. 5.29C compares log transformed data given by the empirical SPM model (on the x axis) versus the inflow values to the seven lakes (on the y axis). These results do not look like an analytical solution as in fig. 5.29A, but more in line with what might hope for and expect ($r^2 = 0.94$, slope = 0.96).

- Fig. 5.29D finally gives a comparison between log transformed data from the empirical SPM model (on the x axis) versus log transformed data from the dynamic river model (on the y axis). The correspondence is very good ($r^2 = 0.94$, slope = 0.95).

Table 5.8. Data on tributary SPM transport into Lakes Batorino, Miastro, Naroch, Kinneret, Erken, Balaton, and Balaton 1 from the empirical river model, the dynamic river model and empirical data used within the lake model (see chapter 4 and table 4.9).

Lake	SPM inflow (tons /yr) calculated from:		
	Empirical river model	Dynmic river model	Empirical data, lake model
Batorino	130	120	160
Miastro	197	160	140
Naroch	510	175	170
Kinneret	20940	39330	40500
Erken	156	150	130
Balaton	41210	14470	15600
Balaton 1	18300	7440	8140

5.4. Concluding remarks

This chapter used several databases to derive operational models to predict concentrations of suspended particulate matter in rivers. The empirical model presented in the first part of this chapter is meant to be used as a submodel in models where the main aim is to predict concentrations of radionuclides, nutrients, or metals in river water and biota. SPM influences the transport of most types of pollutants in most systems and governs both primary phytoplankton and bacterioplankton production and hence also secondary production (e.g., of zooplankton and fish).

An important feature of the requested SPM models for rivers is that all variables needed to run the model should be readily accessible from standard monitoring programs and/or maps. The coefficient of variation (CV) for SPM in rivers sites is very high indeed. The factors influencing the variability of characteristic (median) SPM values among sites in different rivers have been quantified and ranked using the two databases, a European database including data on SPM, water discharge, latitude, longitude, continentality, and altitude, and a Swedish database, which includes a comprehensive set of data on catchment area characteristics. The measured river characteristic

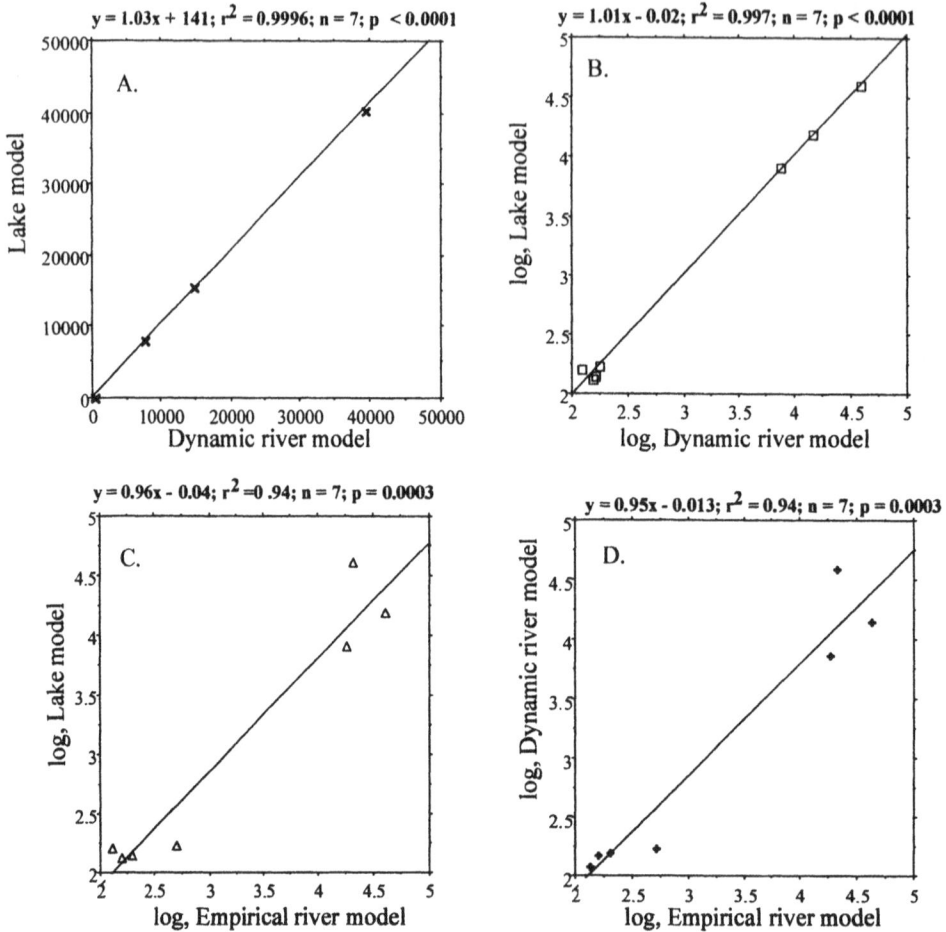

Fig. 5.29. Results of the validation of the dynamic river model for SPM. (A) Gives actual SPM values predicted by the dynamic model on the x axis compared to values calculated from empirical data on the river inflow of SPM to the studied seven lakes on the y axis, (B) gives the same thing using log values on both axes, (C) gives values calculated from the empirical river model on the x axis against SPM values calculated from empirical data on the river inflow of SPM to the studied seven lakes on the y axis and (D) compares the same thing using the empirical SPM model for rivers as compared to SPM values calculated by the dynamic river model.

SPM values can be predicted from the following variables, (1) latitude (as a measure of mean temperature, primary production, and weathering), (2) mean water discharge, (3) continentality (SPM increases with increased continentality), and from a simple system accounting for (4) the percentage of bare rocks, the relief of the catchment, the percentage of upstream lakes, and the forest cover of the catchment area. The latter factors are categorized into three classes (1, 2 and 3) having little, average, or relatively large influences on the transport of SPM from land to water.

It must be stressed that the very good validation results for the dynamic river model should be taken with due reservation since it is based on just 7 cases. These results, however, indicate that the mechanistic principles and assumptions behind the dynamic river model probably represent transport processes in natural ecosystems well. Evidently, it would be very interesting to test the behavior of the dynamic model for more rivers.

6. Coastal Areas

This section first gives a brief introduction to coastal processes and coastal modelling. Section 6.2 presents an empirical/statistical model for SPM in defined coastal areas. Section 6.3 will focus on a dynamic SPM model and section 6.4 on an empirical approach for sites (not areas) in open marine waters. Section 6.1 discusses the fundamental unit, the coastal area and how to define the boundary lines for coastal areas and also two important features of this coastal modelling, (i) the calculation of the water exchange between a coastal area and the open sea and (ii) the oxygen conditions in the deep water areas, which relate to sedimentation of organic particles, bacterial decomposition of organic particles, oxygen consumption, oxygen concentrations and the survival of zoobenthos, a key functional groups in coastal ecosystems.

Due to the complex hydrodynamical conditions characterizing a given coastal area (fig. 6.1), such as the Coriolis-driven coastal jet-zone, tidal effects, effects related to "the zone of maximum turbidity", and winds, it is easy to understand why so much interest and research concern this zone. For basic literature on general sedimentological and hydrodynamical processes in coastal areas, see Muir Wood (1969), Stanley and Swift, (1976), Dyer (1979), McCave (1981), Seibold and Berger (1982), and Postma (1982). Human activities, like trawling, can also influence the sediment resuspension (Floderus, 1989).

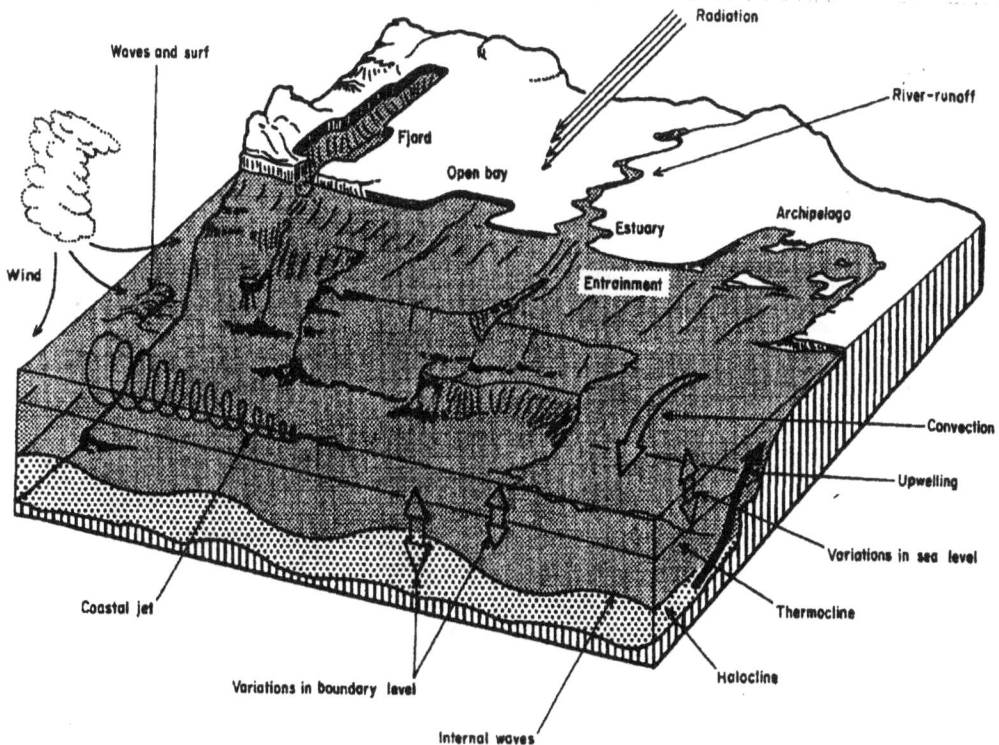

Fig. 6.1. Illustration of key factors regulating water exchange in coastal areas (from Håkanson et al., 1986).

The coastal zone is also a zone of conflicts where many different users, such as professional and leisure-time fishermen and people responsible for recreation, shipping and environmental management and research, place different demands on the coastal waters and apply different criteria to set the value of coastal waters and to define desired conditions (see Wallin et al., 1992; Lundin, 1999, 2000a, 2000b; Wulff et al., 2001). The coastal zone is also the "recipient" of many types of pollutants, such as organic matter, radionuclides, nutrients, metals, and organic toxins (see Pearson and Rosenberg, 1976; Ambio, 1990, 2000; Meeuwig et al., 2000; Aertbjerg, 2001). The coastal zone may be regarded as a "pantry and a nursery" for the sea. It has been demonstrated (fig. 6.2) that shallow coastal areas can have a bioproduction many times higher than the most productive areas on land (Rosenberg, 1985); all three functional groups of primary producers - phytoplankton, benthic algae and macrophytes - are present in coastal areas (but not in open water areas); and where there is a high primary production, there is also a high secondary production of zooplankton, zoobenthos, and fish (Mann, 1982; Sandberg et al., 2000; Håkanson and Boulion, 2002).

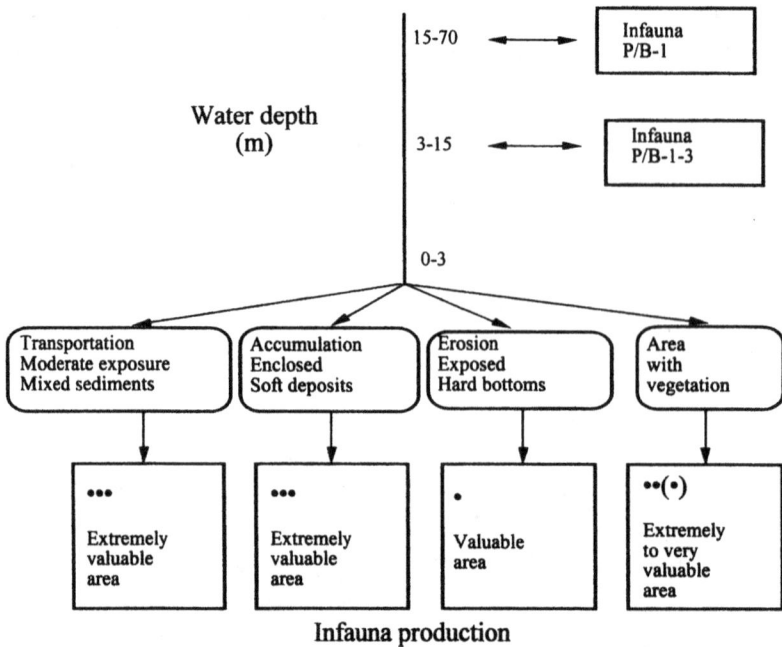

Fig. 6.2. Criteria to estimate the "value" of Baltic coastal areas mainly from a perspective of fishery biology. The infauna is the mobile benthic fauna (in the sediments). Modified from Håkanson and Rosenberg (1985).

From fig. 6.2 shows that at water depths larger than 15 m, one can generally set the production capacity (PR/BM, PR = production in kg/time and BM = biomass in kg) of the infauna (i.e., animals > 1 mm living in the sediments) to be about 1. At water depths of 3 - 15 m, PR/BM is generally between 1 and 3. At water depths smaller than 3 m, it is important also to consider the sediment type, the habitat for the infauna. PR/BM is given on a relative scale of value in terms of fishery biology:

- •   means a valuable coastal area,

- ••   means a very valuable area and

- ••• an extremely valuable area.

This chapter will first discuss criteria to define the fundamental unit, the coastal area. It is, evidently, easy to define what is a lake. A river segment may be defined by the area between two inflowing tributaries. But where and how should one draw the borderlines toward the sea and/or adjacent coastal areas?

The main model in this chapter is the dynamic coastal mass-balance model handling SPM fluxes to, within, and from defined coastal areas. This model is a modified version of the coastal models for SPM and phosphorus presented by Håkanson et al. (2004a) and Håkanson and Karlsson (2004), respectively.

6.1. Fundamental concepts in coastal modelling

6.1.1. Defining coastal area boundaries

In this modelling approach, it is important to define the coastal area. In contexts of process-oriented mass-balance modelling, this should not be a diffuse foodweb structure, but a defined geographical area. The question is where to place the boundaries toward the sea and/or adjacent coastal areas. It is crucial to use a technique that provides an ecologically meaningful and practically useful definition of the coastal ecosystem. How should one define this area so that, e.g., parameters like the mean depth, the volume, and the form factor can be used as relevant model variables (x) for predicting the target variable, the SPM concentration in coastal waters?

Arbitrary borderlines can be drawn in many ways and the morphometric parameters of such areas would be devoid of meaning in contexts of mass-balances and in relation to the target variables one would like to predict and understand. The approach in this work comes from Håkanson et al. (1984) and Pilesjö et al. (1991) and assumes that the borderlines are drawn at the topographical bottlenecks so that the exposure (Ex; defined by the ratio between the section area, At, and the enclosed coastal area, A; fig. 6.3) of the coast from winds and waves from the open sea is minimized. It is always important to define the presuppositions of any model. When and where will it apply? The definition of the coastal boundaries is one crucial aspect for this coastal modelling. Since many scientists working in coastal areas do not have this ecosystem perspective, this method is not well known. If the method were better known, more scientists might see the benefits of applying an ecosystem perspective to coastal studies.

Today, there are also many modern digitized methods applicable in aquatic studies including Geographical Information Systems (GIS). Those methods enable quick and effective estimations of morphometric parameters (see Pilesjö et al., 1991; Gyllenhammar, 2004). In contexts of GIS applications, there are also many commercial software programs available, e.g., Surfer, ArcInfo, MapInfo, and ArcView. Burrough (1986) and Skidmore (1990) discuss GIS methods and also kriging algorithms, which are generally used for interpolations.

Fig. 6.3 illustrates steps to define a coastal area and the boundaries to the open water areas or adjacent coastal areas at the topographical bottlenecks using GIS techniques where the exposure (Ex = 100·At/A) attains minimum values when different alternatives for settling the boundary lines are tested. Once the coastal area is defined, one can also determine important variables for mass-balance calculations, such as the coastal volume (regulating the concentration of any given substance), and important morphometric parameters for internal fluxes, such as the mean depth and the water surface area, and key variables regulating the water exchange between the coast and the sea, such as the section area, the exposure, and the filter factor (fig. 6.4). This method of defining coastal areas also opens a possibility to use empirical models to estimate, e.g., the theoretical water retention times of the surface water and the deep water (table 6.1), and the bottom dynamic conditions (regulating sedimentation, resuspension, and diffusion) from morphometrical parameters (such as area, mean depth, form factor, and section area). This will be outlined in the following sections.

Defining the coastal area

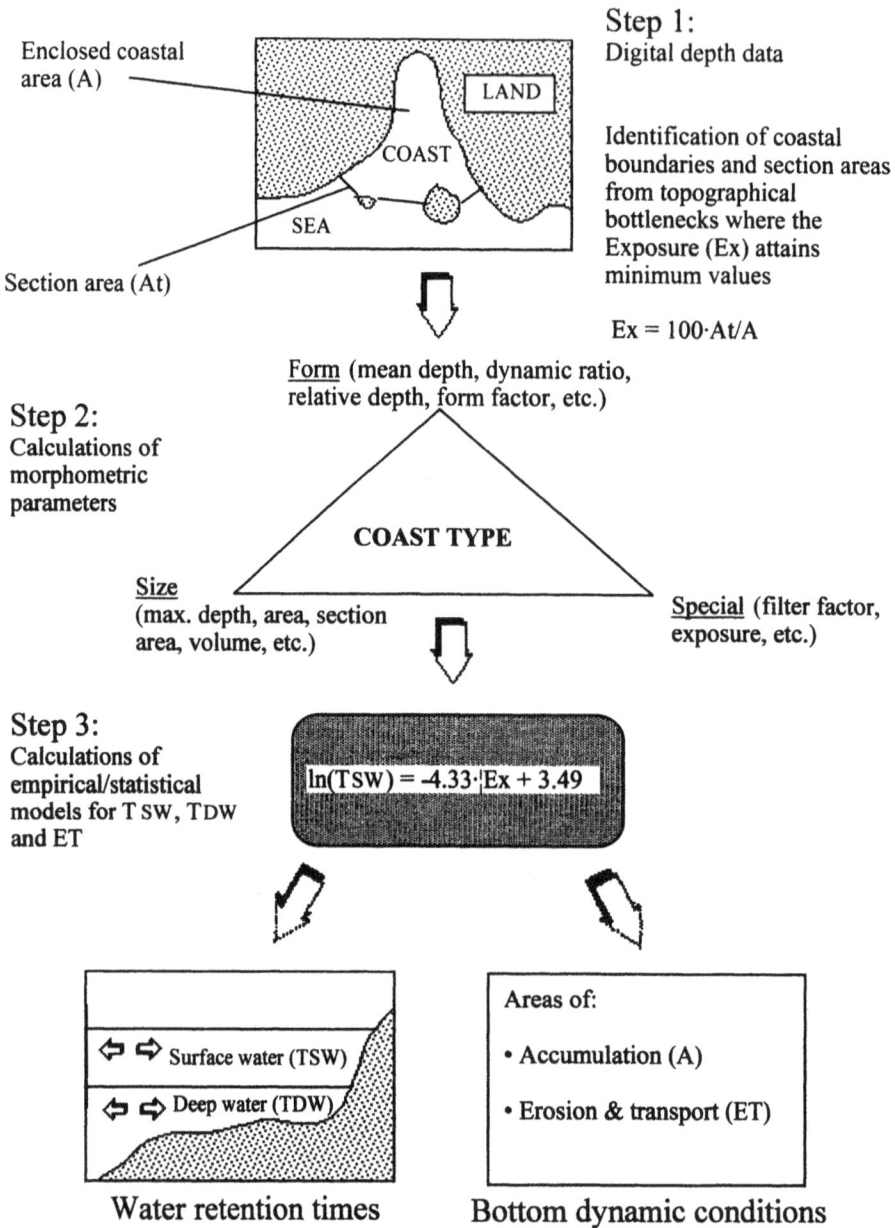

Step 1:
Digital depth data

Identification of coastal
boundaries and section areas
from topographical
bottlenecks where the
Exposure (Ex) attains
minimum values

$$Ex = 100 \cdot At/A$$

Enclosed coastal
area (A)

LAND

COAST

SEA

Section area (At)

Step 2:
Calculations of
morphometric
parameters

Form (mean depth, dynamic ratio,
relative depth, form factor, etc.)

COAST TYPE

Size
(max. depth, area, section
area, volume, etc.)

Special (filter factor,
exposure, etc.)

Step 3:
Calculations of
empirical/statistical
models for T SW, TDW
and ET

$$\ln(TSW) = -4.33 \cdot Ex + 3.49$$

⇐ ⇒ Surface water (TSW)

⇐ ⇒ Deep water (TDW)

Areas of:

• Accumulation (A)

• Erosion & transport (ET)

Water retention times   Bottom dynamic conditions

Fig. 6.3. Illustration of the steps to define the boundary lines for a coastal area (step 1) and how to calculate important coastal parameters in contexts of mass-balance modelling at the ecosystem scale, i.e., for defined entire coastal areas (steps 2 and 3).

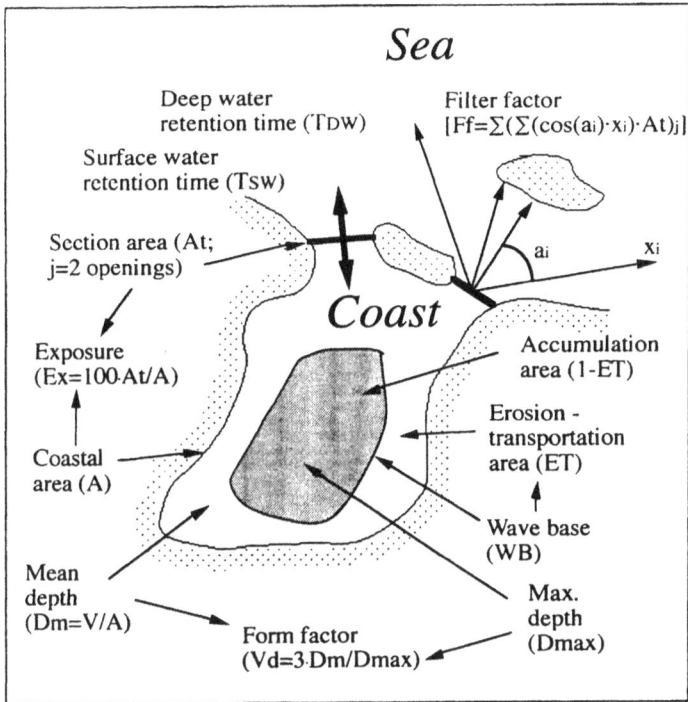

Fig. 6.4. Illustration of key coastal parameters in mass-balance modelling that can be determined quickly from digitized bathymetric maps using GIS methods (GIS = Geographical Information System). The most important criteria in this context is to define the boundary lines, i.e., where the coastal area ends and the sea or adjacent coastal area begins. The approach used in this work is to define the boundary lines so that the topographical openness (the exposure, Ex, defined by the ratio between the section area, At, and the enclosed coastal area, A) attains a minimum value. This is the "topographical bottleneck" method. The filter factor (a measure similar to the fetch) describes how islands and topographical barriers between the defined coastal area and the sea act as energy filter for the wave impact on the coastal area. In this figure, the filter factor is illustrated for one of the two section areas in this coastal area (a_i = the angle between two radials sent out from the opening over open water; x_i = the length of each radial; i = the number of radials at each opening; j = the number of openings; j = 2 in this example).

Table 6.1. Empirical regressions used in this dynamic modelling approach to predict theoretical surface and deep water retention times (T_{SW} and T_{DW} in days) from simple morphometric parameters.

Regression	r^2	n	Reference
$\ln(T_{SW}) = (-4.33 \cdot (\sqrt{Ex}) + 3.49)$	0.95	14	Persson et al. (1994)
$T_{DW} = (-251 - 138 \cdot \log(At) + 269 \cdot \log(Vd))$	0.79	15	Håkanson & Karlsson (2004)

Model domain: $0.002 < Ex < 1.3$; $0.0006 < At < 0.08$; $0.5 < Vd < 1.5$;
note that T_{SW} and T_{DW} are never permitted to be < 1 day and T_{DW} never > 120 days.

$Ex = 100 \cdot At/A$; $Vd = 3 \cdot D_m/D_{max}$; At = section area (km^2); Vd = the form factor (= $3 \cdot D_m/D_{max}$; D_m = mean coastal depth (m), D_{max} = max. coastal depth (m).

6.1.2. Water exchange between the open sea and coastal areas

6.1.2.1. A short background on processes causing coastal water exchange

Many factors influence the water exchange in coastal systems (fig. 6.1) and SPM concentrations in coastal areas cannot be calculated without knowledge of the exchange of water and SPM over the boundary lines between the given coastal area and the open sea. Thus, it is important to discuss some basic concepts concerning the turnover of water in coastal areas. The water exchange varies in time and space. It can be driven by many processes, which also vary in time and space. The importance of the various processes will vary with the topographical characteristics of the coast, which do not vary in time, but vary widely among different coasts. The water exchange sets the framework for the biotic life in a given coastal area and the prerequisites for bioproduction are quite different in coastal waters, since the characteristic retention time may vary from hours to weeks.

Factors influencing the water exchange are (fig. 6.1 and Håkanson, 2000):

- The fresh water discharge (Q is often given in m^3/s) is the amount of water entering the coast from tributaries per time unit. In this modelling, Q can be obtained directly from the river submodel (chapter 5). In marine bays with large tributaries (estuaries), the river discharge may be the most important factor for the water retention time. Criteria for this will be discussed in this chapter.

- Tides. When the tidal variation is larger than about 50 cm, it is often a dominating factor for the surface water retention time. This chapter will mainly discuss data from the Baltic Sea, and in this big estuary, the tidal range is only about 2 cm.

- Water level fluctuations always cause a transport of water. These fluctuations may be measured with simple gauges. They vary with the season of the year and are important for the water retention time of shallow coastal areas.

- Boundary level fluctuations. Fluctuations in the thermocline and the halocline boundary layers may be important for surface and deep water retention times, especially in deep and open coasts.

- Local winds may create a water exchange in all coastal areas, especially in comparatively small and shallow coasts.

- Thermal effects. Heating and cooling, e.g., during warm summer days and nights, may give rise to water level fluctuations which may cause a water exchange. This is especially true in shallow coasts since water level variations in such areas are more linked to temperature alterations in the air than is the case in open water areas.

- Coastal currents are large, often geographically concentrated, shore-parallel movements in the sea close to the coast. They may have an impact on the water retention time, especially in coasts with a great topographical openness.

In theory, it may be possible to distinguish driving processes from mixing processes. In practice, however, this is often impossible. Surface water mixing causes a change in boundary conditions, which causes water exchange, and so on.

The theoretical water retention time (T) for a coastal area is the time it would take to fill a coast of volume V if the water input from rivers is given by Q and the water input from the sea by Q_{sea}, i.e.: $T = V/(Q + Q_{sea})$. This definition does not account for the fact that actual water exchange normally varies temporally, areally, and vertically, e.g., above and beneath the wave base. In the following model, T_{SW} is the theoretical surface water retention time and T_{DW} the theoretical deep water retention time. These concepts evidently only apply to stratified coastal areas.

There are several methods of determining or estimating the water exchange (Dyer, 1979; McCave, 1981; Håkanson et al., 1984, 1986).

- The fresh water input to a bay may be used as a "tracer", and the salinity or the conductivity of the water may be used to determine the water exchange by means of mass-balance calculations. The conductivity may then be measured by CTD-sonds (CTD stands for measurement of Conductivity, Temperature, and Depth). Many kinds of instruments are commercially available. If the CTD-sonds are used to determine the water exchange in an estuary, one must measure salinity or conductivity inside and outside the given coast, as well as the coastal volume and the input of fresh water via tributaries.

- Instead of using the fresh water as a "tracer", one may also use real tracers, like dye tracers (e.g., rhodamine, a red dye). The dye tracer method requires quite a lot of special equipment and trained staff.

- The direction and velocity of water currents may be measured quite simply with inexpensive current meters, which automatically measure the mean direction of the flow and water velocity for the period of registration. If several current meters are placed in a given section, the water exchange can be determined for the coastal area as a whole.

- The water level may be measured by different types of gauges (permanent or mobile, continuously recording or manually handled). The water exchange can be determined from gauge data (differences in water level over time) if the area and volume of the coast are known.

One fundamental abiotic factor that, together with the morphometry (i.e., the size and shape of the coastal area), sets the framework for the biological life is the salinity (fig. 1.11). The salinity in the open water areas outside the coastal zone varies from about 2-4% in estuaries such as the Bothnian Bay, to 6-8% in the Baltic Sea (many data in the following parts of this chapter emanate from the Baltic Sea), and to values in the range 20-30% in the Kattegat, Skagerrak, and the open ocean (fig. 6.5). Beneath the halocline, one finds saltier, denser water with salinities higher than in the surface water. In late summer, one finds warmer, less saline water on top of colder water with approximately the same salinity. These two boundary layers, the thermocline and the halocline, may influence the transport of water, SPM, and pollutants carried by the water and SPM.

In the Baltic coastal zone, the maximum thermocline is generally at a water depth of about 10 m (Persson et al., 1994). The theoretical deep water retention time is generally longer than that of the surface water (this also depends on the volume of the deep water), and the deep water is often exchanged episodically. The mixing between surface water and deep water is generally relatively small during stratified conditions, but efficient during homothermal conditions (see Persson and Håkanson, 1996).

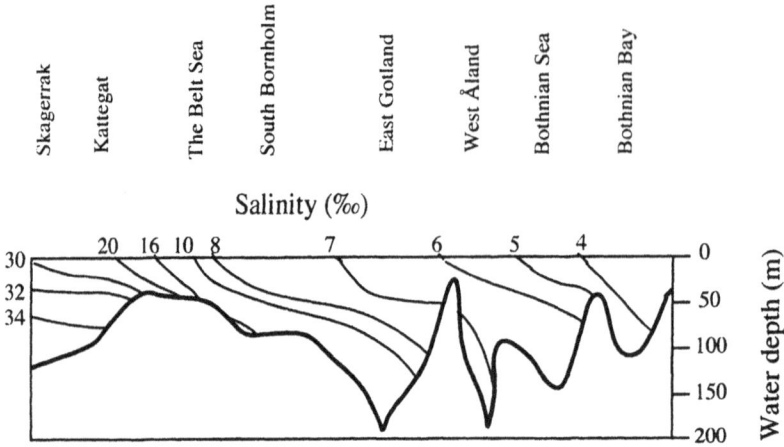

Fig. 6.5. Characteristic salinities in the Baltic Sea (from Håkanson, 1991).

6.1.2.2. Submodels for water exchange

It is very costly and laborious to empirically determine the theoretical (or characteristic) surface or deep water retention time (T_{SW} and T_{DW}) in a given coastal area using any of the discussed methods. The factors regulating inflow, outflow, and retention of substances in coastal areas depend on many more or less stochastic processes. This makes it difficult to give a reliable prediction of T_{SW} or T_{DW} at a given time. In a coastal area, T_{SW} may be indefinitely long on a calm summer day and very short (say a few hours) in connection with a storm or a sudden change in air pressure. T_{SW} or T_{DW} always emanate from frequency distributions (fig. 6.6). This modelling assumes that the characteristic T_{SW} or T_{DW} values, the median value (25 hours as illustrated in fig. 6.6), is used. There is an evident uncertainty associated with such values. Sensitivity and uncertainty tests as well as validations will be performed later in this chapter to see how important this uncertainty is relative to other uncertainties for the target variables in this model (SPM concentrations in water).

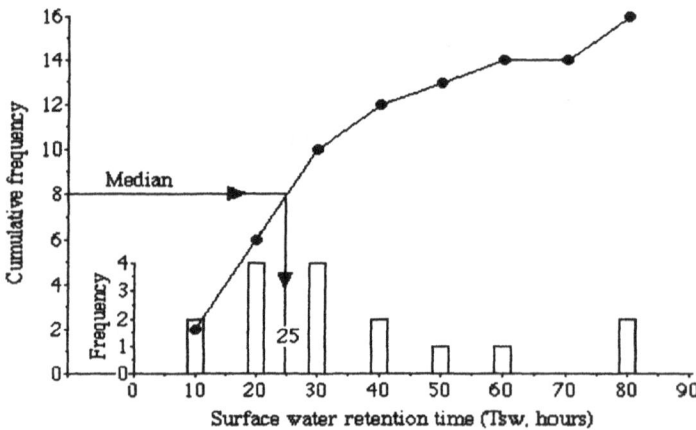

Fig. 6.6. Illustration of empirical data from a dye experiment from a coastal bay to determine the theoretical surface water retention time (T_{SW}). Modified from Håkanson et al. (1984).

As a rule of thumb, one can say that the costs of establishing a frequency distribution such as that in fig. 6.6 from traditional field measurements (using dye, current meters, etc.) is about 20,000 USD for one coastal area. It may not be very meaningful to build a management model if it is a prerequisite that such field work first must be carried out to determine T_{SW} as a driving variable before the model can be used in a given coastal area. This means that it is of major importance that T_{SW} for a given coastal area can be predicted very easily from one coastal morphometric variable, the exposure (fig. 6.3 and table 6.1).

$$T_{SW} = e^{(3.49-4.33\cdot\sqrt{Ex})}/30 \qquad (6.1)$$

The dynamic coastal model for SPM discussed later in this chapter uses predicted T_{SW} values (in months), as given by eq. 6.1. It has been shown (Persson et al., 1994) that T_{SW} can be predicted very well ($r^2 = 0.93$) with this morphometric regression based on the exposure (Ex), which is a function of section area (At) and coastal area (Area) (fig. 6.7). The range of this model for T_{SW} is given by the minimum and maximum values for Ex in table 6.1. That is, Ex should vary from 0.002 to 1.3, and the model should not be used without complementary algorithms if the tidal range is > 10 cm/day or for estuaries, where the fresh water discharge must also be accounted for. For open coasts, i.e., when Ex > 1.3, T_{SW} is calculated not by this equation but from a model based on coastal currents (fig. 6.7) .

It is very difficult and costly to determine T_{DW}, e.g., with the dye method (see Persson and Håkanson, 1996). The costs per area are even higher for T_{DW} than for T_{SW}, but T_{DW} can also be predicted from readily available morphometric parameters ($r^2 = 0.79$; see table 6.1). This empirical model is based on two standard morphometric parameters, which can be determined easily from maps, the section area (At; the larger the section area, the more open the coast for waves and winds from the sea, and the shorter the theoretical deep water retention time) and the form factor (Vd; the larger the deep water volume, the more U shaped the coastal area, the larger the deep water volume, and the longer the theoretical deep water retention time).

$$T_{DW} = (-251 - 138\cdot\log(At) + 269\cdot\log(Vd)) \qquad (6.2)$$

The domain of the model for T_{DW} is given by the minimum and maximum values for the two model variables in table 6.1. This T_{DW} formula is only applicable for stratified coastal areas which are not influenced by tides and or fresh water fluxes since it is based on data from Baltic coastal areas.

For open coasts, T_{SW} (in months) may be estimated from data on the characteristic coastal current (u). This model uses a default value of 2.5 cm/s (= u; see Håkanson et al., 1984). Coastal currents are evidently very important for open coastal areas. The influence of coastal currents is regulated by the exposure (Ex); if Ex > 1.3, the following dimensionless moderator may be applied for the water exchange:

$$Y_{Ex} = (1+0.5\cdot((Ex/10)-1)) \qquad (6.3)$$

Three basic coast types related to water exchange

A. Tidal coast (longitudinal view)

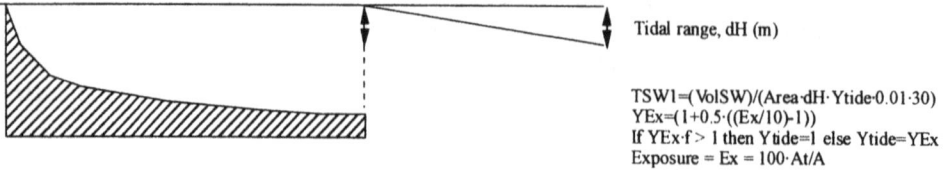

Tidal range, dH (m)

$TSW1 = (VolSW)/(Area \cdot dH \cdot Ytide \cdot 0.01 \cdot 30)$
$YEx = (1 + 0.5 \cdot ((Ex/10)-1))$
If $YEx \cdot f > 1$ then $Ytide = 1$ else $Ytide = YEx$
Exposure $= Ex = 100 \cdot At/A$

B. Open coast (areal view)

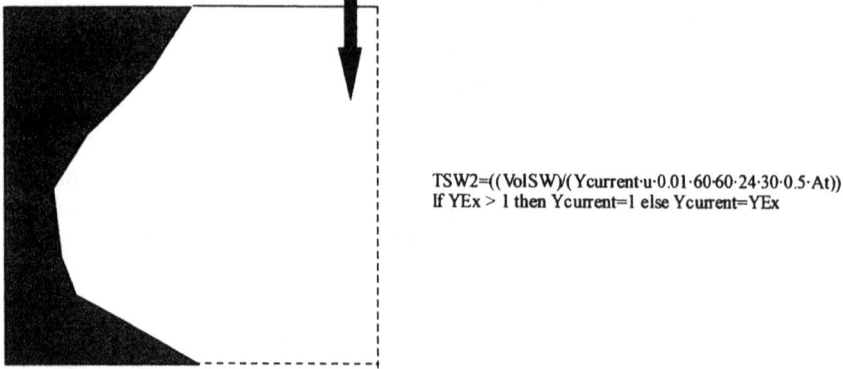

Coastal current (u cm/s)

$TSW2 = ((VolSW)/(Ycurrent \cdot u \cdot 0.01 \cdot 60 \cdot 60 \cdot 24 \cdot 30 \cdot 0.5 \cdot At))$
If $YEx > 1$ then $Ycurrent = 1$ else $Ycurrent = YEx$

C. Archipelago coast (areal view)

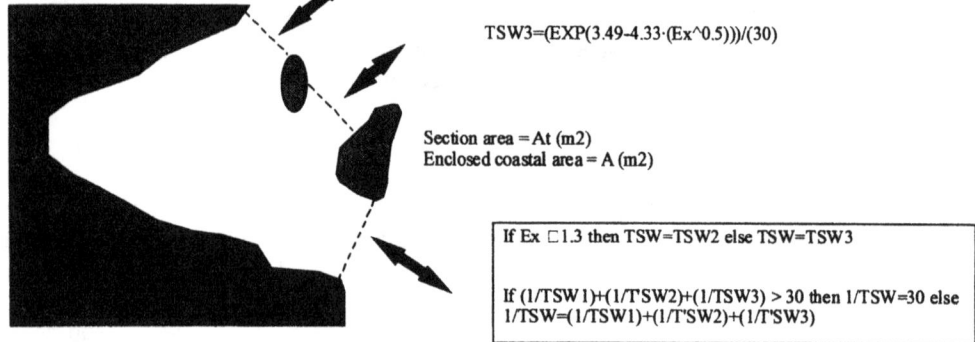

$TSW3 = (EXP(3.49 - 4.33 \cdot (Ex^{0.5})))/(30)$

Section area $= At$ (m2)
Enclosed coastal area $= A$ (m2)

If $Ex \sqsubset 1.3$ then $TSW = TSW2$ else $TSW = TSW3$

If $(1/TSW1) + (1/TSW2) + (1/TSW3) > 30$ then $1/TSW = 30$ else
$1/TSW = (1/TSW1) + (1/TSW2) + (1/TSW3)$

Fig. 6.7. Illustration of the three main coast types and how the theoretical surface water exchange (T_{SW}) may be estimated for each type.

Where 0.5 is the amplitude value; this value regulates the influence that an alteration in the exposure (Ex) would have relative to the norm value (= 10) on the target T_{SW} value. The norm value of 10 means that if the exposure is larger than 10, coastal currents are likely to fully influence the requested T_{SW} value and if the actual Ex value is lower than the norm value, the coastal currents will not influence the T_{SW} value as much. If $Ex = 1$, then $Y_{Ex} = 0.55$ and coastal currents (u) influence T_{SW} 55% less than if $Ex = 10$. If Ex is larger than the norm value of 10, there is an "If then else" statement which sets the moderator to 1.

This means that the theoretical surface water retention time related to coastal currents ($T"_{SW}$ in months) is given by:

$$T''_{SW} = V_{SW}/(Y_{Ex} \cdot u \cdot 0.01 \cdot 60 \cdot 60 \cdot 24 \cdot 30 \cdot 0.5 \cdot At) \qquad (6.4)$$

where

V_{SW} = the volume of the surface water (m^3), which is calculated by the same submodel as already discussed in chapter 4 for the lake model;

u = the characteristic coastal current; u is set to 2.5 cm/s as a default value;

At = the section area in m^2;

0.5 = it is assumed that 50% of the section area is involved in the active water exchange.

For tidal coasts, T_{SW} can be estimated from (see Håkanson, 2000):

$$T'''_{SW} = V_{SW}/(Y_{Ex} \cdot Area \cdot k \cdot 0.01 \cdot dH \cdot 30) \qquad (6.5)$$

where

Area = the coastal area in m^2;

k = a mixing constant; k = 1 means complete mixing. This value can be used as a default assumption if the model is run with dt = 1 month;

0.01 = a calculation constant which changes dH in cm/day to m/day;

dH = the tidal amplitude in m/month (= 30·dH, if dH is given in m/day).

For estuaries, the fresh water inflow (Q) is included in the calculation and the total water flow is given by:

$$Q_{sea} + Q = Vol_{SW}/(T'_{SW} + T''_{SW} + T'''_{SW}) + Vol_{DW}/T_{DW} + Q \qquad (6.6)$$

where the first factor is the surface water flow, and the second factor is the deep water flow, which is calculated from the theoretical deep water retention time (T_{DW}) and the volume of the deep water (V_{DW}).

Fig. 6.8 illustrates how the submodel for water exchange works and how it handles a transition from a situation where the Ex formula is used (i.e., the morphometric equation based on the exposure) to a situation in which the u formula (i.e., coastal currents) is used and when the tidal formula (dH) is applied. Under given conditions for a selected coastal area (Järnavik, table 6.2), the section area, and hence the Ex value, is changed from Ex = 0.23 (default conditions, curve 1) to Ex = 0.023 (curve 2) and Ex = 2.3. The Ex formula should not be used if Ex > 1.3. Under default conditions, the surface water retention time (T_{SW}) is about 4 days and SPM about 3.5 mg/l. The SPM value in the sea outside the coastal area is 4 mg/l. In the next simulation (curve 2), At = 0.1·0.0081 km^2. This means a significantly lower SPM inflow from the sea and that the SPM concentration in the coastal area is lower (between 2 and 3 mg/l); and that T_{SW} value is about 16 days.

If Ex > 1.3, the Ex formula no longer applies and T_{SW} is calculated from the u formula. This situation is shown by curve 3 (At = 10·0.0081 km^2). For such an open coastal area, T_{SW} is about 1 day and the SPM concentration in the coastal areas is very close to the SPM value in the sea outside the given coastal area.

Curve 4 gives results when the tidal amplitude (dH) regulates the water exchange. The tidal amplitude is here set to 100 cm. Then, T_{SW} is about 3 days and the SPM concentration in the coastal areas is slightly lower than the SPM value in the sea outside the given coastal area.

It should be stressed that in many open coastal areas, the water exchange may be driven mainly by winds and in this approach, there are no algorithms for such situations. This is mainly because the data to critically test such algorithms have not been available to the author.

Coastal area Järnavik

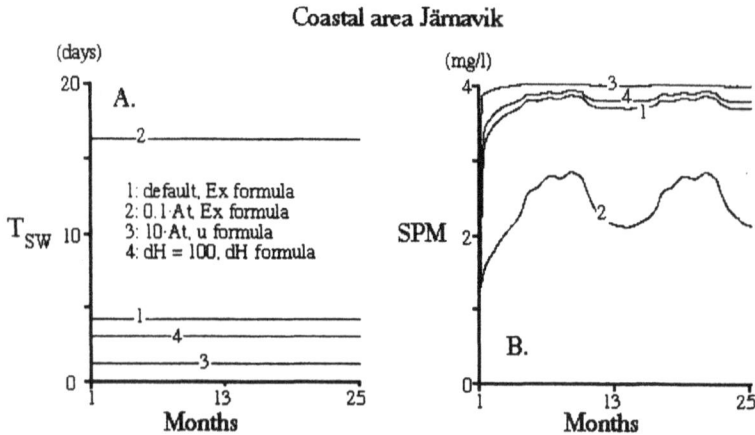

Fig. 6.8. Sensitivity tests illustrating how the dynamic SPM model handles different conditions regulating the water exchange between the coastal area (here Järnavik, Sweden) and the sea. Curve 1 shows the default conditions for this coastal area; curve 2 gives the theoretical surface water retention time (fig. A) and the calculated SPM concentration (fig. B) when the section area is 10 times smaller than under default conditions; curve 3 gives the same results when the section area is 10 times larger, then the water exchange is driven by the coastal current (the u formula); curve 4 shows the results when the tidal amplitude is set to 100 cm.

6.1.3. SPM and oxygen in deep water – an operational ELS model

When the mean O_2 concentration is lower than about 2 mg/l, and the mean oxygen saturation (O_2S in %) lower than about 20%, many key functional benthic groups are extinct (fig. 2.13). It is, naturally, convenient to develop an effect-load-sensitivity model (ELS) based on simple operational effect variables like O_2Sat, but then one must demonstrate the biological/ecological significance of such variables. This is quite clear from fig. 2.13.

Empirical data on the amount of material deposited in deep water sediment traps (1 m above the bottom; Sed_{DW} in g $dw/m^2 \cdot day$) have been used in deriving the ELS model for O_2Sat. This empirical model for O_2Sat will later be put into the dynamic SPM model and then the empirical data on sedimentation in the deep water zone (Sed_{DW}) will be replaced by modelled values of Sed_{DW} from the dynamic model. The values for O_2Sat calculated in this manner will be compared to mean empirical data on O_2Sat from the growing season. The sediment traps were placed at 2-3 sites in each coastal area. They were out for about 7 days at least 2 times during the period July to September in each coastal area (see Håkanson et al., 1989 and Wallin et al., 1992, for further information). Many empirical data expressing the morphometry of the coastal area (fig. 6.4) and the water quality (different forms of N, P, salinity, etc.) that may affect the values for O_2Sat in these areas have been tested using the same statistical procedures (internal correlations, transformations, steps, etc.) as discussed in previous chapters in this book for SPM in lakes and rivers. The data come from 23 Baltic coastal areas. Table 6.2 gives information on these areas. Note that data from adjacent coastal areas have been lumped together in the information given in table 6.2. This is the reason why the empirical model for O_2Sat discussed in this section is based on data from 23 areas and table 6.2 gives only data from 17 areas.

Table 6.2. Data for the 17 studied Baltic coastal areas.

Area	Code	Lati-tude (°N)	Land uplift (mm/yr)	Water area (km²)	D_{max} (m)	D_m (m)	At (km²)	Chl (µg/l)	Sal-ini-ty (%)	Fish prod. (tons/yr)	Sed_{DW} (g dw/m²·d)	Sed_{SW}	Sec_{sea} (m)
Lilla Rimmö	SE1	58	2	2.5	17.6	8.3	0.0172	2.3	6.4	41	20.2	4.2	3
Eknön	SE2	58	2	14.0	19.5	8.5	0.0168	3.5	5.4	32	5.3	2.0	2.5
Lagnöströmmar	SE3	58	2	5.3	20.1	3.8	0.0032	4.6	6.4	125	22.7	11.1	2
Gräsmarö	SE4	58	2	13.8	46.9	13.8	0.0825	2.6	6.6	200	18.3	1.1	3
Ålön	SE5	58	2	6.2	35.2	8.0	0.0162	2.1	6.6	300	9.5	3.7	3.5
Matvik	SS1	56	0	3.1	14.3	5.2	0.0067	1.4	6.5	135	12.7	4.2	4.5
Boköfjärd	SS2	56	0	6.8	21.6	7.1	0.0141	1.4	7.2	70	6.7	1.6	5
Tärnö	SS3	56	0	1.5	11.1	5.1	0.0062	1.6	7.2	50	7.6	3.1	5
Guavik	SS4	56	0	2.7	22.8	5.2	0.0074	2.0	7.3	50	6.4	1.9	5
Järnavik	SS5	56	0	3.4	18.6	5.7	0.0081	1.3	7.3	10	8.1	1.5	5
Spjutsö	SS6	56	0	3.5	15.6	5.8	0.0188	0.9	7.4	50	8.7	4.4	5.5
Ronneby	SS7	56	0	11.0	17.6	4.3	0.0176	2.1	6.5	100	20.2	4.1	4
Käldö	F1	61	5	3.0	16.7	7.6	0.0040	2.7	6.5	50	20.0	11.0	2
Haverö	F2	61	5	2.3	22.5	8.6	0.0172	2.1	6.5	22	38.7	9.2	2
Hämmärösalmi	F3	61	5	2.2	19.3	7.9	0.0114	2.2	6.5	381	13.4	9.1	3
Laitsalmi	F4	61	5	4.2	18.5	7.6	0.0080	2.7	6.5	85	56.4	17.6	1
Kaukolanlahti	F5	61	5	1.4	13.3	4.8	0.0006	9.6	6.5	35	26.3	15.3	1
Min.		56	0	1.4	11.1	3.8	0.0006	0.9	5.4	10	5.3	1.1	1
Max		61	5	14.2	46.9	13.8	0.0825	9.6	7.4	381	56.4	17.6	5.0
Mean (MV)		58	2.1	5.3	20.7	6.9	0.0151	2.65	6.6	102	17.7	6.2	3.3

Of all the many factors that could, potentially, influence O_2Sat, the following have been shown to be most important:

1. Sedimentation in deep water sediment traps (Sed_{DW}); the more oxygen-consuming matter in the deep water zone, the lower O_2Sat $r^2 = 0.43$ at step 1, table 6.3.

2. The prevailing bottom dynamic conditions in the coastal area (ET, i.e., the erosion and transport areas). If variations among coastal areas in ET are accounted for, the r^2 value increases from 0.43 to 0.64. If ET is high (say 0.95), the oxygenation is also likely high and O_2Sat high, and vice versa.

3. The theoretical deep water retention time (T_{DW}); variations in mean O_2Sat among coastal areas can also be statistically related to variations in T_{DW}; the longer T_{DW}, the lower O_2Sat. This is logical and mechanistically understandable. If variations among coastal areas in T_{DW} are accounted for, r^2 increases to 0.74.

4. The mean depth (D_m); the mechanistic reason for this is not so easy to disclose since D_m influences different factors, e.g., (1) resuspension, (2) the volume and hence all SPM

concentrations, (3) stratification and mixing, and (4) the depth of the photic zone and, hence, primary production. However, coastal areas with small mean depths (contrary to lakes with small mean depths; see fig. 1.13) generally have clear water, little SPM, low sedimentation and high O_2Sat. Fine suspended particles in open coastal areas will be transported out of the area and not be entrapped in the same manner as in closed lakes. If variations among coastal areas in D_m are accounted for, the r^2 value increases to 0.80.

Table 6.3. Results of the stepwise multiple regression for the oxygen saturation in the deep water zone (mean O_2Sat in the deep water zone during the growing season in %) using the coastal database. n = 23 Baltic coastal areas; F > 4. y = log(101-O_2Sat). Sed_{DW} = sedimentation in sediment traps placed in the deepest part of the coastal area (g dw/m^2·d); ET = the fraction of ET areas; T_{DW} = the theoretical deep water retention time (days); D_m = the mean depth (m).

Step	r^2	Model variable	Model
1	0.43	$x_1 = \log(Sed_{DW})$	$y = 0.925 \cdot x_1 + 0.132$
2	0.64	$x_2 = \sqrt{ET}$	$y = 0.974 \cdot x_1 - 0.185 \cdot x_2 + 1.72$
3	0.74	$x_3 = \log(1+T_{DW})$	$y = 0.866 \cdot x_1 - 0.151 \cdot x_2 + 0.244 \cdot x_3 + 1.39$
4	0.80	$x_3 = \sqrt{D_m}$	$y = 0.643 \cdot x_1 - 0.118 \cdot x_2 + 0.301 \cdot x_3 + 0.323 \cdot x_4 + 0.470$

From the complex hydrodynamical and sedimentological conditions in coastal areas (fig. 6.1), one might get the impression that the complexity prevents models of high predictive power to be developed. So, it is interesting to conclude that this effect-load-sensitivity (ELS) model, which uses O_2Sat as an operational effect variable, Sed_{DW}, a load factor and ET, D_m and T_{DW} as sensitivity variables, can explain as much as 80% of the variability among the 23 coastal areas in the target y variable. One can also note that the characteristic CV value for O_2Sat is 0.26 (table 3.1), which means that the highest reference r_r^2 is 0.96. This empirical model explains a significant part (83%) of the variation that one can expect to statistically explain.

This O_2Sat model should not be used for coastal areas with characteristics outside the limits given below for the model variables, and it should not be used for coastal areas dominated by tides, or for coastal areas from other coastal types. If the model is used for other coastal areas, then the calculation must be regarded with due reservations, as a hypothesis rather than a prediction.

	D_m (m)	ET (-)	T_{DW} (days)	Sed_{DW} (g/m^2·d)
Minimum	3.8	0.19	1	5.3
Maximum	13.8	0.99	128	82,5

To illustrate how the O_2Sat model works, fig. 6.9 gives three diagrams where the model variables are varied. One can see that differences among coastal areas in mean depth significantly influence the expected value for O_2Sat, while variations in ET generally are not as important.

So, to conclude this section, one can note that many factors could potentially influence O_2Sat variations among these coastal areas. It is easy to speculate about such relationships. With empirical data, it is possible to quantitatively rank such factors and derive a predictive model based on just a few, but the most important, factors influencing O_2Sat (for coasts of the given type). The given model could (statistically) explain about 80% of the variability in the given y variable among the 23 coastal areas. The sedimentation of SPM in the deep water zone, the theoretical deep water retention time, and the mean coastal depth are powerful predictors of O_2Sat, which is also mechanistically understandable.

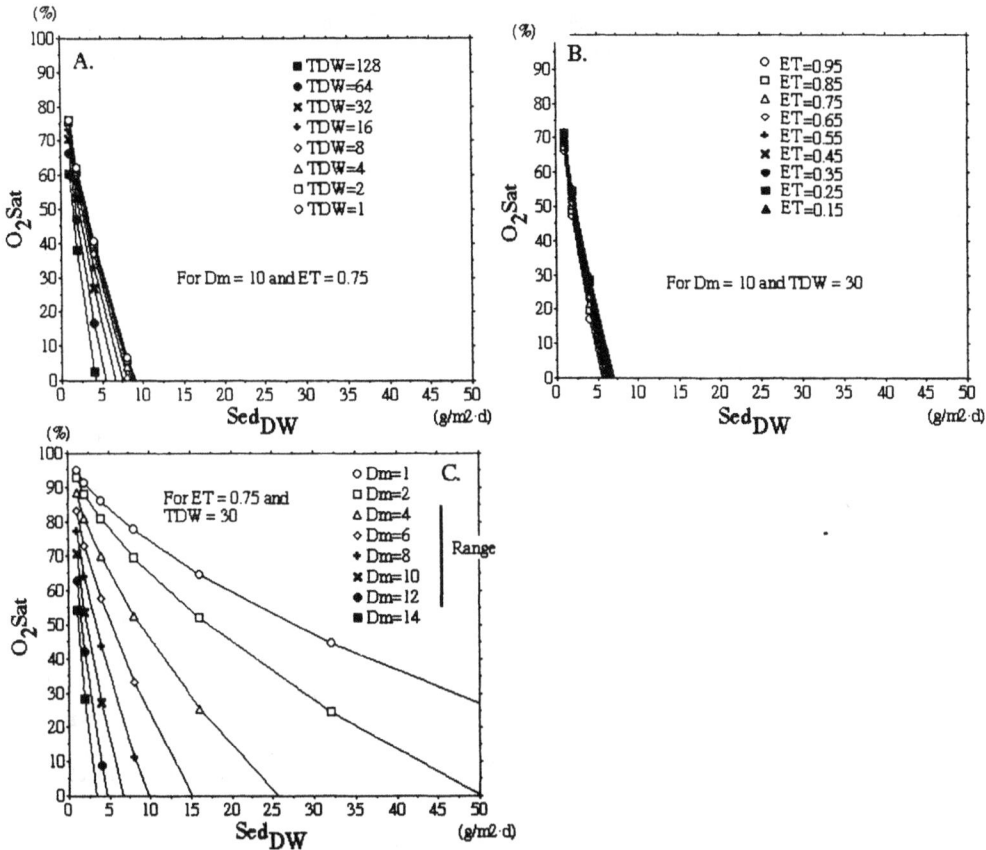

Fig. 6.9. Nomograms illustrating how different model variables influence the oxygen saturation in the deep water zone: (A) gives the information when sedimentation of SPM in deep water sediment traps and theoretical deep water retention time have been varied and mean depth and ET held constant, (B) gives the same when the ET areas have been varied and (C) when the mean depths have been varied and all else kept constant.

6.2. Empirically based coastal models for SPM

This section will present an empirical model for SPM in coastal areas. This model has been developed in two steps:

• First, the empirical database for Baltic coastal areas (appendix 9.1) will be used to derive empirical models for Secchi depth, since there are no data on SPM from these coastal areas for which there is comprehensive information on coastal morphometry, water chemistry, Secchi depth and sedimentation (table 6.2). The aim is to quantify and rank the factors influencing the variability in mean Secchi depths (mean values for the growing season) among these 23 coastal areas in the same manner as already discussed for the deep water oxygen saturation.

• Then, a model for SPM will be presented. It relates SPM to Secchi depth and salinity. In a following section, SPM values predicted by the dynamic model will replace empirical SPM values so that Secchi depths are predicted from the dynamic SPM model.

6.2.1. Empirical models for Secchi depth

Table 6.4 gives three models for the Secchi depth (mean values for the growing season) using the same database from 23 Baltic coastal areas as before. Fig. 6.10 first gives scatter plots showing statistically significant relationships between empirical data on Secchi depths versus tested coastal variables.

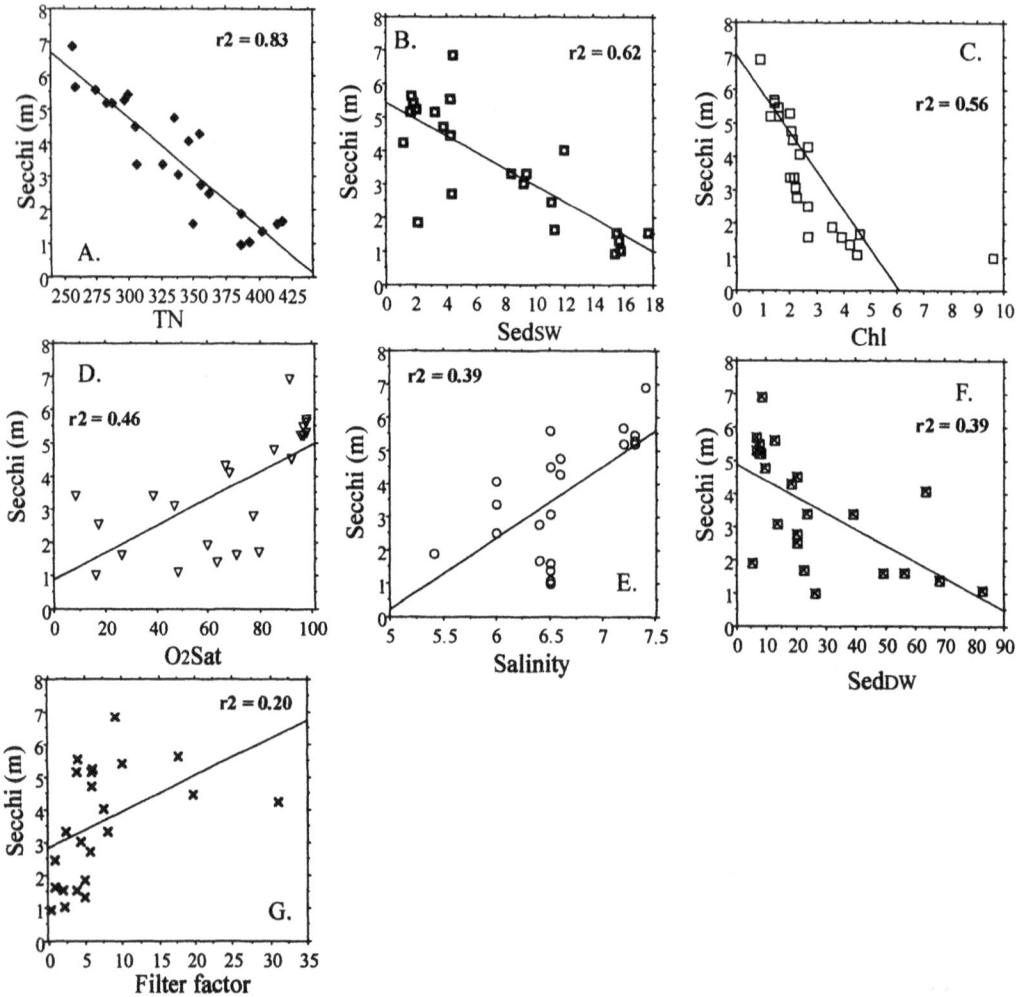

Fig. 6.10. Scatter plots, regression lines, and r^2 values for Secchi depth versus total N concentrations (TN in µg/l), sedimentation in surface water sediment traps (Sed$_{SW}$ in g/m^2·d), chlorophyll-a concentrations (Chl in µg/l), oxygen saturation in the deep water zone (O$_2$Sat in %), salinity (%), sedimentation in deep water sediment traps (Sed$_{DW}$ in g/m^2·d) and the filter factor (Ff in km^3).

Table 6.4. Results of the stepwise multiple regression for Secchi depth (Sec; mean value for the growing season) using the coastal database. n = 23 Baltic coastal areas; F > 4.

Step	r^2	Model variable	Model
A. Actual data on y = Sec and model variables			
1	0.83	$x_1 = TN$	$y = -0.033 \cdot x_1 + 14.5$
2	0.89	$x_2 = Vd$	$y = -0.033 \cdot x_1 - 1.65 \cdot x_2 + 16.3$
3	0.91	$x_3 = At$	$y = -0.033 \cdot x_1 - 1.51 \cdot x_2 + 18.3 \cdot x_3 + 15.9$
B. Transformed data; y = log(Sec)			
1	0.83	$x_1 = \log(Chl)$	$y = -1.03 \cdot x_1 + 0.88$
2	0.88	$x_2 = Ff$	$y = -0.96 \cdot x_1 + 0.0078 \cdot x_2 + 0.80$
3	0.91	$x_3 = \log(D_m)$	$y = -0.090 \cdot x_1 - 0.0095 \cdot x_2 - 0.30 \cdot x_3 + 1.03$
C. Transformed data; y = log(Sec); Ff = the filter factor omitted			
1	0.83	$x_1 = \log(Chl)$	$y = -1.03 \cdot x_1 + 0.88$
2	0.88	$x_2 = \log(Vd)$	$y = -1.036 \cdot x_1 - 0.445 \cdot x_2 + 0.887$

One can note the significant logical and negative relationships between Secchi depth and total N, chlorophyll, and sedimentation (the higher the production, the more SPM, the higher the sedimentation and the lower the Secchi depth). Even in this relatively narrow range, there is also a significant positive relationship between salinity and Secchi depth – the higher the salinity, the clearer the water. The filter factor (fig. 6.4) is an expression for how an archipelago between the given coastal area and the sea may act as an energy filter – the denser the topographical filter effect, the lower the Secchi depth.

The first ladder in table 6.4 gives Secchi depth in three steps:

- First from mean total N concentrations ($r^2 = 0.83$); the higher the TN concentration, the higher the primary production, the more turbid the water, and the lower the Secchi depth.

- Then from the form factor (Vd; $r^2 = 0.89$); the role of the form factor has been discussed on several occasions; Vd relates to resuspension; if the coastal area is very shallow, i.e., if Vd attains a small value, then the coastal area will be dominated by ET areas with hard bottoms and relatively coarse sediments and little fine sediments and little resuspension of SPM. So, the Secchi depth will be high. This is in contrast to the general conditions in lakes. The Secchi depth is often lower in shallow lakes with low form factors (fig. 1.13). As discussed, the main reason for this seems to be that for open coastal areas, the fine suspended particles will be transported out of the area and not be trapped in the same manner as in lakes.

- The last step gives the section area (At in km^2) as a model variable; then r^2 reached 0.91. If At is large, the water exchange between the coastal area and the sea will be very important and one should expect that the Secchi depth in the coastal area would be close to the Secchi depth in the sea.

The next ladder is based not on TN concentration but on chlorophyll as a more direct measure of primary production. Then the steps are:

- Chlorophyll enters at the first step ($r^2 = 0.83$); the higher the primary production, the lower the Secchi depth.

- The filter factor appears at the second step (fig. 6.4 for definition; Ff; $r^2 = 0.88$). If Ff is large, the coast is open to wind/wave influences from the sea. Then the Secchi depth in the coastal area would be similar to the Secchi depth in the sea.

- At the third and last step, one finds the mean depth (D_m; $r^2 = 0.91$). The larger D_m, the more SPM will be entrapped and retained in the coastal area and the smaller the Secchi depth.

In the third ladder, the filter factor has been omitted because Ff may not always be easily accessed. Then the steps are:

- Chlorophyll enters at the first step ($r^2 = 0.83$); as before.

- The form factor, as before ($r^2 = 0.88$).

These empirical regressions for O_2Sat and Secchi depth demonstrate that there are some morphometric parameters (Vd, At, and D_m) that seem to be very important in understanding and predicting how variables like Secchi depth, SPM, sedimentation, and oxygen vary among coastal areas. These morphometric variables are incorporated in the dynamic model in a mechanistic way where they regulate transport fluxes of SPM.

6.2.2. SPM versus Secchi depth and salinity

It has already been stressed that there exists a relationship between SPM, Secchi depth, and salinity – the higher the salinity, the higher the aggregation of suspended particles, the larger the particles, and the higher the water clarity. An SPM concentration of 10 mg/l, would imply turbid conditions in a freshwater system, but relatively clearer water in a saline system. This is shown again by the same data as in fig. 2.4 in fig. 6.11. In this case, however, the figure does not give regressions but deterministic relationships based on boundary requirements. First, that the maximum average SPM in surface waters in lakes and marine coastal areas (but not in rivers) should approach 50 mg/l. If the SPM value is very low (0.5 mg/l), the Secchi depth in lakes (with salinity = 0%) should be about 10 m; the corresponding Secchi depth in coastal areas with a salinity of 6.5% should be about 50 m; and the Secchi depth should approach 200 m if the salinity approaches 30%. There are many data in fig. 6.11 supporting the relationship between SPM, Secchi depths and salinity for lakes, less data for Baltic coastal areas, and no data for coastal areas with higher salinities. So, the relationships outlined in fig. 6.11 should be regarded as a working hypothesis, nothing more. But this working hypothesis will be elaborated and tested in the next section.

From the boundary conditions defined for the three lines in fig. 6.11, one can see that:

- for salinity 0%, log(Sec) = 1 for $SPM_{SW} = 0.5$ mg/l; this defines z = 1;

- for salinity 6.5%, log(Sec) = 1.7 for $SPM_{SW} = 0.5$ mg/l; this defines z = 1.7; and

- for salinity 30%, log(Sec) = 2.3 for $SPM_{SW} = 0.5$ mg/l; this defines z = 2.3.

It is assumed that the Secchi depth can never be higher than 200 m because even distilled water will scatter light and set a limit to the Secchi depth.

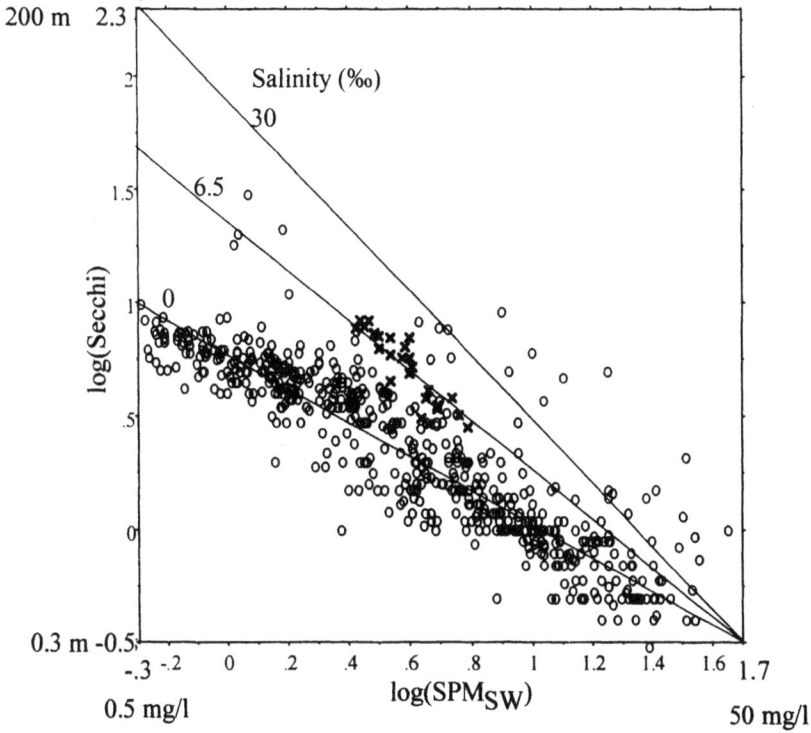

Fig. 6.11. Illustration of the relationship between log(SPM$_{SW}$) and log(Secchi) using the data for freshwater systems (surface water salinity 0%) and Baltic areas (mean surface water salinity 6.5%) given in fig. 2.4. Note that these are not regression lines but deterministically drawn lines from the end points on the x axis and the starting points on the y axis, where the values are given for the actual data for the surface water SPM value and Secchi depth. o = lake data; x = coastal data.

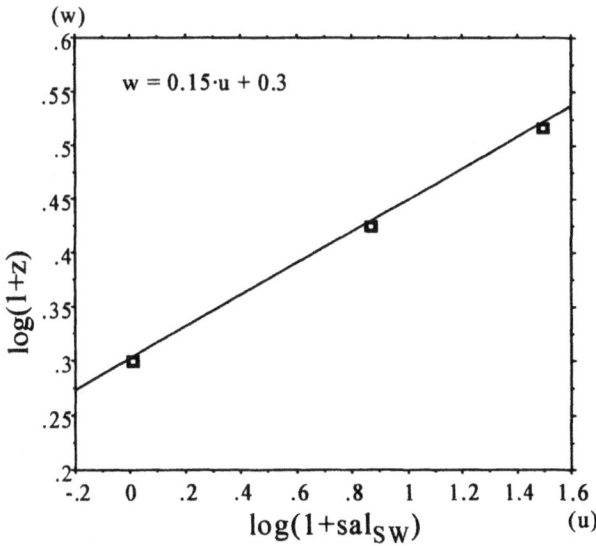

Fig. 6.12. The relationship between log(1+z) and log(1+sal$_{SW}$). See the text for explanations to abbreviations.

Fig. 6.12 shows the relationship between z and the surface water salinity (salSW). One can see that:

$$w = 0.15u + 0.3 \qquad\qquad (6.7)$$

where $w = \log(1+z)$ and $u = \log(1+sal_{SW})$; and hence:

$$(y-z) = (z+0.5)/(-0.3-1.7)\cdot(x+0.3) \qquad\qquad (6.8)$$

or

$$y = (z+0.5)\cdot(x+0.3)/(-2) + z \qquad\qquad (6.9)$$

where $y = \log(Sec)$ and $x = \log(SPM_{SW})$

$$z = (10^{\wedge}(0.15\cdot\log(1+sal_{SW})+0.3)-1)) \qquad\qquad (6.10)$$

This means that the requested algorithm to estimate Secchi depth from SPM_{SW} and salinity may be given by:

$$Sec=10^{\wedge}(-((10^{\wedge}(0.15\cdot\log(1+sal_{SW})+0.3)-$$
$$1))+0.5)\cdot(\log(SPM_{SW})+0.3)/2+(10^{\wedge}(0.15\cdot\log(1+sal_{SW})+0.3)-1))) \quad (6.11)$$

SPM_{SW} as a function of Secchi depth and salinity may be expressed by:

$$SPM_{SW}=10^{\wedge}(-0.3-2\cdot(\log(Sec)-(10^{\wedge}(0.15\cdot\log(1+sal_{SW})+0.3)-$$
$$1))/((10^{\wedge}(0.15\cdot\log(1+sal_{SW})+0.3)-1)+0.5)) \qquad\qquad (6.12)$$

Fig. 6.13 gives two nomograms illustrating the relationship between Secchi depth, surface water salinity (salSW) and surface water SPM (SPMSW). One can note that a SPM value of 10 mg/l corresponds to a Secchi depth of about 1.5 m in a lake and a Secchi depth of about 3.5 m if the salinity is 30%. Evidently, SPM values as high as 10 mg/l are probably rare in the ocean where also the Secchi depths should be high.

This empirically based deterministic model linking SPM, Secchi depth, and salinity will be tested in the following section, where SPM values will not come from measurements but from the dynamic SPM model.

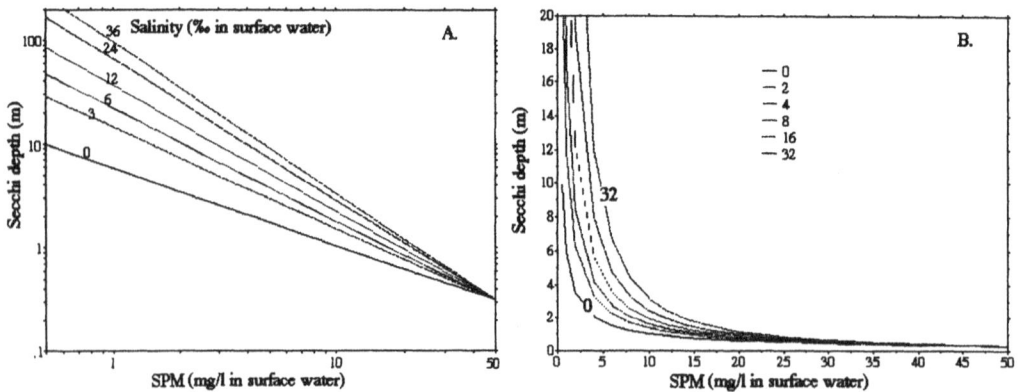

Fig. 6.13. Illustration of the relationship between Secchi depth, SPM in surface water, and salinity in surface water; (A) gives the nomogram using log data and (B) the same thing using actual data.

6.3. Dynamic modelling

6.3.1. Introduction, aim and working hypotheses

The dynamic coastal model discussed in this chapter draws much from the lake model and the main text in the following section will minimize repetitions and focus on the new parts that are specific for the coastal model. In short, these new parts concern:

1. The relationship between SPM, salinity, and Secchi depth (and hence also the depth of the photic zone and primary production).

2. The fact that coastal areas are more topographically open than lakes and the influence this has on water dynamics and bottom dynamic conditions. The SPM transport via surface water and deep water to and from the open sea will be discussed. There are also new algorithms to predict: (1) sedimentological effects when fresh water meets salt water in estuaries ("the zone of maximum turbidity", see Dyer, 1979; Håkanson et al., 1984); (2) how turbulence in the deep water zone in coastal areas influence sedimentation; and (3) areas of fine sediment erosion and transport (ET) for open coastal areas.

3. The studied coastal areas in the Baltic Sea (see appendix 9.1.3 and table 6.2) have two specific features related to land uplift. Since land uplift is a rather specific process adding SPM to some of the studied coastal areas, it will not be presented in the main text, but those equations are given and explained in appendix 9.7. In all these coastal areas, there are also direct emissions of SPM from a point source, namely from fish cage farms. This may also be regarded as a special case and the equations quantifying how much SPM comes from the fish farms are given in appendix 9.6.

Basically all other internal processes are those given by the dynamic lake model. This means that:

- Production of SPM in coastal areas is calculated in the same manner from chlorophyll, area, Secchi depth, and surface water temperature, as explained in section 4.3.3.2.

- Sedimentation is given in a modified manner for coastal areas compared to lakes (see section 4.3.3.3.1.).

- Resuspension is basically given in the same way as for lakes (section 4.3.3.3.2.).

- Mixing in coastal areas is given by the equations in section 4.3.2.3.

- Mineralization in surface water, deep water and on ET areas is given by the equations in section 4.3.3.3.4.

- River inflow to coastal areas is given by the river model (see chapter 5).

This section first presents the new parts of the dynamic SPM model for coastal areas, mainly inflow and outflow between the sea and the complementary features related to the bottom dynamic submodel.

It should be stressed that this dynamic model for SPM and sedimentation in coastal areas meet the criteria for useful predictive models discussed by Håkanson and Peters (1995):

1. Such models should predict important target variables for water management and science. Sedimentation and SPM belong to this category since sedimentation and SPM in

the water must be known to calculate the flux of matter, nutrients, metals, radionuclides, or organic toxins to, within, and from coastal areas. Sedimentation must also be known to calculate the age of sediments and hence historical and time-related aspects of sediment contamination. Sedimentation also regulates the oxygen concentration, which in turn, influences the survival of zoobenthos, a key functional group, which in turn is of paramount importance as food for fish.

2. Such models should also yield high predictive power when validated against independent empirical data. The validation of the model presented here is a central task of this chapter. The quality of models is not governed by statements or arguments but by performance in blind tests.

3. Such models should also apply over a wide domain, which should be clearly defined. This dynamic model for SPM will be validated using data from Baltic coastal areas. However, the structure of the model is such that it should apply more generally, although this remains to be demonstrated. The only reason why this model has not been tested for other parts of the world is that relevant data from such areas have not been accessible to the author.

4. The variables needed to run the model (the obligatory driving variables) should be easily accessed, e.g., from standard monitoring programs or maps.

There are some hydraulic coastal models, e.g., Threetox and Coasttox. Unlike the model discussed in this work, those are spatially distributed 2D or 3D models based on partial differential equations; the model POSEIDON is based on several interlinked boxes. These models are used in the RODOS DSS (see http://www.rodos.fzk.de/) and they are mainly designed to handle short-term (hours to days) spatial variations. They are driven by online meteorological data (winds, temperature, and precipitation) and hence cannot be used for predictive purposes over longer time periods than 2 to 3 days since it is not possible to forecast weather conditions for longer periods. Such models may be excellent tools in science and may provide descriptive power rather than long-term predictive power.

There are also different types of ecosystem-oriented models and modelling approaches for sedimentation and variables influencing sedimentation in coastal areas (see, e.g., Wulff et al., 2001). However, there are major differences among the model discussed here and other models related to differences in target variables (from conditions at individual sites to mean values over larger areas), modelling scales (daily to annual predictions), modelling structures (from using empirical/regression models to the use of ordinary or partial differential equations) and driving variables (whether accessed from standard monitoring programs, climatological measurements, or specific studies). To make meaningful model comparisons is not a simple matter, and this is not the focus of this section. As far as the author is aware, there are no mass-balance models for SPM and coastal sedimentation of the type presented here accounting for total primary production, point source emissions, fresh water input, surface and deep water exchange processes, land uplift, internal loading, mixing and mineralization in a general manner designed to achieve practical utility and monthly variations. Also the fundamental unit, the defined coastal area, is determined in a way that, to the best of the author's knowledge, has not been used before in dynamic modelling of sedimentological processes; no comparable models use the topographical bottleneck approach to define the coastal area. As discussed, this approach also makes it possible to estimate the theoretical surface water and deep water retention times (which are fundamental components in coastal mass-balance modelling) from bathymetric map data.

The following specific questions and working hypotheses will be addressed in this section:

•   Which SPM fluxes (g/month) to, within, and from a given coastal area can be expected to be large and small? Evidently, it is more important to predict large fluxes with a smaller uncertainty than smaller fluxes. The answer to this question is also important for sensitivity and uncertainty analyses (see chapter 3).

- The exchange processes between the coast and the sea are of fundamental importance for the SPM concentrations (and other substances as well) in coastal water - but is it possible to find operational criteria and guidelines for this? For which coast types will the SPM concentration in the coastal water be close to the concentrations in the sea, and when and why will there be differences?

- In models of this kind (i.e., process-oriented mass-balance models for coastal ecosystems based on ordinary differential equations and time-scales of months) there are always uncertainties related to the model structure and uncertainties in driving variables – but which uncertainties related to fluxes and driving variables are generally small, and which are important and need to be reduced to increase the predictive power of the model?

- In this model, there are generic parts which apply for all substances and substance-specific parts. In applying this model for different contaminants, it is essential to stress this point.

- When is it important to account for river inflow and when not? Is it possible to find simple operational criteria for this?

- How will coastal morphometry (coastal size and depth conditions and the openness towards the sea) influence SPM concentrations in water?

- How would tidal effects influence SPM concentrations in coastal areas?

It should be noted that it would be extremely difficult and costly to scrutinize questions like these using more traditional approaches based on field studies. This is an area where process-based modelling can provide an important input to clarify what is happening in natural coastal areas.

This dynamic model for SPM is also an ELS model (Effect-Load-Sensitivity model; see chapter 1). ELS models are designed for practical utility and are often based on a dynamic (mass-balance) model yielding concentrations of substances in the water and then complemented with regressions between such concentrations and operational variables of ecosystem effect. This dynamic model gives SPM concentrations in coastal waters and the operational effect variables are the oxygen saturation in the deep water (O_2Sat in %) and the Secchi depth (in m).

The accuracy of a model prediction is strongly influenced by the uncertainty in the empirical data used to run and validate the model (chapter 3). Sedimentation is known to display considerable natural variations between years, seasons and/or even between closely located stations (Matteucci and Frascari, 1997; Blomqvist, 1992; Heiskanen and Tallberg, 1999). Douglas et al. (2003) have shown that there are large variations in sedimentation even during 36-48 hour periods. The empirical sediment trap data used to validate this model have coefficients of variation (CV) of 0.58 for the surface water and 0.50 for the deep water compartment (Wallin et al., 1992).

In models for lakes and/or coastal areas, the surface water compartment is often separated from the deep water compartment by the thermocline (Carlsson et al., 1999), the pycnocline (Abdel-Moati, 1997), or the halocline (Andreev et al., 2002). However, the classic effect-load-sensitivity model by Vollenweider (1968) for lakes does not separate surface and deep water at all and neither does De Schmedt et al. (1998) when modelling suspended sediments and heavy metals in the Scheldt estuary. The thermocline, halocline, and pycnocline are all gradients, meaning gradual changes, and they can be found over wide ranges of water depths (chapter 4). This means that it is often difficult to find a relevant value separating the surface water compartment from the deep water compartment using, e.g., temperature data. In this section, the separation is not done in the traditional way using temperature data, but by the wave base (the "critical water depth"), i.e., the depth below which fine cohesive particles following Stokes's law are continuously being deposited (see chapter 4). This gives one defined critical water depth for each coastal area. From this water depth, it is easy to calculate requested water volumes, sedimentation, resuspension, mixing, mineralization, and outflow. This also leads to a relatively simple model structure since the sedimentation of SPM from the deep water, and the deep water alone, ends up on areas of continuous sedimentation (the accumulation areas).

This chapter is structured in such a way that after this introduction, the studied coastal areas and the methods and data used to build and test the model will be briefly presented (see also appendix 9.1.3). A fundamental requirement for the model is the definition of a coastal area – i.e., how and where to draw the boundary lines toward the sea and/or adjacent coastal areas. For the studied coastal areas, the presented "topographical bottleneck" approach has been used to define each coastal area (fig. 6.3). This gives the basic unit to determine water volume, mean depth, maximum depth, inflow, outflow, and internal fluxes of SPM. For other types of coastal areas (such as tidal areas, open coastal areas, and estuaries), suggestions how to modify this modelling approach have already been given (fig. 6.6).

Modelled data on sedimentation and Secchi depths will be compared to empirical data. The sedimentation data come from measurements using sediment traps placed at two to four sites in each coastal area (in the surface water and in the deep water, and for two periods of about one week each during the growing season). The fieldwork and the sediment trap data have been presented in detail by Wallin et al. (1992). This chapter will also present validation results of the dynamic model and sensitivity and uncertainty tests using Monte Carlo techniques according to procedures discussed in chapter 3. Finally, a case study on the practical use of the dynamic SPM model in coastal management will be presented.

6.3.2. Methods and data

This work is based on data from 17 coastal areas located in three different archipelagos in the Baltic Sea (table 6.2). Five of the areas are off the Swedish east coast, 7 areas are in the south of Sweden, and the remaining 5 areas southwest of Finland (see fig 9.1 in the appendix).

Table 6.2 gives a compilation of data:

- Area code.

- Latitude; used to calculate surface and deep water temperatures; the temperature submodel is presented in appendix 9.4.

- The morphometric parameters, coastal area (water surface area), maximum depth (D_{max}), mean depth (D_m), section area (At); area and D_m are used to calculate coastal volume (V) and concentrations of SPM; the morphometric parameters are also used to determine the coastal form, which influences internal fluxes of SPM.

- The measured mean summer chlorophyll-a concentrations (Chl in µg/l) are used to calculate primary production.

- The mean surface water salinity, which influences aggregation of suspended particles and sedimentation.

- "Fish production" in table 6.2 relates to point source emissions of SPM from fish cage farms. All these data were collected in a project where the environmental impacts of emissions from fish cage farms were studied (Wallin et al., 1992). There are no other major point sources of nutrients or SPM to these areas. All of them are also little influenced by tributaries, which is indicated by the salinities in table 6.2.

- The data used to validate the model are given by the columns Sed_{DW} and Sed_{SW} (sedimentation in sediment traps placed in the deep water and the surface water); Wallin et al. (1992) and Wallin and Håkanson (1991) have given reports on how, when, where, and for how long the sediment traps were deployed. The data used here are the mean values for July, August, and September.

- The last column in table 6.2 gives the median Secchi depths at the sample sites in the section areas closest to the sea (Sec_{sea} in m). These data are used to estimate the inflow of SPM from the sea and/or adjacent coastal areas to the given coastal area [$Q_{SW} \cdot SPM_{sea}$ in (m³/month)·(g/m³) = g/month]. SPM_{sea} is calculated from Sec_{sea} from eq. 6.3.

Table 6.2 also gives the ranges (minimum and maximum values) for the data and these ranges give important information about the model domain.

6.3.3. The dynamic SPM model

6.3.3.1. Basic structure

The structure of the dynamic model is shown in fig. 6.14. As in the lake model, there are three main compartments: surface water, deep water, areas where processes of erosion and transport dominate the bottom dynamic conditions (ET areas). The volumes of the surface and deep water are calculated from the water depth separating transportation areas for fine particles from accumulation areas, the critical depth (D_{crit}, see chapter 4).

There are six inflows:

1. Primary production (F_{prod}) which includes all types of plankton (phytoplankton, bacterioplankton and zooplankton) influencing SPM in the water.

2. Inflow of SPM to coastal surface water from the sea (F_{inSW}).

3. Inflow of SPM to the deep water from the sea (F_{inDW}).

4. Land uplift (F_{LU}). Land uplift is a special case and the submodel for F_{LU} is given in appendix 9.7.

5. Emissions of SPM from point sources (F_{PSSW}), in this case from fish cage farms. The submodel for these emissions of SPM is given in appendix 9.6.

6. Tributary inflow (F_Q), see chapter 5.

The structure of the dynamic SPM model for coastal areas

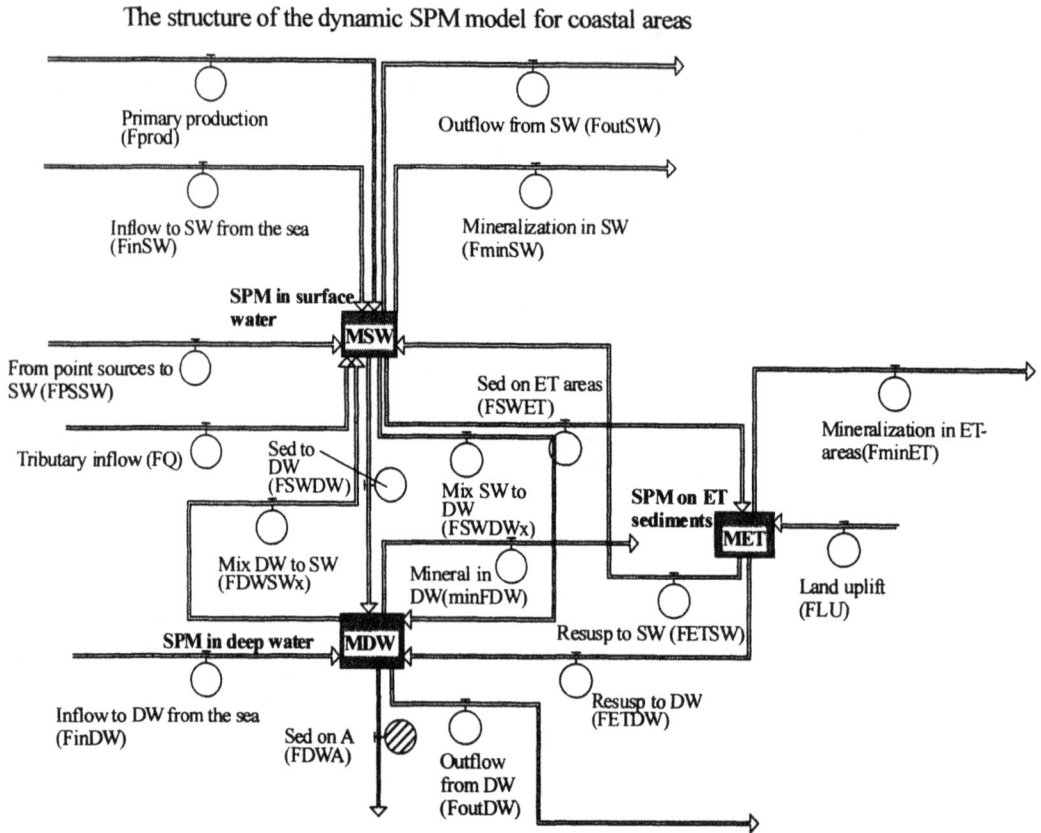

Fig. 6.14. A general outline of the structure of the coastal model. Note that for simplicity point source emissions to the deep water compartment have been omitted in this figure.

The amount of matter deposited on ET areas may be resuspended by wind/wave action or slope processes, so resuspension is an important internal process influencing the SPM flux in coastal areas. The resuspended matter can be transported either back to the surface water (F_{ETSW}) or to the deep water (F_{ETDW}). How much that will go in either direction is regulated by a distribution coefficient calculated from the form (Vd = the form factor) of the coastal area. Other internal processes are mineralization, i.e., the bacterial decomposition of organic SPM particles in surface water, the deep water and the ET compartments (F_{minSW}, F_{minDW} and F_{minET}) and mixing, i.e., the transport from deep water to surface water (F_{DWSWx}) or from surface water to deep water (F_{SWDWx}).

All basic equations are compiled in table 6.5.

Table 6.5.   A compilation of the differential equations for the coastal model.

Surface water (SW):

$M_{SW}(t) = M_{SW}(t - dt) + (F_{inSW} + F_{DWSWx} + F_{ETSW} + F_{prod} + F_{PSSW} + F_{LU} - F_{outSW} - F_{SWDW} - F_{SWET} - F_{minSW} - F_{SWDWx}) \cdot dt$

$M_{SW}(t)$	= Mass (amount) in the SW compartment at time t (g)
F_{inSW}	= Flow into the SW compartment from the sea (g/month); see text
F_{DWSWx}	= Flow from deep water to surface water (upward mixing; g/month); see below
F_{ETSW}	= Flow (resuspension) from ET areas to the SW compartment (g/month); see below
F_{prod}	= Flow into the SW compartment from primary production (g/month); see text
F_{PSSW}	= Flow into the SW compartment from point source emissions (g/month; see Håkanson et al., 2004a)
F_{outSW}	= Flow from the SW compartment and out of the coastal area (g/month); see text
F_{SWDW}	= Flow (sedimentation) from the SW compartment to deep water compartment (g/month); see below
F_{SWET}	= Flow (sedimentation) from the SW compartment to ET areas (g/month); see below
F_{minSW}	= Flow (mineralization) from the SW compartment (g/month); see below
F_{DWSWx}	= Flow from surface water to deep water (downward mixing; g/month); see below

ET areas (ET):
$M_{ET}(t) = M_{ET}(t - dt) + (F_{LU} + F_{SWET} - F_{ETDW} - F_{ETSW} - F_{minET}) \cdot dt$

$M_{ET}(t)$	= Mass (amount) in the ET compartment at time t (g)
F_{LU}	= Flow into the SW compartment from land uplift (g/month; see Håkanson et al., 2004a)
F_{ETDW}	= Flow (resuspension) from ET areas to the DW compartment (g/month); see below
F_{minET}	= Flow (mineralization) from the ET areas (g/month); see below

Deep water (DW):
$M_{DW}(t) = M_{DW}(t - dt) + (F_{SWDW} + F_{ETDW} + F_{SWDWx} + F_{inDW} + F_{PSDW} - F_{DWSWx} - F_{DWA} - F_{outDW} - F_{minDW}) \cdot dt$

$M_{DW}(t)$	= Mass (amount) in the DW compartment at time t (g)
F_{inDW}	= Flow into the DW compartment (g/month); see text
F_{PSDW}	= Flow into the DW compartment from point source emissions (g/month; see Håkanson et al., 2004a)
F_{DWA}	= Flow (sedimentation) from the DW compartment to A areas (g/month); see below
F_{outDW}	= Flow from the DW compartment and out of the coastal area (g/month); see text
F_{minDW}	= Flow (mineralization) from the DW compartment (g/month); see below

Other important algorithms

F_{DWSWx}	= $M_{DW} \cdot R_{mix} \cdot (V_{SW}/V_{DW})$
F_{ETSW}	= $M_{ET} \cdot (1 - Vd/3) \cdot 1/T_{ET}$ [T_{ET} = 1 month]
F_{SWDW}	= $M_{SW} \cdot (1 - ET) \cdot (v_{def}/D_{SW}) \cdot Y_{ZMT} \cdot Y_{SPMSW} \cdot Y_{salSW} \cdot Y_{DR} \cdot ((1 - DC_{resSW})) + Y_{res} \cdot DC_{resSW})$
v_{def}	= 6 m/month
Y_{ZMT}	= If $Q > Q_{sea}$ then $Y_{ZMT} = (sal_{sea}/sal_{SW}) \cdot (Q_{sea} + Q)/Q$ else $Y_{ZMT} = (sal_{sea}/sal_{SW}) \cdot (Q_{sea} + Q)/Q_{sea}$ [Q values in m^3/month; calculates sedimentation effects related to the "zone of maximum turbidity"]
Y_{SPMSW}	= $(1 + 0.75 \cdot (C_{SW}/50 - 1))$ [calculates how changes in SPM (C_{SW}) influences sedimentation]
Y_{SPMDW}	= $(1 + 0.75 \cdot (C_{DW}/50 - 1))$ [calculates how changes in SPM (C_{DW}) influences sedimentation]
Y_{salSW}	= $(1 + 1 \cdot (Sal/1 - 1) = 1 \cdot Sal/1$ [calculates how changes in salinities > 1‰ influence sedimentation]
Y_{DR}	= If DR < 0.26 then 1 else 0.26/DR [calculates how changes in DR and turbulence influence sedimentation]
DC_{resSW}	= $F_{ETSW}/(F_{ETSW} + F_{in} + F_{prod})$ [the resuspended fraction of SPM]
Y_{res}	= $(((T_{ET}/1) + 1)^{0.5}$ [calculates how much faster resuspended sediments settle out]

F_{SWET} $= M_{SW} \cdot ET \cdot (v_{def}/D_{SW}) \cdot Y_{SPMSW} \cdot Y_{salSW} \cdot Y_{DR} \cdot ((1-DC_{resSW})) + Y_{res} \cdot DC_{resSW}))$ $[v_{def} = 6 \text{ m/month}]$

F_{minSW} $= M_{SW} \cdot R_{min} \cdot Y_{ET} \cdot (SWT/9)^{1.2}$ $[R_{min} = 0.125]$

Y_{ET} $= 0.99/ET$ [calculates how changes in ET among systems influence mineralization]

F_{minDW} $= M_{DW} \cdot R_{min} \cdot Y_{ET} \cdot (DWT/9)^{1.2}$ $[R_{min} = 0.125]$

F_{DWA} $= M_{DW} \cdot R_{DW}$

R_{DW} $= v_{DW}/D_{DW}$

v_{DW} $= (v_{def}/12) \cdot Y_{SPMDW} \cdot Y_{salDW} \cdot Y_{DR} \cdot Y_{DW} \cdot ((1 - DC_{resDW}) + Y_{res} \cdot DC_{resDW})$

Y_{DW} $=$ If $T_{DW} < 7$ (days) then $Y_{DW} = 1$ else $Y_{DW} = (T_{DW}/7)^{0.5}$ [calculates how changes in T and turbulence influence deep water sedimentation]

6.3.3.2. Primary production of SPM

Primary production is basically calculated from chlorophyll-a in the same way as for lakes (see eq. 4.22). One can note that the surface water temperatures (SWT) needed to calculate F_{prod} can either come from measurements, climatological tables or from the temperature submodel given in appendix 9.4. Since this is a model for coastal water temperatures, altitude and continentality have been omitted as driving variables. To predict SWT and deep water temperatures (DWT), and also mixing, one needs data on latitude, coastal area, and coastal mean depth (such data are given in table 6.2). As discussed in chapter 4, there are several boundary conditions for the temperature submodel. The coastal waters are not likely stratified if the mean depth is lower than 2 m and if the dynamic ratio (DR = $\sqrt{Area}/D_m$; Area in km^2 and D_m in m) is larger than 3.8.

Primary production of SPM is basically given by the same equations as in the lake model, but with one modification related to the fact that there are no mean monthly empirical data on chlorophyll available for these coastal areas, only a mean value for the growing season. This means that F_{prod} is given by:

$$F_{prod} = ((Chl_{MV}/Chl_{MVref}) \cdot 30.6 \cdot Chl_{ref}^{0.927}) \cdot 0.45 \cdot 30 \cdot Area \cdot Sec \cdot 0.001((SWT+0.1)/9) \cdot (BM_{PL}/BM_{PH})$$
(6.13)

Chl_{ref} = The mean monthly chlorophyll concentrations for a reference system (here mean monthly chlorophyll data for Lake Miastro, Belarus, fig. 4.18B):

Jan	Feb	Mar	Apr	May	Jun	Jul	Aug	Sep	Oct	Nov	Dec
0.8	1.3	4.3	6.2	13.1	9.6	11.3	10.9	16.3	15.0	4.3	3.3

These values are assumed to give a characteristic season pattern for chlorophyll in µg/l; the expression ($30.6 \cdot Chl^{0.927}$) transforms Chl into phytoplankton production (in µg C/l·d). The factor 0.45 is a standard transformation factor to change g C to g dw. Multiplication with 0.001, 30 days, coastal area (Area) and the mean monthly value of the effective depth of the photic zone (Secchi depth) gives the biomass of phytoplankton produced per month (g dw per month).

- Chl_{MV} = The mean empirical Chl value for the given coastal area for the growing season (µg/l).

- Chl_{MVref} = The mean Chl value for the given reference system for the growing season (11.2 µg/l).

- BM_{PL}/BM_{PH} = The ratio between the biomass of all sorts of plankton (phytoplankton, bacterioplankton zooplankton, etc.; BM_{PL}) to the calculated biomass of phytoplankton (BM_{PH}) comes from the foodweb model (LakeWeb, see appendix 9.2). This foodweb model has been used to get a simple general calculation constant for this ratio along a trophic state gradient (fig. 4.14). This ratio is on average about 2.5 which indicates that the total biomass of bacterioplankton plus zooplankton plus phytoplankton is a factor of about 2.5 higher than the phytoplankton biomass. The dimensionless moderator for SWT gives a seasonal pattern to this related to the seasonal changes in mean monthly surface water temperatures.

- SWT = Mean monthly surface water temperatures (°C). By dividing SWT with a reference temperature of 9°C (related to the duration of the growing season), this approach accounts for seasonal variations in SWT in a dimensionless manner. The moderator is (SWT+0.1)/9. The constant 0.1 is used since SWT may approach 0°C during the winter and since there is also production under the ice.

- Sec = Secchi depth (in m).

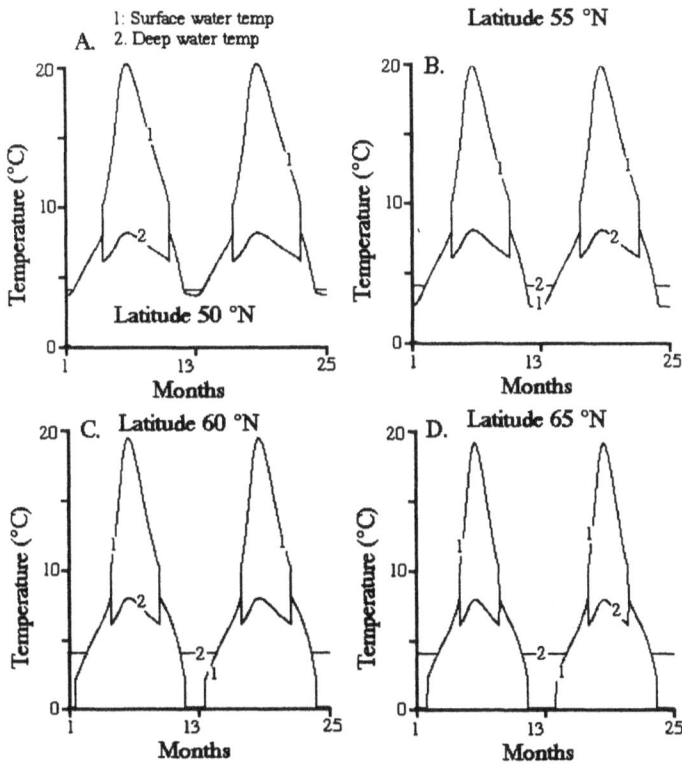

Fig. 6.15. Simulations using the water temperature submodel to estimate surface water temperatures and deep water temperatures in four coastal areas (using data for Guavik, southern Sweden, table 6.2) placed at latitude 50°N, 55°N, 60°N and 65°N.

Fig. 6.15 shows how the temperature submodel predicts SWT and DWT at four latitudes using data for the coastal area Guavik (SS4, table 6.2). One can note that there would be a likely ice cover at latitudes 60°N and 65°N, but not at latitude 55°N.

6.3.3.3. Inflow of SPM from the sea

The inflow of SPM to the surface water from the sea or adjacent coastal areas (F_{inSW}) is calculated from the surface water flow (Q_{SW}), which is calculated from the theoretical surface water retention time (T_{SW}, fig. 6.6) and the concentration of SPM outside the coast (SPM_{sea}), which is calculated here from measured median values of Secchi depth close to the section areas (Sec_{sea}, table 6.2). Evidently, it would have been preferable to have access to reliable empirical data both on T_{SW} and SPM_{sea} but such data are very hard to get. This means that the inflow of SPM from the sea is given by:

$$F_{inSW} = Q_{SW} \cdot SPM_{sea} = (V_{SW}/T_{SW}) \cdot SPM_{sea} \tag{6.14}$$

where V_{SW} is the surface water volume (in m^3). Fig. 6.16 gives a compilation of the submodel to calculated surface water and deep water volumes.

Submodel for volumes

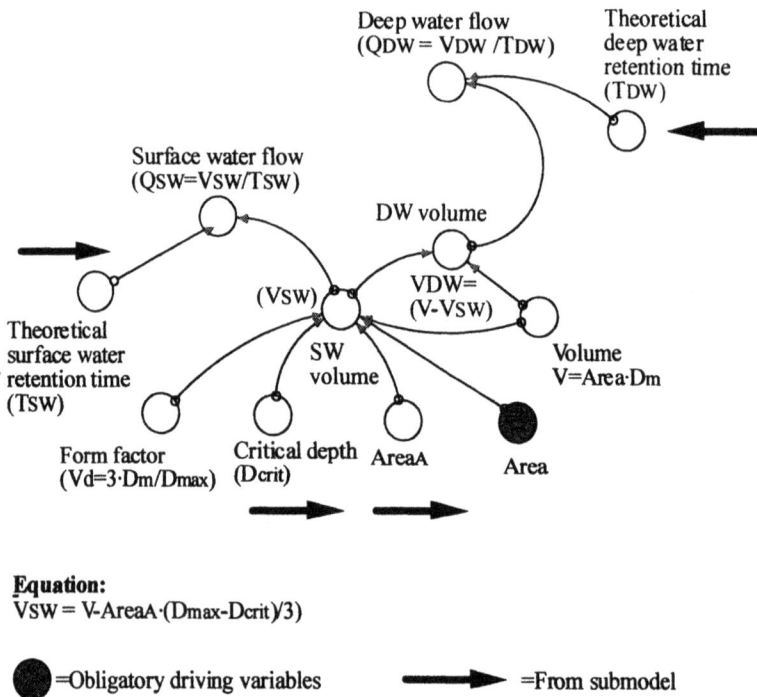

Equation:
$VSW = V\text{-}AreaA \cdot (Dmax\text{-}Dcrit)/3$

● =Obligatory driving variables   ➤ =From submodel

Fig. 6.16. The submodel for surface water and deep water volumes.

The deep water inflow of SPM from the sea (F_{inDW}) is quantified in the following way:

$$F_{inDW} = Q_{DW} \cdot (SPM_{sea}+2) = (V_{DW}/T_{DW}) \cdot (SPM_{sea}+2) \tag{6.15}$$

where

Q_{DW} = The deep water flow (m^3/month), which is given by the ratio between the volume of the deep water (V_{DW} in m^3, fig. 6.16) and the theoretical deep water retention time (T_{DW} in months; table 6.1), i.e., V_{DW}/T_{DW}.

SPM_{sea} = The SPM concentration in the surface water in the sea outside the coast. 2 mg/l are added to get a better estimate of the SPM concentration in the deep water value because the SPM concentration is likely higher close to the bottom than in the surface water zone (see section 6.4).

6.3.3.4. Sedimentation, ET, and resuspension in coastal areas

This section will discuss three modifications for sedimentation and resuspension, as compared to the dynamic lake model. The first modification concerns the conditions in estuaries where salt water meets fresh water and there is increased flocculation and sedimentation of SPM. This is quantified by the following algorithm:

If $Q > Q_{sea}$ (values in m^3/month) then $Y_{ZMT} = (sal_{sea}/sal_{SW}) \cdot (Q_{sea}+Q)/Q$

$$\text{else } Y_{ZMT} = (sal_{sea}/sal_{SW}) \cdot (Q_{sea}+Q)/Q_{sea} \qquad (6.16)$$

Where Y_{ZMT} is dimensional moderator quantifying how the default fall velocity v_{def} (eq. 4.26) for SPM increases in estuaries depending on the fresh water inflow (Q), the salt water inflow (Q_{sea}), the salinity in the coastal area (sal_{SW}, the surface water salinity in %), and the mean salinity in the sea outside the given coastal area (sal_{sea} in %). The greater the difference in salinity between the sea and the coast, the higher the ratio sal_{sea}/sal_{SW} and the higher the flocculation and sedimentation. The higher the fresh water inflow relative to the salt-water inflow, the higher the effects related to the "zone of maximum turbidity", ZMT), and the higher the value of Y_{ZMT}. $(Q_{sea}+Q)/Q$ or $(Q_{sea}+Q)/Q_{sea}$ attain values between 1 and 2 and sal_{sea}/sal_{SW} may attain different values depending on the prevailing conditions; if sal_{sea} is 10% and sal_{SW} 5% and if the $(Q_{sea}+Q)/Q$ ratio is 2, then Y_{ZMT} is 4 and the fall velocity is 4 times higher than the default value (6 m/month).

This approach (called the ZMT correction) has been tested using the dynamic SPM model with the dynamic river model (from chapter 5) using data from the only studied coastal area in this work with a significant fresh water inflow, Ronneby (S. Sweden, which has a tributary with a catchment area of 633 km^2). The approach given in eq. 6.16 has also been tested for predictions of SPM, Secchi depth, sedimentation, and phosphorus concentrations in water and sediments for other estuaries using the dynamic mass-balance model for phosphorus and coastal data presented by Håkanson and Karlsson (2004). The results are encouraging, which is exemplified in fig. 6.17 for the Ronneby coastal area. The predictions are improved using the ZMT correction, although not so much in the Ronneby estuary because the difference between the salinity in the sea and the coast is small (7.4% and 6.5%, respectively for the surface water data). The fresh water inflow is also logically small compared to the inflow from the sea (on average about 6%), but there is still a clear improvement in predicting sedimentation (and also Secchi depth and oxygen saturation) in this coastal area.

Coastal area, Ronneby, S. Sweden

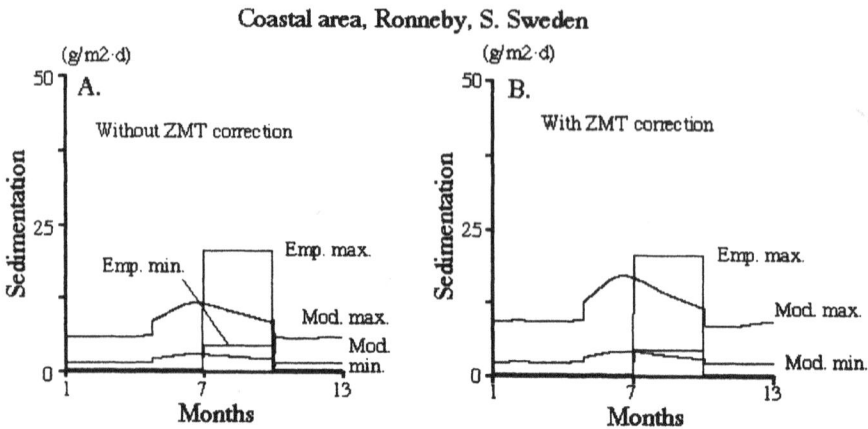

Fig. 6.17. Illustration of how the ZMT correction (the algorithm to account for sedimentation/flocculation in the "zone of maximum turbidity") works for predicting sedimentation in the coastal area Ronneby (S. Sweden). This is an estuary with a tributary with a catchment area of 633 km^2. For comparative purposes, the figures also give empirical data on sedimentation in surface water and deep water sediment traps (Emp. min. and Emp. max. values, respectively).

The next change as compared to the lake model concerns the algorithm quantifying how the degree of turbulence in the deep water zone influence sedimentation of SPM. For lakes, there are no practically useful and well tested algorithms to estimate the theoretical deep water retention time (T_{DW}), as far as the author is aware. So, in the lake model, there is a simple approach (eq. 4.34) to estimate the effect of turbulence from to the theoretical lake water retention time (T). However, for the studied coastal areas, there is, as discussed, a simple method to estimate T_{DW} (table 6.1) and this means that eq. 4.34 will be replaced by the following expression, which is meant to apply for coastal areas where T_{DW} can be estimated in a simple, operational manner:

$$\text{If } T_{DW} < 7 \text{ (days) then } Y_{DW} = 1 \text{ else } Y_{DW} = (T_{DW}/7)^{0.5} \qquad (6.17)$$

This means that if T_{DW} is smaller than 7 days, the turbulence in the deep water zone corresponds to what one might generally expect in the surface water zone where the theoretical water retention time (T_{SW}) is generally between 2 and 7 days in these and similar coastal areas. If T_{DW} is longer than that, the turbulence is likely lower and the possibilities for the suspended particles to settle out higher. If T_{DW} is 120 days, which is generally the boundary value in dimictic coastal areas, i.e., in coastal areas which mix in spring and fall, Y_{DW} is 4.1 and the settling velocity in the deep water zone is 4.1 times higher compared to situations when T_{DW} is smaller than 7 days. One can also assume that Y_{DW} should never be higher than, say 7, corresponding to a T_{DW} value of 1 year. The calculations shown in fig. 6.17, and in all the following presentations, utilize these two algorithms (equations 6.16 and 6.17) for sedimentation in coastal areas.

There are several models to predict resuspension and the ET areas for coastal systems (Persson and Håkanson, 1995), but those approaches generally require data on the filter factor (Ff) or the mean filter factor, i.e., how the conditions outside the defined coastal area (islands, etc.) work as an energy filter and reduce the impact of the waves from the sea. In this modelling, the filter factor is omitted since it may be difficult to access. This approach predicts the critical depth (D_{crit}; the wave base) from the information given in table 6.2.

To estimate ET (dimensionless) for lakes, one accounts for the form, as given by the form factor (Vd). The same approach is used for coastal areas, but modified for the topographical openness (Ex). So, shallow coasts have larger areas above the critical depth than deep coastal areas, if all else is constant. This calculation of ET is related to the hypsographic form (depth/area form as given by the form factor, $Vd = 3 \cdot D_m/D_{max}$; D_m = the mean depth; D_{max} = the max. depth; fig. 4.11).

The larger the exposure ($Ex = 100 \cdot At/Area$), the larger the potential energy impact from the sea (Persson and Håkanson, 1995) and the deeper the wave base (= the critical depth, D_{crit}). This is accounted for by the following approach:

$$\text{If } Ex < 0.003 \text{ then } Y_{Ex1} = 1 \text{ else } Y_{Ex1} = (Ex/0.003)^{0.25} \tag{6.18}$$

and the other boundary condition for very open coastal areas is:

$$\text{If } Ex > 10 \text{ then } Y_{Ex2} = 10 \text{ else } Y_{Ex2} = Y_{Ex1} \tag{6.19}$$

The value for Y_{Ex2} is multiplied by the value for D_{crit}. This means that when the exposure varies between 0.003 and 10, D_{crit} is increased by the factor $(Ex/0.003)^{0.25}$. That is, if $Ex = 0.1$, the factor is 2.4 and D_{crit} likely at a water depth of $2.4 \cdot D_{crit}$ m rather than at D_{crit}.

The ET value is used as a dimensionless distribution coefficient. It regulates the sedimentation of SPM either to deep water areas or to ET areas and hence also the amount of matter available for resuspension on the ET areas. ET generally varies from 0.15, since there must always be a shallow shore zone where processes of erosion and transport dominate the bottom dynamic conditions, to 1 in large and shallow areas totally dominated by ET areas. In this modelling, ET is, however, never permitted to become higher than 0.99, since one can assume that in most coastal areas there are deep holes, sheltered areas, or macrophyte beds that would function as A areas. For simplicity, this approach is used also when there is an ice cover (if SWT = 0°C), because the stratification is weak during the winter, primary production low, and the error in predicting sedimentation with this simplification small. The algorithms to calculate ET and the critical depth are shown in fig. 6.18.

6.3.3.5. SPM outflow

The outflow from the surface water is given by a flux called F_{outSW}, where $F_{outSW} = M_{SW}/T_{SW}$; T_{SW} is the theoretical surface water retention time (months) and M_{SW} is the modelled amount of SPM in the surface water compartment (g).

The deep water outflow is calculated in the same way as $F_{outDW} = M_{DW}/T_{DW}$, where $1/T_{DW}$ is the outflow rate (1/month); T_{DW} the theoretical deep water retention time; and M_{DW} is the modelled amount of SPM in the deep water (g).

Bottom dynamic conditions submodel

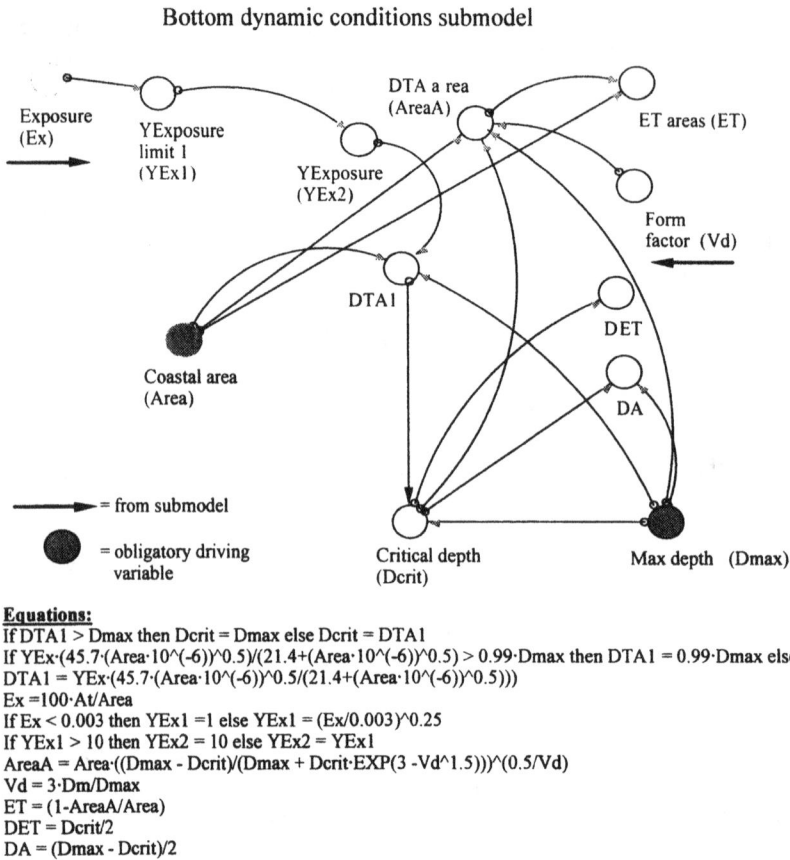

Equations:
If DTA1 > Dmax then Dcrit = Dmax else Dcrit = DTA1
If YEx·(45.7·(Area·10^(-6))^0.5)/(21.4+(Area·10^(-6))^0.5) > 0.99·Dmax then DTA1 = 0.99·Dmax else
DTA1 = YEx·(45.7·(Area·10^(-6))^0.5/(21.4+(Area·10^(-6))^0.5)))
Ex =100·At/Area
If Ex < 0.003 then YEx1 =1 else YEx1 = (Ex/0.003)^0.25
If YEx1 > 10 then YEx2 = 10 else YEx2 = YEx1
AreaA = Area·((Dmax - Dcrit)/(Dmax + Dcrit·EXP(3 -Vd^1.5)))^(0.5/Vd)
Vd = 3·Dm/Dmax
ET = (1-AreaA/Area)
DET = Dcrit/2
DA = (Dmax - Dcrit)/2

Fig. 6.18. The submodel to calculate the critical water depth separating T areas and A areas (D_{crit}), the area above D_{crit} (the ET areas), the area below D_{crit} (the A areas); operational boundary conditions and algorithms.

6.3.3.6. The panel of driving variables

Table 6.6 gives the panel of driving variables. These are the coastal area specific variables needed to run the dynamic SPM model. No other parts of the model should be changed unless there are good reasons to do so.

6.3.4. Results – blind tests

Note that there have been no calibrations of the dynamic SPM model for coastal areas. All equations have been motivated by empirical data or results based on empirical data and many parts of the model are similar to what was described in chapter 4 for the lake model.

The results of the validations will be presented in the following way. First, comparisons between empirical data, uncertainties in empirical data, and model-predicted values will be given for sedimentation, Secchi depths and oxygen saturation in three coastal areas, one from southern Sweden, one from eastern Sweden and one from Finland. Then, the modelling results for all 17 coastal areas will be directly compared to empirical data. The basic question is: How well does the model predict considering all the uncertainties in the empirical data used to run the model and the assumptions behind the given algorithms?

Table 6.6. Panel of driving variables.

A. Morphometric parameters:

1. Coastal area
2. Mean depth
3. Maximum depth
4. Section area
5. Latitude

B. Chemical variables:

6. Characteristic mean salinity in the coastal area
7. Characteristic SPM concentration in the sea outside the given coastal area or adjacent coastal areas (here predicted from the corresponding Secchi depth values)
8. Characteristic concentrations of chlorophyll for the growing season.

C. Inflow variables:

1. Monthly emissions from point sources or tributaries
2. Land rise

The first results are given in fig. 6.19 for coastal area Järnavik, southern Sweden. Fig. 6.19A gives the empirical values for sedimentation in surface water sediment traps (called empirical minimum values) and deep water sediment traps (empirical maximum values). The minimum modelled values are calculated from sedimentation of SPM on accumulation areas (F_{DWA}). Sedimentation on A areas should vary from zero at the wave base to maximum values in the deepest part of the coastal area (sediment focusing, fig. 1.6). In these calculations, a correction factor has also been applied to the model-predicted values of F_{DWA} based on this knowledge. The predicted values for F_{DWA} are assumed to be directly comparable to the values from deep water sediment traps for U-shaped basins, with a form factor of 3, and too low for V-shaped basins with a form factor smaller than 3. Thus, the ratio $Vd/3$ is used as a correction factor. This means that the values from the deep water sediment traps (Sed_{DW}) should be compared to modelled values given by $(Vd/3) \cdot F_{DWA}$ and values from the surface water sediment traps (Sed_{SW}) to F_{DWA} (after dimensional adjustments so that the values are expressed in $g/m^2 \cdot d$). From fig. 6.19A, one can note the good correspondence between empirical and modelled minimum and maximum values (= SW and DW values, respectively) for sedimentation in this coastal area.

Coastal area Järnavik, southern Sweden

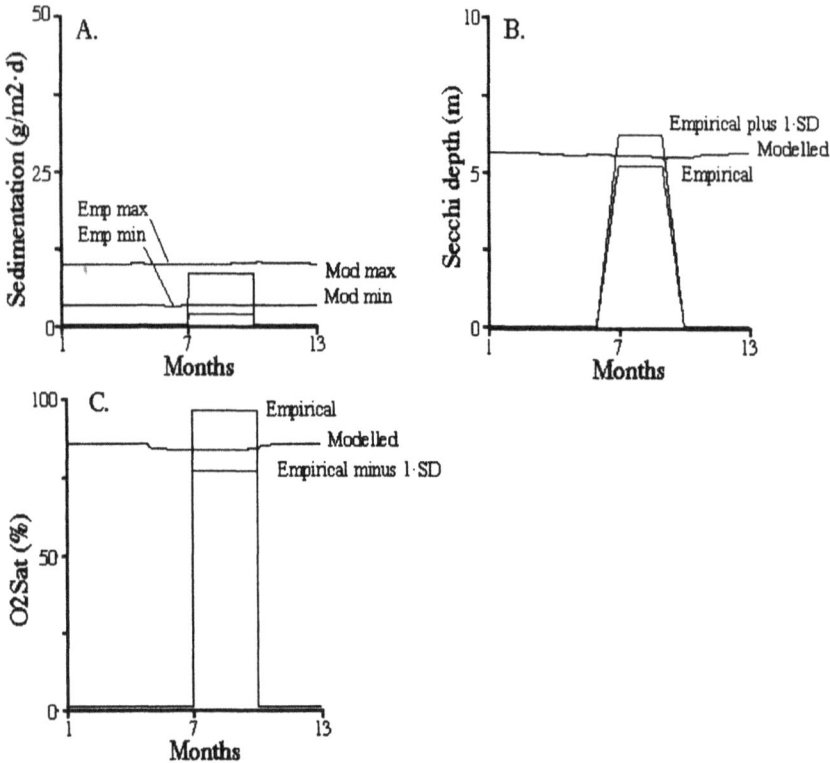

Fig. 6.19. Validations in the Järnavik area (table 6.2) (A) for empirical and modelled minimum and maximum values for sedimentation, (B) for modelled Secchi depths versus empirical data and uncertainty in empirical data (MV plus 1·SD) and (C) for modelled values of the oxygen saturation in the deep water and empirical data and uncertainty in empirical data (MV minus 1·SD).

Fig. 6.19B gives similar results for Secchi depths. The empirical data in this figure are the mean measured Secchi depth for the growing season and the mean value plus one standard deviation (SD) as a measure of the uncertainty in the mean value for this coastal area. One can see that the modelled Secchi depth (using eq. 6.12) is a little higher than the measured mean value but well within one standard deviation of the mean.

The results for the oxygen saturation in the deep water zone in coastal area Järnavik are given in fig. 6.19C. The figure gives the mean O_2Sat value for the growing season and also the empirical mean value minus one standard deviation. The modelled O_2Sat value is slightly lower than the mean empirical value, but well within the uncertainty of the mean.

From the very good results in fig. 6.19, one can ask: How will the model predict in the other 16 coastal areas?

Coastal area Gräsmarö, Swedish east coast

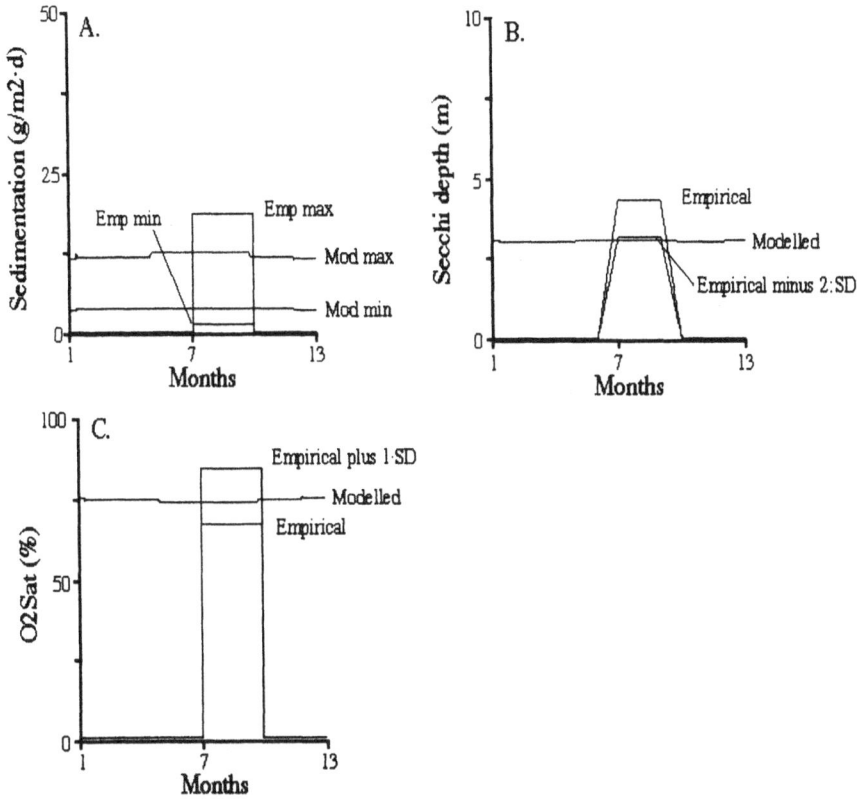

Fig. 6.20. Validations in the Gräsmarö coastal area (table 6.2) (A) for empirical and modelled minimum and maximum values for sedimentation, (B) for modelled Secchi depths versus empirical data and uncertainty in empirical data (MV minus 2·SD) and (C) for modelled values of the oxygen saturation in the deep water and empirical data and uncertainty in empirical data (MV plus 1·SD).

Fig. 6.20 gives the results for coastal area Gräsmarö off the Swedish east coast.

- The most important part in validating the dynamic SPM model is how well the model predicts measured values of sedimentation in sediment traps. From fig. 6.20A, one can see that also in this case there is a very good correspondence between modelled and measured data.

- The modelled Secchi depths are lower than the measured mean value, but at the 95% confidence interval of the mean (as given by MV- 2·SD).

- The modelled O_2Sat is between the mean measured value and the value for the mean plus 1·SD.

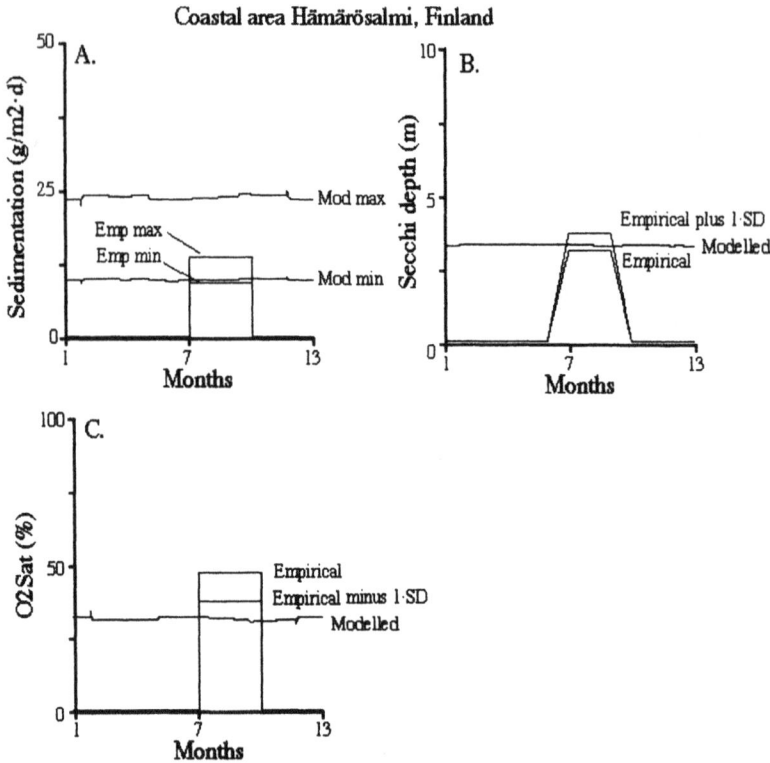

Fig. 6.21. Validations in the Hämärösalmi coastal area (table 6.2) (A) for empirical and modelled minimum and maximum values for sedimentation, (B) for modelled Secchi depths versus empirical data and uncertainty in empirical data (MV plus 1·SD) and (C) for modelled values of O₂Sat and empirical data and uncertainty in empirical data (MV plus 1·SD).

Fig. 6.21 gives similar results for a Finnish coastal area, Hämärösalmi.

- Modelled values of sedimentation are close to the corresponding values measured in sediment traps (fig. 6.21A)

- The modelled Secchi depth are slightly higher than the measured mean value (fig. 6.21B).

- The modelled O₂Sat is within the 95% confidence interval of the mean (as given by MV-2·SD).

The data for all 17 coastal areas are compiled in fig. 6.22 for sedimentation. In this case, modelled mean values $[((Vd/3) \cdot F_{DWA} + F_{DWA})/2]$ are compared to empirical mean values $[(Sed_{SW} + Sed_{DW})/2]$. Fig. 6.22 gives the regression line for actual data. The r^2 value is 0.89 and the slope 1.17. Fig. 6.22B gives the corresponding error function.

$$y = 1.17x - 1.63; \quad r^2 = 0.89; \quad n = 17; \quad p < 0.0001$$

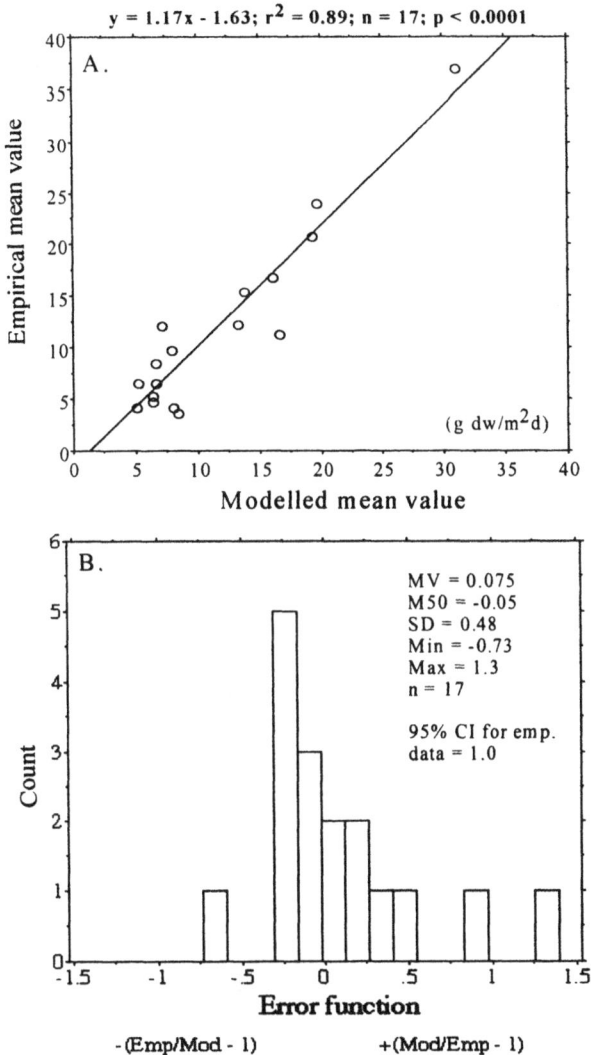

Fig. 6.22. Compilation of validation results for mean sedimentation for the 17 coastal areas, (A) gives the regression between empirical data and modelled values (regression line, r^2, n, and p) and (B) gives the corresponding error function and statistics.

One can see that the mean error is close to zero (= 0.075) and that the modelled values generally are within the 95% uncertainty interval for the empirical data (± 1). These are good results and it is probably not possible to obtain better results with these data. That is, the limiting factor for the predictive power is access to reliable empirical data rather than uncertainties in model structures. Evidently, it would be interesting to test this model also for other coastal areas.

Fig. 6.23 gives the corresponding information for Secchi depth. Here, modelled values for the growing season are compared to empirical mean data. The r^2 value is 0.84 and the slope 1.08. The error function is shown in fig. 6.23B. The mean error is close to zero (= 0.086) and most modelled values are within the 95% uncertainty interval for the empirical data (± 0.38). These are also very good results.

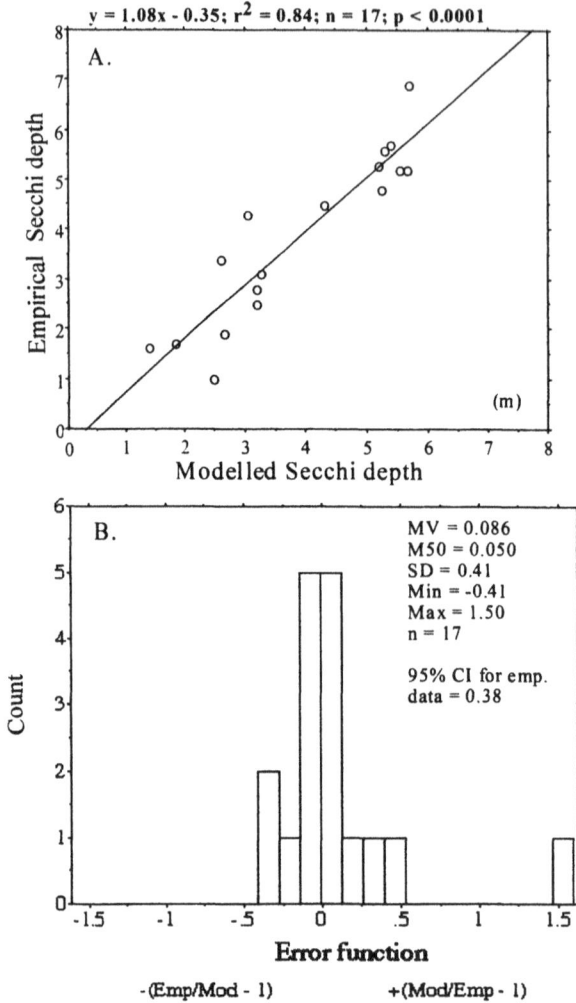

Fig. 6.23. Validation results for Secchi depths for the 17 coastal areas, (A) gives the regression between empirical data and modelled values (regression line, r^2, n, and p) and (B) gives the corresponding error function and statistics.

The results for the oxygen saturation are compiled in fig. 6.24, where modelled values for the growing season are compared to empirical mean values. The r^2 value is 0.86 and the slope 1.01. The error function is given in fig. 6.24B. The mean error is -0.079 and most modelled values are within the 95% uncertainty interval for the empirical data (± 0.45).

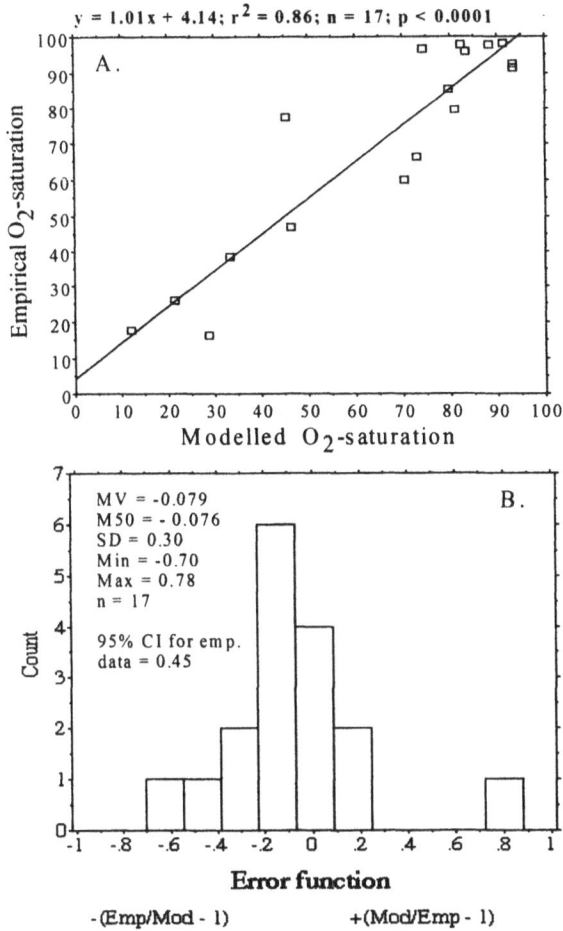

Fig. 6.24. Validation results for the oxygen saturation in the deep water zone, (A) gives the regression between empirical data and modelled values (regression line, r^2, n, and p) and (B) gives the corresponding error function and statistics.

One should note that this is a blind test in the sense that there have been no changes in the model variables; only the obligatory coast-specific driving variables listed in table 6.2 have been changed. The results will be further elaborated in the next section.

6.3.5. Calculation and ranking of fluxes

This model may be used to address several interesting issues, e.g., (1) to calculate and compare sedimentation in different coastal areas, which is essential in studies concerning the age of sediments and sediments as a historical archive, (2) to test hypotheses about the relative role of various processes determining the value of the target variable, SPM in coastal water, and (3) to calculate and rank SPM fluxes. These issues will be discussed in this section. It is very important to identify small and large fluxes of substances, toxins, nutrients, and SPM in remedial contexts so that realistic expectations can be obtained for various remedial measures intended to reduce SPM and the associated risk of low oxygen concentrations and, hence, the risks to key functional benthic species.

The next figure (fig. 6.25) compares the fluxes in three coastal areas, one with direct contact with the sea, Gräsmarö, on the Swedish east coast, one area situated deep in the Finnish archipelago, Hämärösalmi, and one area from S. Sweden, Järnavik. The aim is to see which are the dominating fluxes in the three cases. The working hypothesis is that there may be major differences in the ranking of the fluxes between the areas, that the fluxes from the sea dominate in area Gräsmarö and not in area Hämärösalmi, where other fluxes, maybe autochthonous production (called Prim Prod in fig. 6.25), may dominate.

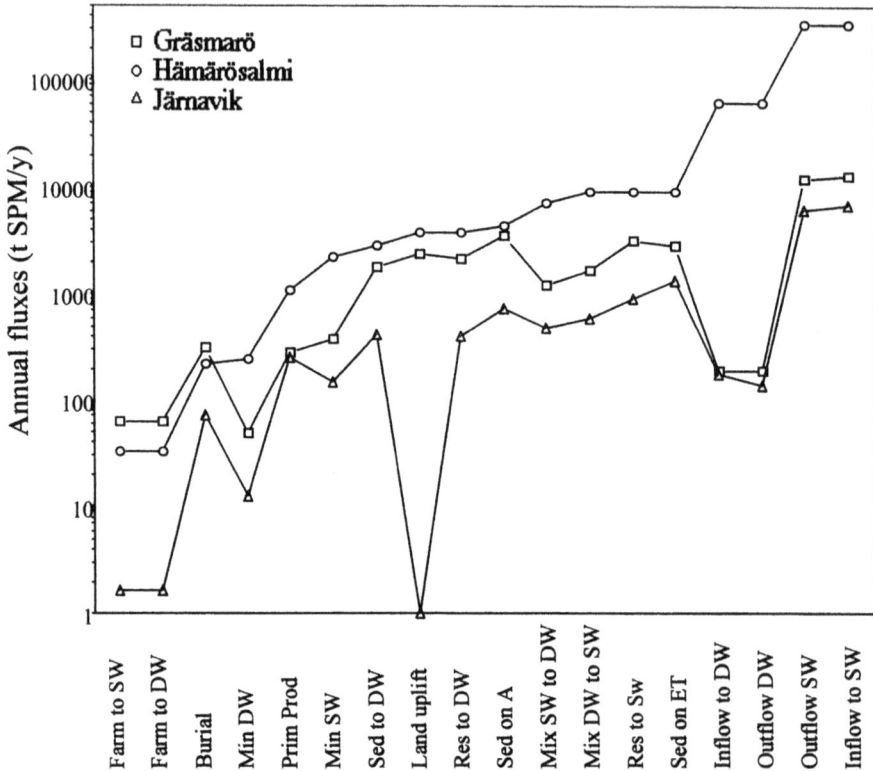

Fig. 6.25. Simulations giving a ranking of all the SPM fluxes (transport processes) for area Gräsmarö, Sweden, which has direct contact with the Sea), area Järnavik, which is also exposed to the sea, and area Hämärösalmi, which is deep in the Finnish archipelago.

From fig. 6.25, one can note:

- There are no major differences between the three coastal areas. The surface water fluxes of SPM dominate in all three areas.

- There is a logical difference in the ranking of the fluxes in the three cases, especially concerning land uplift, which is more important in area Hämärösalmi, which is part of a relatively shallow archipelago system. Land uplift is zero in coastal area Järnavik.

- The working hypothesis concerning the role of the primary production in the coastal areas must be rejected. Autochthonous production is one of the smallest of all the given SPM fluxes in the three areas.

- Most importantly, the main working hypothesis, that there is a major difference in the ranking between the areas must also be rejected. There are no major, only minor differences between the three areas. The surface water fluxes dominate in these cases.

This means that the conditions in the sea play an important role for the transport of SPM, and hence also for all pollutants associated with carrier particles included in the SPM group. This is also a logical consequence of the fact that the characteristic theoretical surface water retention times in these coastal areas are generally short (in the order of days, or in exceptional cases, of weeks).

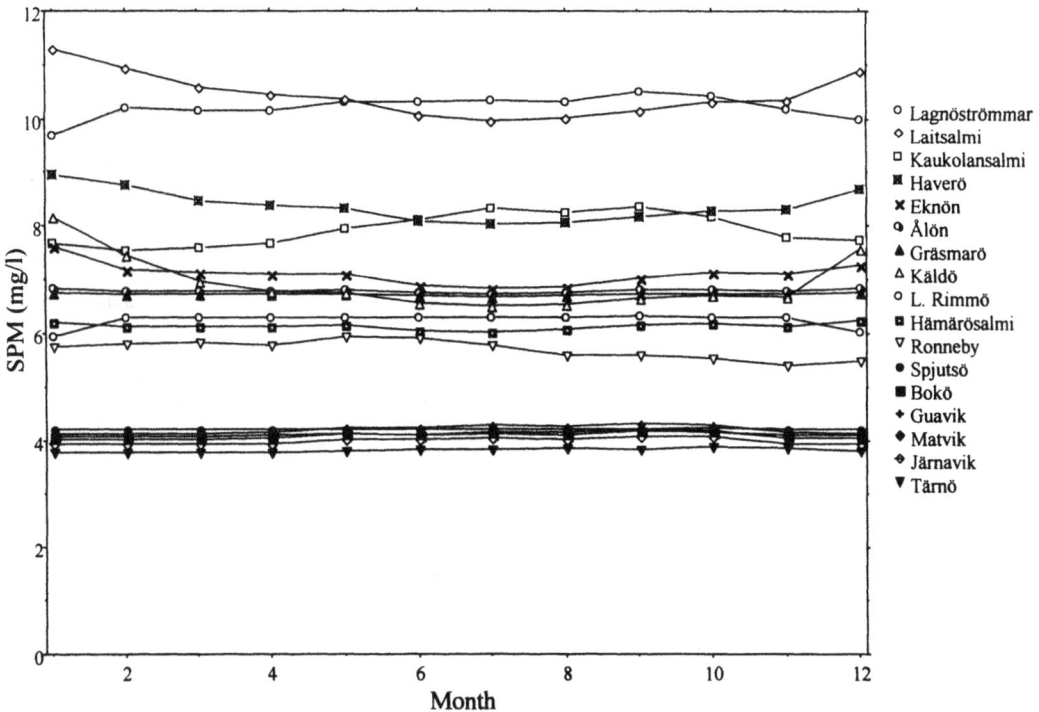

Fig. 6.26. Simulations to show how SPM concentrations vary seasonally (monthly) in the studied coastal areas.

Fig. 6.26 gives a compilation of seasonal (monthly) SPM values in these 17 coastal areas and fig. 6.27 shows how sedimentation in cm/yr varies on a monthly basis in the areas. It is interesting to note:

- That area characteristic SPM values vary from about 4 mg/l to 11 mg/l.

- That there are coastal areas without any clear seasonal SPM patterns, and also areas with higher SPM concentrations in the summer and fall and also areas with lower values in the summer, depending on the specific characteristics of the coast.

- That the mean net sedimentation on accumulation areas varies between 0.1 and 1.5 cm/yr and that also for sedimentation there is no clear seasonal pattern prevailing in all coastal areas. One should also note that these are mean values for the entire A area and that the net sedimentation varies from zero at sites close to the critical water depth to maximum values at sites in the deepest parts of the coastal areas.

- This means that the age of the bioactive A sediments (often set to 0-10 cm) is between 7 and 100 year in these coastal areas.

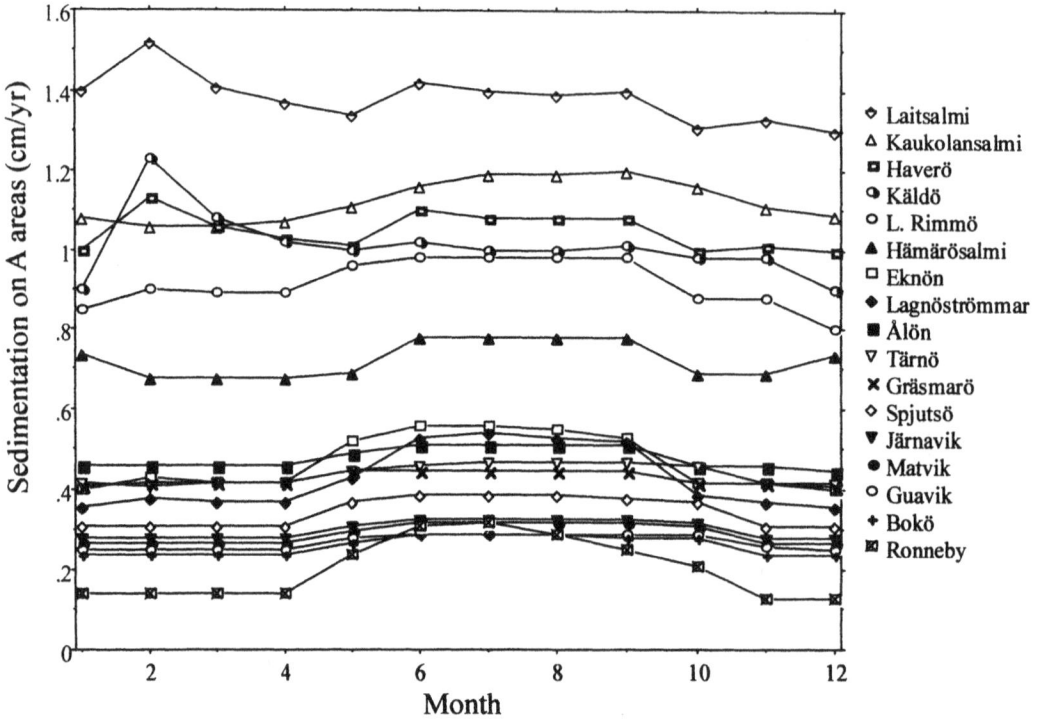

Fig. 6.27. Simulations to show how sedimentation in cm/yr varies on a monthly basis in the studied coastal areas.

It is also interesting to look at differences in mean SPM values calculated for the surface water compartment (SW), the deep water compartment (DW), for the entire water volume (Tot in fig. 6.28) as compared to SPM in the sea just outside the coastal areas (Sea).

Fig. 6.28 gives three contrasting examples on this matter.

- Coastal area Langnöströmmar (SE Sweden) has higher SPM concentrations compared to the outside sea. The SPM concentration in the deep water compartment is higher than in the surface water compartment. The mean SPM concentration in the entire coastal area (Tot) is logically between the values in the surface and deep water compartments.

- The conditions in coastal area Käldö (Finland) are quite different. The SPM concentrations in the outside sea are generally higher than in the coastal area, even higher than in the deep water compartment. This means that this coastal area works as a sediment trap. It is comparatively small and deep and has a very small section area (At).

- The SPM conditions in coastal area Tärnö are relatively close to the conditions in the sea, because this coastal area is small and exposed to winds and waves from the sea.

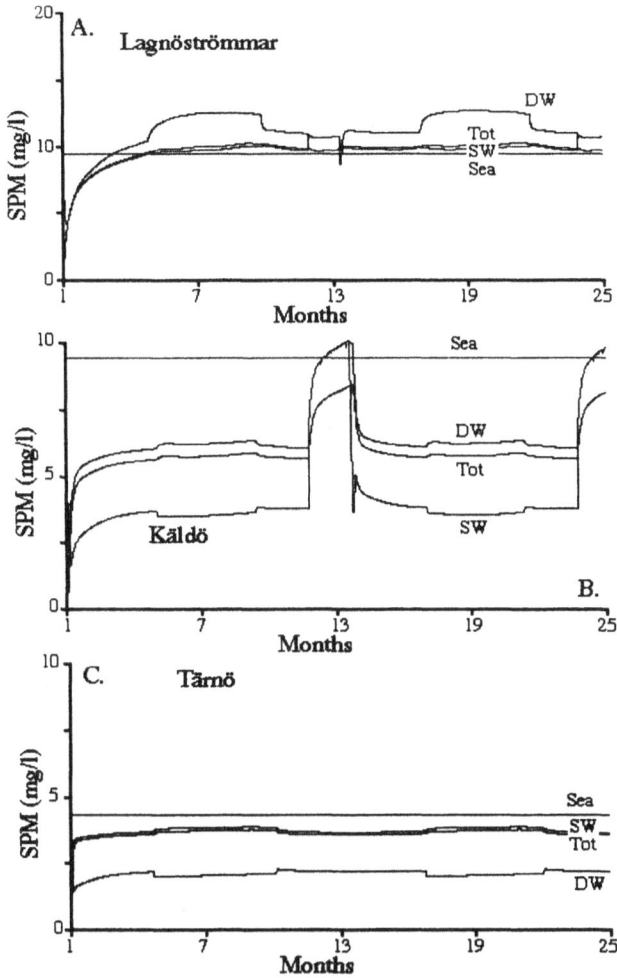

Fig. 6.28. SPM concentrations in surface water (SW), deep water (DW), total SPM concentrations in coastal water (Tot) and SPM concentrations in the sea outside the three coastal areas (A) Lagnöströmmar, Swedish east coast), (B) Käldö, Finland, and (C) Tärnö, southern Sweden.

6.3.6. SPM in different coast types

This section will give several sensitivity tests. The aim has been to try to clarify important factors regulating SPM concentrations in coastal areas. Fig. 6.29 first shows results from the coastal area Ronneby, southern Sweden, when the salinity has been set to 0, 6.5 (the actual value for this coastal area) and 30% while all else have been kept constant, including the SPM concentration in the sea outside the given coastal area (which has been set to 5 mg/l = the default value (see fig. 6.29B). Under these conditions, one can see no major differences in the SPM concentration in the total water volume (fig. 6.29A), but sedimentation is much higher if the salinity is high (fig. 6.29C) and the water clarity is also much higher at higher salinities (fig. 6.29). There are interesting compensatory effects in this example: A higher Secchi depth means a deeper photic zone and a higher bioproduction; a higher salinity also means a higher flocculation and aggregation, so sedimentation becomes higher. The model quantifies such dependencies and the net result is shown in fig. 6.29.

Coastal area, Ronneby, Sweden

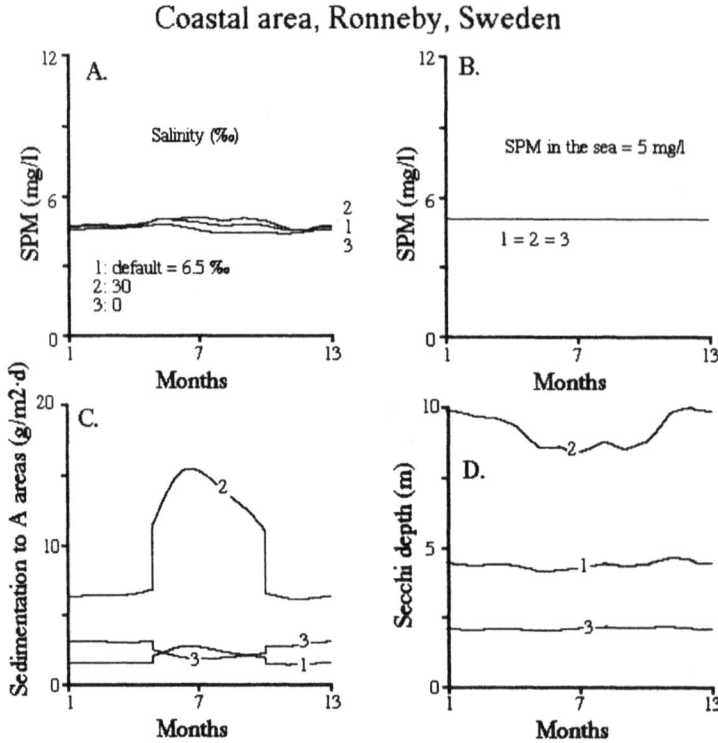

Fig. 6.29. Sensitivity analyses illustrating how different salinities (0, 6.5, and 30%) would influence (A) total SPM concentrations in water, (C) sedimentation on accumulation areas, and (D) Secchi depths if all else is constant for coastal area Ronneby, southern Sweden, including the SPM concentration in the sea outside the coastal area (fig. B).

Fig. 6.30 gives results from nine similar sensitivity tests:

- Figures A and B address the problem when it becomes important to account for river inflow of SPM to a coastal area. The river model (from chapter 5) has been added to the coastal model. From fig. 6.30A, one can see that the ratio between the area of the catchment (A_{DA} = 633 km^2) and the given coastal area (11 km^2) is 58. Had this ratio been lower (curve 1 gives the results if the ratio is 10), then less allochthonous SPM would be transported into the coastal area and the SPM concentration would be lower, but only slightly lower since river inflow is not a major SPM flux to this coastal area. If the ratio had been 500, then river inflow would have been a dominating factor for SPM in this coastal area, which has a section area of 0.0176 km^2. Had the section area been smaller, such a high river inflow of SPM would have been even more important. This is shown in fig. 6.30B where the section area is 10 times smaller and all else the same as in fig. 6.30A (also note the change in scale on the y axis).

- Fig. 6.30C gives results from a sensitivity test where the area of the coast has been varied (set to 11, 110, and 1.1 km^2). There are no major differences in the three curves shown in the figure but the highest SPM values appear during the summer in the largest coastal areas, with the highest bioproduction (which by definition is directly related to area).

- The role of different mean depths is illustrated in fig. 6.30D. The shallower the coast, the higher the SPM concentration – if all else is constant. This is related to both resuspension and the fact that the SPM concentration by definition is amount of SPM per volume of water; a smaller mean depth means a smaller volume of water.

- Fig. 6.30E shows results when the maximum depth has been altered; curve 1 gives the default value, curve 2 results when the D_{max} value is increased by a factor of 3, and curve 3 the results when the maximum depth is 3 times smaller than the default value. The coast with the smaller maximum depth relative to the mean depth will have the most U-shaped basin with the least resuspension, so the SPM concentrations would be lower.

- The section area (At) is generally very important in regulating water and SPM transport between the coast and the sea. In this case, the default At has been multiplied by a factor of 3 and divided by a factor of 3. This has no major impact in this coastal area (fig. 6.30F) which is quite open and where the SPM concentration in the coast is close to the SPM concentration in the sea (5 mg/l).

- The latitude of the coastal area influences water temperatures and bioproduction. From fig. 6.30G, a coastal area situated at latitude 40°N would have higher SPM concentrations than a similar coastal area at latitude 70°N.

- The mean annual precipitation influences the river water discharge. Had the precipitation been 3 times higher than the default value of 650 mm/yr, the river inflow would have been a dominating source of SPM to this coastal area. This is shown in fig. 6.30H.

- Fig. 6.30I gives results from three simulations where the tidal rage has been varied (and all else kept constant). In this relatively open coastal area, the SPM concentration in the coast is already close to the SPM concentration in the sea and had the tidal effect been higher, the SPM concentration in the coast would be even closer to the value in the outside sea.

The main message from these tests is that different fluxes dominate different coastal areas of different topographical characteristics and this means that in order to predict SPM in a given coastal area, one needs a model that has been validated and shown to be able to handle and quantify such differences. Such a model should not be too complicated and the necessary driving variables should be easily accessed. This dynamic model is meant to meet such demands. Although the model is based on general principles and processes that appear in all/most coastal areas, it still needs to be further tested, e.g., for tidal coasts, for more open coasts, and coasts with higher salinities.

Coastal area, Ronneby, Sweden

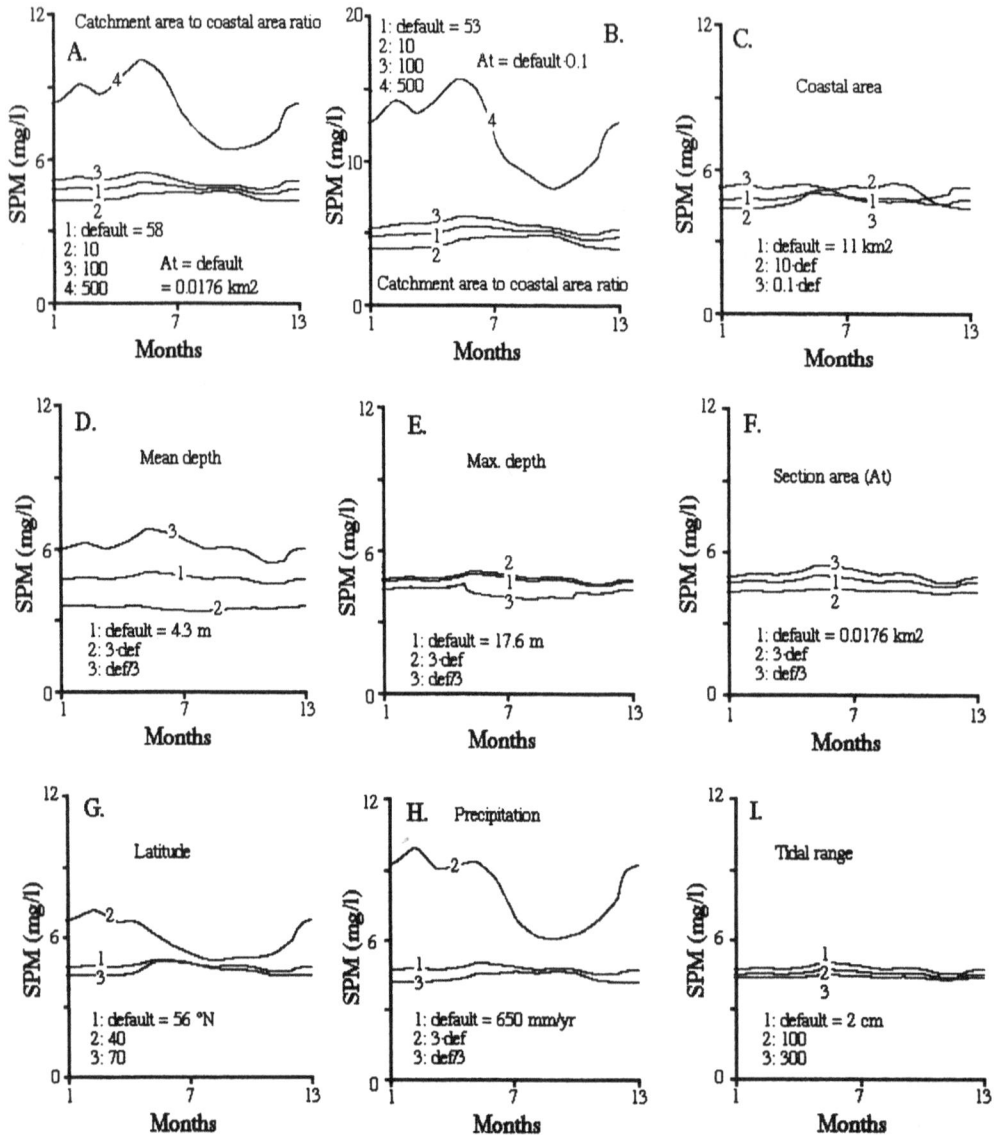

Fig. 6.30. Results from nine sensitivity analyses illustrating how total SPM concentrations in the water of Ronneby coastal areas depend on (A) the size of the catchment areas and river inflow of SPM (A_{DA}/A ratios of 10, 53, 100, and 500 have been tested) for the default values for this coastal area that has a section area, At, of 0.0176 km²; (B) gives the results as in fig. A had the section area been 10 times smaller; (C) illustrates the role of different values for the coastal area; (D) the role of different mean depths; (E) the role of different maximum depths; (F) how different section areas influence SPM; (G) how SPM relates to different latitudes; (H) how different mean annual precipitations likely influence SPM; and (I) how different values for the tidal range influence the conditions – if all else is constant.

6.3.7. Sensitivity and uncertainty analyses using Monte Carlo simulations

An initial sensitivity analysis for one of the studied coastal areas, Järnavik, southern Sweden using a uniform coefficient of variation (CV) of 0.5 is given in fig. 6.31. The results demonstrate how the uncertainty in one flux, the inflow of SPM to surface water from the sea, influences the uncertainty in the SPM concentration in the entire coastal volume, the target y variable (SPM). The idea is to get a quantification of the role of this flux for the value of the target variable, while all else is constant. One hundred runs have been simulated and a normal frequency distribution has been applied for the uncertainty in the x variable. Then, the uncertainty in the target y variable is considerable (CV for y = 0.45) (see fig. 6.31B, which also gives a box-and-whisker plot using the data from month 13). In the following, many results like this will be given and the aim is to see how the uncertainties in the y variable depend on the uncertainty in the given fluxes.

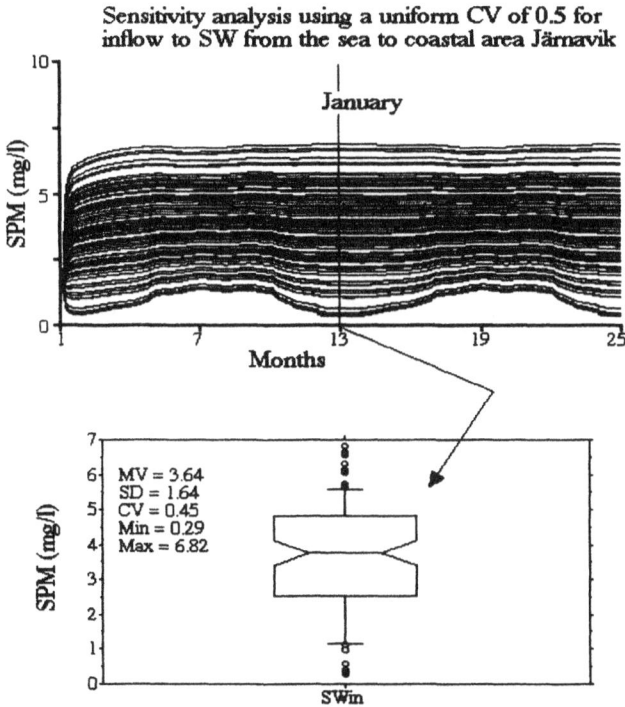

Fig. 6.31. Results from a sensitivity test (100 runs) using data for coastal area Järnavik, southern Sweden. The figure illustrates the role of uncertainty for one of the fluxes, inflow of SPM to the surface water compartment, assuming a CV of 0.5, and a normal frequency distribution around the mean flux on modelled SPM concentrations in the coastal water. For the following comparative tests, data from January (month 13), have been selected as indicated in the figure. Figure B gives the box-and-whisker plot for these data and basic statistics.

Fig. 6.32 gives the results for all the fluxes in the model. This figure also ranks the importance of the various uncertainties for the predictions of SPM in this coastal area (Järnavik). From these presuppositions, one can note that the three most important uncertainties concern the SPM fluxes to and out of the surface water compartment and mineralization in the surface water compartment. All other uncertainties in the SPM fluxes are of less importance and the uncertainty in land uplift (LU) is of no significance since land uplift is zero in this region. It may be realistic to use a uniform uncertainty for the given fluxes, but it is not realistic to assume a uniform uncertainty for the different driving variables (see Håkanson, 1999).

In the next model test, uncertainty analyses (Monte Carlo techniques) will be carried out to study how characteristic uncertainties in the driving variables influence the uncertainty in the y variable (SPM) in another coastal area Ronneby, southern Sweden (fig. 6.33).

Sensitivity analyses using a uniform CV pf 0.5 for all SPM-fluxes to, within and from coastal area Järnavik, Sweden

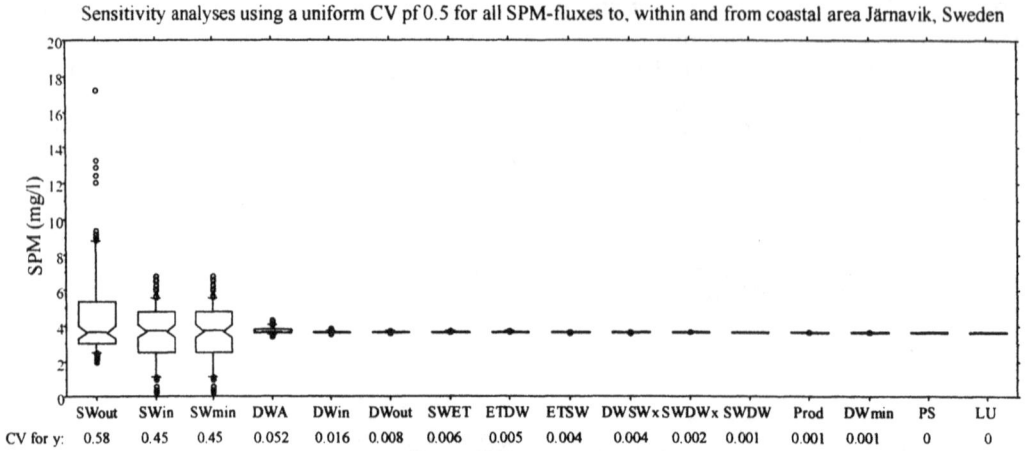

	SWout	SWin	SWmin	DWA	DWin	DWout	SWET	ETDW	ETSW	DWSWx	SWDWx	SWDW	Prod	DWmin	PS	LU
CV for y:	0.58	0.45	0.45	0.052	0.016	0.008	0.006	0.005	0.004	0.004	0.002	0.001	0.001	0.001	0	0

Fig. 6.32. Sensitivity tests according to the procedure shown in fig. 6.31 where all fluxes in the model are accounted for, one by one, and all else kept constant. A uniform uncertainty for all the fluxes has been used (a CV of 0.5 and a normal frequency distribution around the mean value). The figure also ranks the importance of the fluxes in relation to the prediction of the target variable, SPM in the Järnavik coastal area, under these presuppositions. The figure gives the box-and-whisker plots (median, 25 and 75 quartiles, 10 and 90 percentiles and outliers) as well as the CV for the y variable (e.g., 0.58 related to the uncertainty in the flux of SPM from surface water to the sea in January).

Uncertainty analyses using characteristics CVs for the driving variables for coastal area Ronneby, southern Sweden

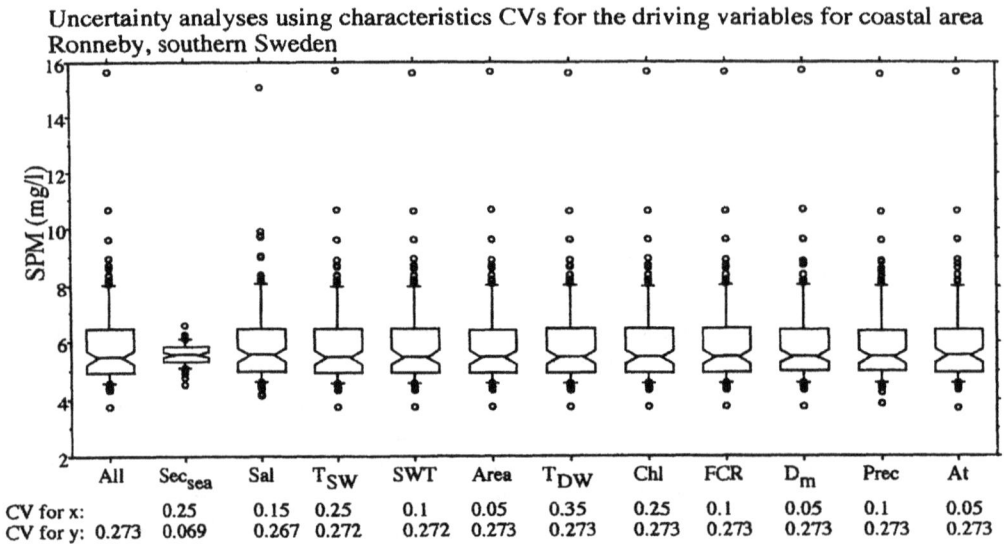

	All	Sec$_{sea}$	Sal	T_{SW}	SWT	Area	T_{DW}	Chl	FCR	D_m	Prec	At
CV for x:		0.25	0.15	0.25	0.1	0.05	0.35	0.25	0.1	0.05	0.1	0.05
CV for y:	0.273	0.069	0.267	0.272	0.272	0.273	0.273	0.273	0.273	0.273	0.273	0.273

Fig. 6.33. Uncertainty analyses using Monte Carlo techniques with data for area Ronneby, southern Sweden for the target variable, total SPM concentration in water. The figure gives characteristic CVs for the eleven selected driving variables (e.g., 0.15 for salinity and 0.05 for coastal area), calculated CVs for SPM (e.g., 0.273 when all these uncertainties are accounted for at the same time, and 0.069 when the uncertainty for the Secchi depth in the sea just outside the coastal area is neglected and all other uncertainties are accounted for), and a ranking based on the calculated CVs of how uncertainties in these driving variables influence the uncertainty for SPM.

The tests have been done for the following driving variables:

1. The concentration of SPM in the surface water outside the coast, as calculated from eq. 6.13 and from regional data on the Secchi depth close to the section area (table 6.2). This is, as the previous sensitivity analyses indicated, a very important variable. The characteristic CV for Sec_{sea} is set to 0.25.

2. The chlorophyll-a (Chl) concentration regulating primary production; CV = 0.25 (Wallin et al., 1992).

3. Surface water salinity (Sal). There are comparatively reliable data on the salinity; CV = 0.15.

4. The feed conversion ratio (FCR) influencing the point source emissions of SPM (see appendix 9.6). A standard FCR value of 1.5 has been used in this modelling, which may not be true for the fish farm in this coastal area. CV is set to 0.1.

5. Surface water temperatures (SWT) are predicted from latitude. There is no uncertainty in latitude but the predicted values for SWT are uncertain and the CV value is set to 0.1.

6. The coastal area. This value can be determined quite well but there may be uncertainties where the boundary lines defining the coastal area are drawn. This CV is set to 0.05.

7. This uncertainty (CV = 0.05) will also be used for the section area (At).

8. And also for the mean depth (D_m) for the same reason.

9. This coastal area is an estuary and there are uncertainties in the value used for the mean annual precipitation (650 mm/yr), which influence the river water discharge and the river transport of SPM to this coastal area. The CV for precipitation is set to 0.1. The theoretical surface water retention time (T_{SW}) and the theoretical deep water retention time (T_{DW}) are not obligatory driving variables since for this coastal area, they are predicted from morphometric parameters. But there are uncertainties in the empirical submodels to predict T_{SW} and T_{DW}, so they have been included in this test.

10. The theoretical surface water retention time (T_{SW}). This value is important because it regulates the fluxes of SPM to and from the sea and/or adjacent coastal areas. T_{SW} is calculated from an empirical model that has given an r^2 value of 0.95 (table 6.1), but the predicted T_{SW} value is uncertain and CV is set to 0.25.

11. The theoretical deep water retention time (T_{DW}). T_{DW} is also calculated from an empirical model (table 6.1), which gave an r^2 value of 0.79; CV is set to 0.35.

The aim now is to produce a ranking of these uncertainties for the target variable. The results are given in fig. 6.33. Note:

•   The total calculated CV for SPM according to this testing procedure is 0.27.

•   The most important factor is the uncertainty associated with the value used for the value used for the Secchi depth just outside the coastal area, which is used to calculate the SPM concentration in the surface water outside the coastal areas, which regulates the inflow of SPM from the sea, which is the dominating inflow to this coastal area.

- If this uncertainty is omitted, CV for SPM decreases the most, from 0.27 to 0.07. This means that future work should concentrate on getting more reliable data and/or submodels for this driving variable in this (and most) coastal areas. This is often (but not always) the best way to reduce the uncertainties in predictions of SPM in coastal areas.

- The model is not so well balanced since the model predictions depend so much on the uncertainty in one single driving variable.

6.3.8. Scenario illustrating the practical use of the dynamic SPM model

This section has been included to exemplify how this dynamic SPM model can be used in practical contexts in coastal water management. The questions asked here are: How large point source emissions of SPM can be accepted in a given coastal area? And why?

We have selected coastal area Hämärösalmi (Finland) with the largest fish cage farm. It produced 381 tons/yr of rainbow trout during the fieldwork. In this coastal area, there are no other point source emissions of SPM beside those from the fish cage farm. It has been assumed that the fish farm emissions also influence the total nitrogen (TN) concentration in the water. It would have been best if this influence could have been predicted by a validated mass-balance model for nitrogen, which meet the discussed criteria for a useful model. Unfortunately, to the best of the author's knowledge no such model is available, so in this scenario the second best has been done. It will be assumed that the fish farm emissions influence the TN concentration in the same proportional manner as they influence the SPM value in the coastal area. It will also be assumed (from data given by Wallin et al., 1992) that the regional reference value for the TN concentration in this coastal area is 310 µg/l. From these presuppositions, the changes in TN concentrations may be estimated (fig. 6.33A). These changes in TN (in µg/l) will influence the chlorophyll concentration (in µg/l). This is calculated from the following empirical model (see Håkanson, 1999):

$$\log(\text{Chl}) = 2.78 \cdot \log(\text{TN}) - 6.66 \qquad (6.20)$$
$$(n = 22; r^2 = 0.91)$$

From chlorophyll, one can estimate primary production of SPM and apply the dynamic coastal model. The target variables in this scenario are: (i) total SPM in coastal water, (ii) sedimentation in the deep water (modelled maximum values), and (iii) oxygen saturation in the deep water (O_2Sat), as defined previously in this chapter.

The annual fish production in the farm is set to 0, 381 (the actual value), 1000, and 3000 tons/yr. How will this influence the given variables? From fig. 6.34, one can note:

- If 3000 tons/yr fish were to be produced in this coastal area, the TN concentrations would likely increase very much, which would reduce the Secchi depth, increase SPM (fig. 6.34B), increase sedimentation (fig. 6.34C) and decrease O_2Sat (fig. 6.34D).

- The changes are likely small if less than 1000 tons/yr are produced, so this should probably be the maximum production that the regional environmental authorities would permit.

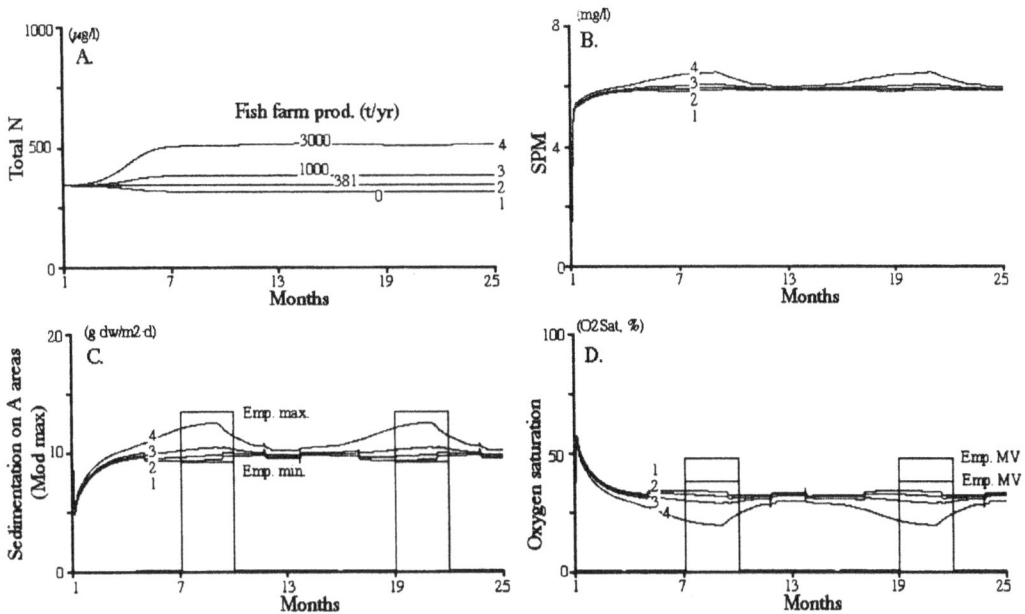

Fig. 6.34. A scenario (and a sensitivity test) illustrating the practical use of the model in coastal management. The annual fish production in the fish cage farm in this coastal area (Hämärösalmi, Finland) from 0 tons/yr to 3000 tons/yr (the actual value in this area in this study was 381 tons/yr) and the likely influence this would have on (A) total N concentrations, (B) total SPM in coastal water, (C) sedimentation on accumulation areas, and (D) oxygen saturation in the deep water. The figure also gives the corresponding empirical data and uncertainties in the empirical data.

Note that the reason for this result is that for this coastal area (and for most but not all coastal areas), there is a very dynamic exchange of water and SPM between the coast and surrounding waters. The theoretical surface water retention time (T_{SW}) is 5 days in this coastal area, which means that the total surface water volume is exchanged 6 times each month. This also means that if the emissions from the point source will not stay in the coastal area very long and that higher SPM concentrations than in the surrounding waters can not easily be maintained.

6.3.9. Conclusions

This section has presented a dynamic mass-balance model for SPM and sedimentation in coastal areas handling all important fluxes of SPM to, from, and within coastal areas, as defined according to the topographical bottleneck method. The model is based on ordinary differential equations and the calculation time (dt) is one month to reflect seasonal variations. An important demand, related to the practical utility of the model, is that it should be driven by variables readily accessed from standard monitoring programs or maps. Added to the dynamic core model are several (static) empirical regressions for standard operational effect variables used in coastal management, such as the Secchi depth and the oxygen saturation in the deep water. The obligatory driving variables include four morphometric parameters (coastal area, section area, mean, and maximum depth), latitude (to predict surface water and deep water temperatures, stratification, and mixing), salinity, chlorophyll concentrations, and the Secchi depth, or SPM concentration in the sea outside the given coastal area. The model is based on three compartments: Two water compartments (surface water and deep water; the separation between these two compartments is done not in the traditional manner from temperatures but from sedimentological criteria, as the water depth separating transportation areas from accumulation areas) and a sediment compartment (ET areas, i.e., erosion and transportation areas where fine sediments are discontinuously being deposited). The processes accounted for include inflow and outflow via surface and deep water, input from point sources, from primary production, from land uplift, sedimentation, resuspension, mixing, and mineralization. The model has been validated with good results (the predictions of sedimentation, Secchi depth, and oxygen saturation are generally within the 95% uncertainty limits of the empirical data. The section has also presented sensitivity and uncertainty tests of the model. Many of the structures in the model are general and have also been used with similar success for other types of aquatic systems (lakes and rivers) and for other substances (mainly phosphorus and radionuclides; see chapter 1). The model could potentially be used for coastal areas other than those included in this study, e.g., for open coasts, estuaries, or areas influenced by tidal variations.

Finally, it may be said that the only way to derive a generic model yielding perfect predictions for the entire domain of coastal areas on earth, would be to account for all processes. This would be like map making in scale 1:1! Simplifications are always needed, and the main challenge is too find the simplest and mechanistically best model structure yielding the highest possible predictive power using the smallest number of driving variables in blind tests.

6.4. SPM in open water areas – empirical modelling

6.4.1. Introduction and working hypotheses

To the best of the author's knowledge, there exist no useful SPM models for open marine areas/sites that may be used in practice and are based on few and readily available driving variables (x). One reason for this may be the difficulty in collecting SPM data in a coherent manner from defined sites in a gradient from calm to stormy conditions. It is hoped that the SPM models presented in this section can be used to get a better understanding of SPM and the factors regulating SPM in open marine areas. They could also be as submodels in wider ecosystem contexts where the aim can be to predict other important variables, such as toxins in fish or production and biomasses of key functional groups of organisms.

This section concerns SPM data from the open Baltic Sea, and for this marine system, there has been very limited knowledge of the SPM concentration, its variation and the factors influencing variation among and within sites. A default value of 3 mg/l (Pustelnikov, 1977) has often been used for SPM in the Baltic Sea. From data given by HELCOM (1998), approximately $7500 \cdot 10^3$ tons SPM per year are transported into the Baltic Sea by rivers, most coming from the former Soviet Union (USSR), Poland, and Sweden.

Besides allochthonous inflow and autochthonous production, there is also, as discussed in chapter 1, another source of SPM in the Baltic Sea related to land uplift.

In Baltic coastal areas, one can generally expect a high sedimentation (Håkanson, 1999; Persson and Jonsson, 2000). The existence of the strong Coriolis-driven currents implies that less material is transported to the deep, open parts of the Baltic Sea since these currents move the suspended particulates along and into the coasts. This also means that allochthonous matter from the large rivers entering the Baltic Sea, the autochthonous production in the coastal areas, and the resuspension in the shallow coastal areas together with the dominating coastal currents create an environment for high sedimentation within coastal areas.

The general working hypotheses for this work are:

- No single factor is likely to explain all or very much of the variability among and within sample sites in SPM; the requested empirical model for SPM as y variable is likely to include several x variables.

- The SPM concentrations depend on the distance to the bottom. Close to the bottom, one should expect higher SPM concentrations (Håkanson et al., 1989; Floderus and

Håkanson, 1989; Nõges et al., 1999). The water depth also determines whether a bottom is dominated by erosion, transport, or accumulation processes for fine particles. To test this, this study includes data on the water depth at the sample site as well as data on the distance from the bottom to the sample site.

- Shallow and wind/wave exposed sites/areas are generally dominated by processes of fine particles erosion and transport. The relationships between winds, waves, and SPM concentrations have been investigated in several studies (Floderus and Håkanson, 1989; Hellström, 1991; Kristensen et al., 1992). A high wind speed might imply an increased SPM concentration. However, not only the prevailing wind situation is of importance, but also the frequency of resuspensions (fig. 6.35).

- If there are many resuspensions per month, it is likely that there is less material on the bottom to be resuspended, compared to a situation with few resuspension events. Also, the wind direction should be of interest. If there is a large fetch, the wave base (i.e., the water depth down to which the wave orbitals can resuspend fine particles) can be deep.

- Studies by Andersson (2000) demonstrate that resuspension correlates with winds higher than 7 m/s in four different archipelago areas in the Baltic proper, whereas studies by Eckhéll et al. (2000) indicate that wind speeds higher than 14 m/s correlate best with resuspension. In this study, we have used wind data to calculate the number of days with winds lower than 10, 7, 5, and 3 m/s. Fig. 6.36 illustrates wind speeds during two sampling events at two stations. This sampling was carried out before and after storms up to 20 m/s. Evidently, this kind of sampling is not very common since in practice it is very difficult to collect and prepare samples (filtration, measurements of temperature, salinity, etc.) under stormy conditions. The basic assumption is that if there is a long period of calm wind conditions, a fraction of the suspended particulate matter will settle out and the SPM concentration in the water will be reduced – if all else is constant. Burban et al. (1989, 1990) have demonstrated that changes in water turbulence, SPM, and salinity are key regulatory factors for the aggregation and flocculation of suspended particles and hence also for the fall velocity. A typical fall velocity for SPM in lakes is 10-100 m/yr depending on the material and this can increase significantly (by a factor of about 20) along a salinity gradient (see also Kranck, 1973, 1979; Lick et al., 1992).

- In the summer when the water temperature is high, the biological production is also high and this affects the SPM concentration. One can assume that the SPM values will increase with water temperature. Temperatures have also been measured at all sites and vertical levels.

- One of the main sources for SPM is allochthonous input from rivers. In this work, it will be tested if the salinity can be used as a predictor of SPM and the salinity has been measured at all sites and verticals. However, this is not a simple matter since, as mentioned, it is also well known that the salinity will increase the flocculation of suspended particles and enhance sedimentation of SPM causing lower values of SPM in the water. A further complication regarding the relationship between salinity and temperature in relation to SPM variations concerns stratification and the fact that saltier and colder water masses are heavier than less salty and warmer water, so there is a general correlation between salinity, temperature, and water depth.

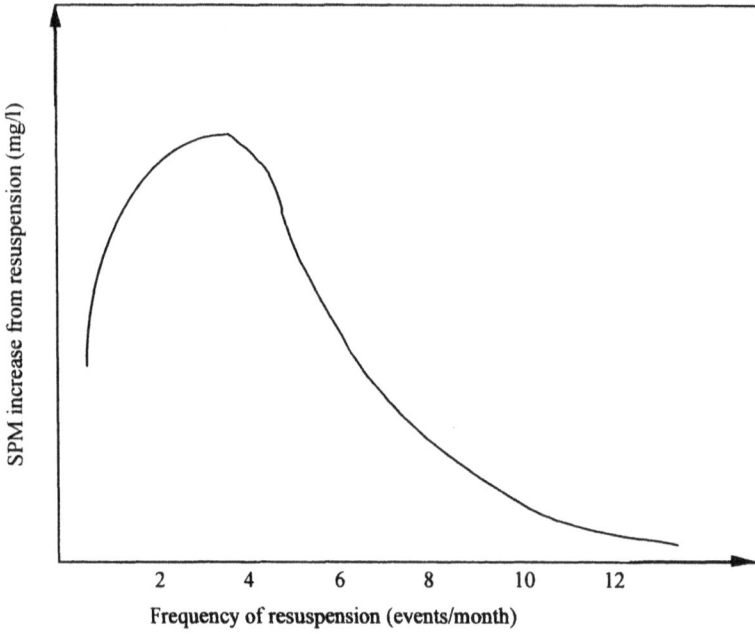

Fig. 6.35. Schematical illustration of the conceptual relationship between the frequency of resuspension events and increase in suspended particulate matter (SPM; note that there is no scale on the y axis). Few resuspension events per time unit would imply a small increase in SPM in the water; many resuspensions would also mean a small increase in SPM because the amount of resuspendable matter would be small.

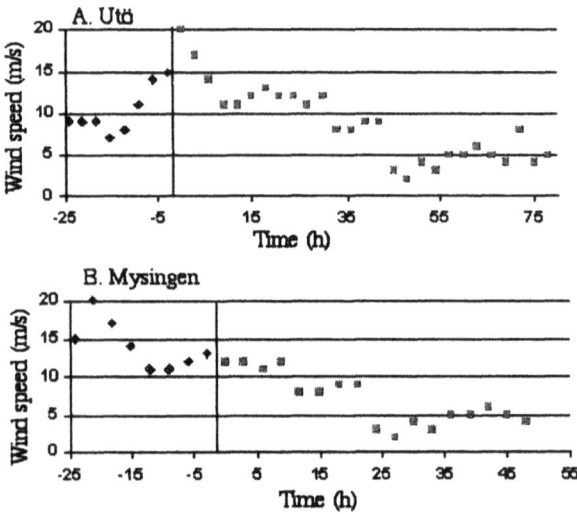

Fig. 6.36. Measured wind speeds at stations Utö and Mysingen before and after the sampling.

6.4.2. Methods and data

The data used in this section are described in appendix 9.3. The basic aim of this section is to carry out a statistical analysis, which aims to quantify and rank the factors causing variations within and among the studied sites in the Baltic Sea. The data are compiled in table 6.7. One can see that there are data from five stations on SPM, fetch (eq. 6.21), days with wind speed higher than 10, 7, and 5 m/s, salinity, temperature, and water depth.

Table 6.7. Compilation of data (mean values for vertical profiles) from field work between 1999 and 2001 in the open Baltic Sea.

Site	Fetch km	Days wind speed (m/s) <10	<7	<5	SPM mg/l	Depth m	Salinity %	Temp. °C
Utö	1694	24	8	3	2.45	40.0	5.52	14.17
	1694	23	7	2	2.59	40.0	5.64	12.92
	1694	22	6	1	2.60	36.5	5.67	12.66
	1694	23	6	1	2.51	35.0	5.65	12.73
Mysingen	757	23	7	2	3.52	18.0	5.92	9.60
	757	22	6	1	3.85	16.9	5.83	10.38
	757	23	6	1	2.93	18.2	5.73	11.18
Gälnan	208	18	6	1	4.86	15.3	5.35	13.69
	208	11	2	0	3.96	13.4	5.40	9.76
Arholma	5000	10	2	0	2.46	22.0	5.43	8.44
	5000	11	2	0	1.77	24.8	5.51	8.47
Laxen	5000	30	12	4	0.67	30.0	5.83	4.47
	5000	30	12	4	0.88	50.9	5.90	5.26
	5000	30	12	4	1.05	48.5	6.94	5.21
	5000	30	11	2	1.01	49.8	6.05	6.72
	5000	30	11	2	2.91	35.6	5.55	6.06
	5000	30	11	2	1.99	15.5	4.90	6.40

It should be stressed that the values given in table 6.7 for temperature and salinity are mean values for each vertical profile and sampling event. These values correspond to values for mixed conditions (= unstratified) when there is no vertical gradient.

Wind/wave influences on SPM in open water areas may be calculated from an equation based on the effective fetch (Beach Erosion Board, 1972; Håkanson and Jansson, 1983). This study includes data on the fetch at each site, which is a modification (from Pilesjö et al., 1991; fig. 6.37 for illustration) of the effective fetch and gives the wind/wave impact as:

$$\text{Fetch} = \sum(\cos(a_i) \cdot x_i) \tag{6.21}$$

where a_i and x_i are angles in radians and distances to land in km, respectively, for a number of radians starting from a defined site and going in many directions (i).

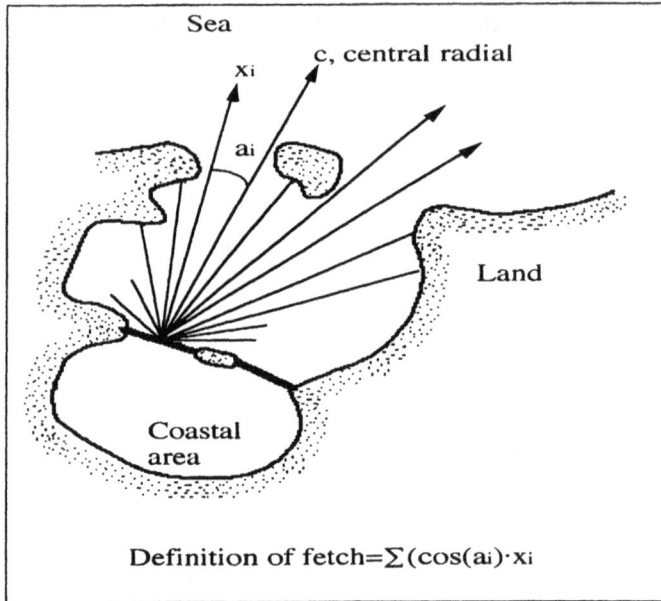

Fig. 6.37. Definition of the fetch for a site (from Pilesjö et al., 1991).

This is a measure of the potential wind/wave exposure at a given site. This measure is easy to determine if one has access to a GIS program (GIS = Geographical Information System). Without this program, it is also possible to determine the fetch, but it is more demanding. So, in the following, results/models will be given both including and excluding the fetch. For this work, the maximum value for the fetch is set to 5000 km since the wave base changes very little for higher values (fig. 2.18).

To evaluate results of regression analyses based on empirical data, it is important to recognize that such data are always more or less uncertain due to problems related to sampling, transport, storage, analytical procedures, natural variations, etc. This will restrict the descriptive or predictive power of any model. The theoretically highest reference coefficient of determination (r_r^2; see chapter 3) of a model is related to the characteristic CV of the y variable (CV = coefficient of variation; CV = SD/MV; SD = standard deviation; MV = mean value).

As a background to the following regressions, table 3.1 gave a compilation of characteristic CV values for within-site variability related to individual samples for three variables of interest in this context, SPM, water temperature, and salinity. If modelled values are compared to empirical data, one can not expect to obtain r^2 values higher than the r_r^2 values, so these r_r^2 values should be used as reference values for the following models. A characteristic CV for SPM in this dataset is 0.67 and $r_r^2 = 0.70$. This very high CV for individual SPM values may be reduced if one calculates mean values based on a certain number of data and then determines the CV for the mean values. The high CV value for SPM implies that many samples are required in order to determine a mean value with a high certainty (see eq. 3.2, the sampling formula in chapter 3).

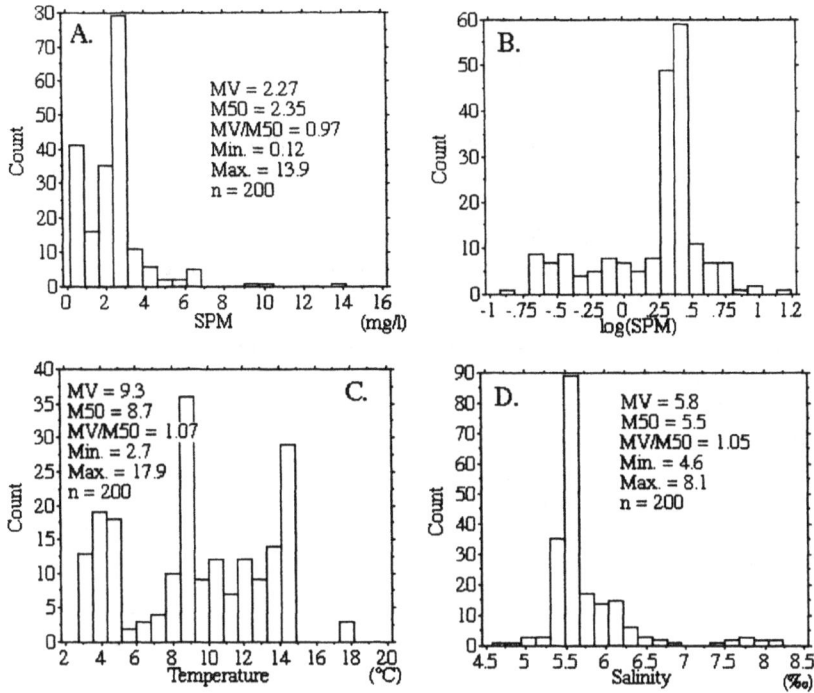

Fig. 6.38. Frequency distributions and basic statistics for (A) suspended particulate matter (SPM), (B) log(SPM), (C) water temperature, and (D) salinity.

6.4.3. Statistical modelling

The entire data set indicates that SPM has a positively skewed frequency distribution (fig. 6.38A) and the recommended transformation (Håkanson and Peters, 1995) is log(SPM), fig. 6.38B. All variables assumed to influence the SPM variation have been transformed so that the best possible normal frequency distributions are obtained for the given x variables for the statistical analyses. Fig. 6.38C shows the frequency distribution for water temperature and fig. 6.38D the distribution for salinity. From fig. 6.38, the range in the individual SPM values is from 0.12 to 13.9 mg/l; the mean and the median values are close, at about 2.3 mg/l. The temperature range in the individual values is from 2.7 to 17.9°C and the range in salinity from 4.6 to 8.1%. Note that the range in salinity is relatively small.

The model only applies in the domain given by the range of the model variables. If the model is used outside this domain (fig. 6.38), this should be done with caution, as a working hypothesis only.

6.4.3.1. Variations among sites/events

The basic aim here is to rank the factors influencing variations in mean SPM values among sites when the water mass is homothermal (i.e., for mixed conditions). Data for SPM exist for 17 sites/events (table 6.7). The mean SPM varies from 0.67 to 4.86 mg/l. The question is: What factors can statistically explain this variability?

1. **Correlation matrix.** The correlation matrix (giving correlation coefficients, r; table 6.8) for untransformed values of mean SPM and the x variables given gives a first scanning of the relationships. From this table, one can note that the best r values appear between mean SPM and:

    (1) The fetch (r = - 0.81; p = 0.0001; p = the statistical certainty) – the larger the fetch, the lower the SPM value; the more enclosed the site is, the higher the SPM value. This is logical since exposed sites are often dominated by coarse deposits (sand, etc.), which are unlikely to influence SPM in water. However, this type of relationship is often difficult to explain in a mechanistic manner since the fetch covaries with other variables.

    (2) A significant relationship exists between the fetch and mean temperature (r = -0.66; p = 0.0001) – the water temperature is generally higher in shallow coastal areas. This is certainly nothing new; it only shows that these empirical data provide logical results. There are also strong causal as well as statistical relationships between depth and mean salinity, depth and mean temperature, and mean temperature and mean salinity because saltier and colder water with a higher density prevail at greater water depths. There are also strong correlations between the number of days with winds speeds lower than 10, 7, and 5 days.

    (3) There are also strong and interesting relationships between the mean SPM and the water depth at the site (r = -0.72); deep locations generally have lower SPM values than shallower sites near the coast. There is also a clear connection between SPM and days with wind speeds lower than 5 m/s (r = -0.62), the more calm days the lower the mean SPM values. These results are in good agreement with the working hypotheses.

    (4) All other relationships are weaker. These relationships will be elaborated in more detail in the following parts.

2. **Stepwise multiple regressions.** Table 6.9A gives the results for stepwise regressions of log(SPM) vs the given variables and table 6.9B gives similar results when the fetch has been omitted. One can note:

    - After three steps, the r^2 value in table 6.9A is 0.81. That is, 81% of the variations in log(SPM) among these 17 sites/events may be statistically explained by variations in the three given parameters. The fetch is the most powerful predictor (r^2 = 0.61), which shows that wind/wave exposure affects SPM considerably – the higher the fetch, the coarser/harder the sediments, the lower the mean SPM value at the site.

    - The next most important factor is number of days with wind speeds lower than 5 m/s (r^2 increases from 0.61 to 0.79).

    - The third factor, mean salinity, increases r^2 to 0.81, but this is not a statistically significant relationship (F set to 1).

    - If the fetch is omitted from the analysis, the mean water temperature at the site emerges as the most important factor explaining mean SPM variations among sites (r^2 = 0.59) - the higher the mean temperature, the higher the SPM value. This is logical and in good agreement with the working hypothesis, although the causal reason may be obscured since temperature influences both production and stratification/mixing.

- The second variable (if fetch is omitted) is the water depth at the site. Accounting for the water depth increases r^2 from 0.59 to 0.78; this is also logical and significant.

- The third factor is days with wind speeds lower than 5 m/s (r^2 increases from 0.78 to 0.80, but this is obtained by lowering F to 1). The result is logical, but not statistically significant.

Table 6.8. Correlations for mean values among sites/samplings (5 sites and 17 sampling situations). The correlation matrix shows linear correlation coefficients (r) and statistical certainties (p) for the actual values (untransformed values).

p \r	SPM	Fetch	Depth	Salinity	Temp	Days,10	Days,7	Days,5
SPM	1	-0.81	-0.72	-0.41	0.66	-0.51	-0.58	-0.62
Fetch	0.0001	1	0.54	0.17	-0.66	0.13	0.24	0.16
Depth	0.001	0.024	1	0.73	-0.39	0.52	0.57	0.65
Salinity	0.1	0.51	0.0009	1	-0.32	0.38	0.40	0.52
Temp	0.004	0.0001	0.12	0.21	1	-0.44	-0.56	-0.47
Days,10	0.036	0.16	0.034	0.14	0.080	1	0.97	0.82
Days,7	0.015	0.04	0.016	0.12	0.019	0.0001	1	0.89
Days,5	0.008	0.12	0.0047	0.031	0.058	0.0001	0.0001	1.00

From this, it may be concluded that the variables causing a low mean SPM value (negative correlations) appear with the strongest correlations rather than the factors that one would assume would create high SPM values. The exception from this is temperature in model B (table 6.9).

3. **Highest r^2 and unexplained residual (R).** The models based on mean SPM values can statistically explain about 80% of the variability among these sites/events. This means that the unexplained residual term of 20% ($R = 1 - r^2$) includes all other factors that could potentially influence the mean SPM value including the inherent uncertainty in the mean SPM values. These results confirm the working hypothesis: "It is likely that no single factor could explain the variation in SPM" among the sites.

4. **Scatter plots.** Fig. 6.39 gives pair wise scatter plots (and statistics) for selected relationships. There is a negative relationship between mean SPM and water depth (fig. 6.39A; $r^2 = 0.50$, p = 0.0015), SPM and salinity ($r^2 = 0.22$, p = 0.059), SPM and fetch ($r^2 = 0.61$, p = 0.0002) and SPM vs days with wind speeds lower than 10 m/s and 5 m/s (figures E and F; $r^2 = 0.29$ and 0.53, respectively). Figure C illustrates the positive relationship between mean SPM and mean temperature (for each site/event; $r^2 = 0.59$, p = 0.0003). Evidently, there is a considerable scatter around all the regression lines since no single variable explains more than 61% of the variability. As already stressed, there are also marked correlations among several of the x variables (table 6.8).

Table 6.9. Regression models for variations in mean values for suspended particulate matter (SPM) among sites (n = 17 sampling situations for 5 sites). y variable = log(SPM). F > 4 for the two first steps and F = 1 for the third steps; F = degrees of statistical freedom.

A. Equations when the fetch is included

Step	x variable	r^2	Regression
1	Fetch^0.5	0.61	$y=-0.0088 \cdot x_1+0.77$
2	(Days,5)^1.2	0.79	$y=-0.0064 \cdot x_1-0.0645 \cdot x_2+0.79$
3	log(Salinity)	0.81	$y=-0.0066 \cdot x_1-0.0521 \cdot x_2-1.294 \cdot x_3+1.75$

B. Equations when the fetch is omitted

Step	x variable	r^2	Regression
1	log(Temp)	0.59	$y=1.17 \cdot x_1-0.77$
2	log(Depth)	0.78	$y=0.89 \cdot x_1-0.636 \cdot x_2-0.59$
3	(Days,5)^1.3	0.80	$y=-0.764 \cdot x_1-0.524 \cdot x_2-0.0221 \cdot x_3+0.569$

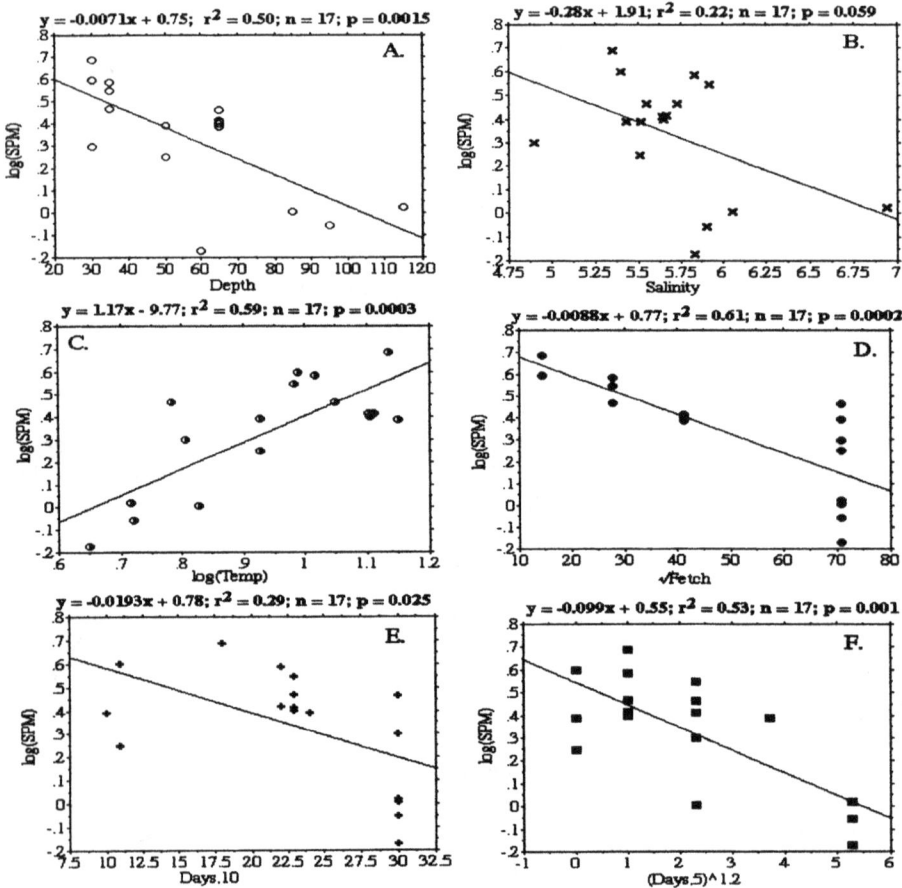

Fig. 6.39. Regressions (regression line, r^2 value, n = number of data, and p = statistical certainty) between mean SPM [log(SPM) on the y axis] and all the x variables describing the variability among the given sampling sites, (A) water depth, (B) mean salinity, (C) mean water temperature [log(Temp)], (D) fetch, (E) number of days with wind velocities lower than 10 m/s prior to the sampling and (F) number of days with wind speeds lower than 5 m/s (transformed as [Days,5)^1.2].

6.4.3.2. Variations within sites/events

The same statistical procedures will be used in this section to analyze the variations for individual data within the sites, i.e., SPM vs temperature, salinity, and distance from the bottom under stratified conditions.

1. **Correlation matrix**. The correlation matrix for untransformed values of mean SPM and the three x variables is given in table 6.10. One can note that the highest r values appear between SPM and

    (1) the distance from the bottom ($r = -0.30$; $p = 0.0001$) – the SPM values are higher close to the bottom within the sites. It is interesting to note that in a given vertical profile, the highest SPM values are likely to appear close to the bottom and not near the surface and, on the other hand, that samples from shallow sites close to the coast are likely to have higher mean SPM values than samples from deep pelagic sites;

    (2) there is also a positive ($r = 0.26$, $p = 0.0002$) relationships between temperature and SPM - the warmer the water the lower SPM;

    (3) within the sites, there is also a weak negative relationship between salinity and SPM – the saltier the water the higher the SPM values;

    (4) there is a rather strong and expected negative relationship between temperature and salinity ($r = -0.45$).

2. **Stepwise multiple regressions**. Table 6.11 gives the results for log(SPM) vs the given three variables. One can note:

    - After three steps, the r^2 value is 0.53. That is, 76% of the variation that one can expect to explain ($r^2/r_r^2 = 0.53/0.70$) is explained by this model for the individual SPM data. The steps are, water temperature ($r^2 = 0.39$), distance from the bottom ($r^2 = 0.50$), and salinity ($r^2 = 0.53$). The third factor, salinity, increases r^2, but not very much.

The results for the variations within the sites have been used to derive a model based on dimensionless moderators for how these three factors are likely to affect SPM values at different sites/events individually (see Håkanson and Peters, 1995, for more information about modelling using dimensionless moderators) and to expand the model for the mean values ($\approx$ mixed conditions) so that it applies also for stratified conditions. The scatter plots in fig. 6.40 have been used to derive the dimensionless moderators.

Table 6.10. Correlations within sites (n = 200). The correlation matrix shows linear correlation coefficients (r) and statistical certainties (p) for the actual values (untransformed values).

p \r	SPM	Distance	Salinity	Temp
SPM	1	-0.30	-0.12	0.26
Distance	0.0001	1	-0.29	0.14
Salinity	0.09	0.0001	1	-0.45
Temp	0.0002	0.041	0.0001	1

Table 6.11. Regression models for variations in individual values within sites for SPM within sites (n = 200). y variable = log(SPM); F > 4.

Step	x variable	r^2	Regression
1	log(Temp)	0.39	$y=1.097 \cdot x_1 - 0.782$
2	log(1+Distance)	0.50	$y=1.181 \cdot x_1 - 0.335 \cdot x_2 - 0.389$
3	Salinity^1.25	0.53	$y=1.039 \cdot x_1 - 0.384 \cdot x_2 - 0.067 \cdot x_3 + 0.409$

Table 6.12. Definition of dimensionless moderators quantifying how variations within sites in temperature, salinity, and distance from the bottom are likely to influence individual SPM values. The calibration constant for the moderators is 0.45.

A. Temperature

		SPM regression	SPM moderator	
Temp. max.:17.9		4.0	1.83	SPM regression = $10^{\wedge}(1.1 \cdot \log(\text{Temp})-0.78)$
Temp. min.: 2.7		0.50	0.51	Moderator for temperature:
	SPM factor 8		$3.57=0.45 \cdot 8$	$Y_{Temp}=(1+1.9 \cdot (\log(2+\text{Temp})/\log(8)-1)$

B. Salinity

		SPM regression	SPM moderator	
Salinity, max.:	4.6	3.0	2.70	SPM regression = $10^{\wedge}(-0.115 \cdot \text{Sal}^{\wedge}1.25+1.25)$
Salinity, min.:	8.1	0.47	0.95	Moderator for salinity: If salinity < 8 then
	SPM factor 6.38		$2.85 = 0.45 \cdot 6.38$	$Y_{Sal}=(1-3.4 \cdot (\text{Sal}^{\wedge}1.25/8^{\wedge}1.25-1))$ else
				$Y_{Sal}=(1-0.4 \cdot (\text{Sal}^{\wedge}1.25/8^{\wedge}1.25-1))$

C. Distance from the bottom

		SPM regression	SPM moderator	
Distance, max.:	1	3.15	1.12	SPM regression = $10^{\wedge}(-0.24 \cdot \log(1+D)+0.57)$
Distance, min.:	120	1.18	0.91	Moderator for distance (D):
	SPM factor 2.68		$1.22=0.45 \cdot 2.68$	$Y_D=(1-0.15 \cdot (\log(1+D)/\log(1+20)-1))$

Table 6.13. Empirical data (mean values from 4 stations/events) used to calibrate the model (high and low SPM values).

	High SPM = 11 (mg/l)	Low SPM=(mg/l)
Fetch (km)	391	5000
Water depth (m)	32	108
Days, 5	1	4
Salinity (%)	5.8	7.3
Temperature (°C)	8.9	4.7

Fig. 6.40A illustrates the positive relationship between SPM and temperature ($r^2 = 0.39$, $p < 0.0001$). From this figure, one can note that the range in the individual temperatures values is from 2.7 to 17.9°C. In this range, the regression line shows that, on average, SPM varies from 0.5 to 4 mg/l, i.e., with a factor of 8. However, as stressed before, temperature covaries with several variables, and so does the distance from the bottom, which also influences SPM (fig. 6.40B) and salinity (fig. 6.40C). For this model derivation, it has been assumed that the relative range in how temperature, distance from the bottom and salinity influence SPM is given by the regression lines in fig. 6.40. A calibration constant has been sought, which, in a simple manner, accounts for the fact that there is a significant covariation among these three variables and the variables already accounted for in the models in table 6.11 (for mixed conditions). To find the best possible value for this calibration constant, an analysis related to the boundary conditions of the model has been carried out. This means that the empirical data have been used to determine mean values of the fetch, water depth, days with wind speeds lower than 5 m/s (= Days,5), salinity, temperature, and SPM for a situation with high SPM values and for a situation with low SPM values. Different values for the calibration constant have been tested. The constant operates on the factor expressing how a temperature change from 2.7 to 17.9°C influences SPM. If the temperature did not covary with any variable, this factor is 8 (table 6.12). If the calibration constant is set to 0.5, the factor is 0.5·8 = 4, and changes in temperature influence SPM values less because temperature influences have already been incorporated in the model for mixed conditions since temperature covaries both with the fetch ($r = -0.66$; table 6.8), Days,5 ($r = -0.47$), salinity ($r = -0.32$), and water depth ($r = -0.39$). Temperature is also the first model variable in the second model in table 6.12B.

Different values for the calibration constant (from 0.1 to 1) have been tested (table 6.13 gives data used in the calibrations). If the constant is set to 0.45, and the dimensionless moderators duly adjusted, one gets the best fit between the empirical data, i.e., SPM = 11 mg/l for the conditions causing a high SPM values and SPM = 1 mg/l for the conditions causing a low SPM value, and the model predictions. This is shown in fig. 6.41. The predictions should be valid for surface water conditions during stratified periods. There are two basic models for mixed conditions (table 6.12), the alternative with fetch as x variable and the alternative without fetch when the mean temperature is the first model variable (see also fig. 6.42, which gives an outline of the model). These two models do not produce identical SPM values (which should not be expected), but the values are fairly close (fig. 6.41) and if one uses a calibration constant of 0.45, the empirical data are in between the values predicted by these two approaches, both for the high SPM value (fig. 6.41A) and the low (fig. 6.41B). This also means that one can define the amplitude values of the three requested dimensionless moderators (table 6.12).

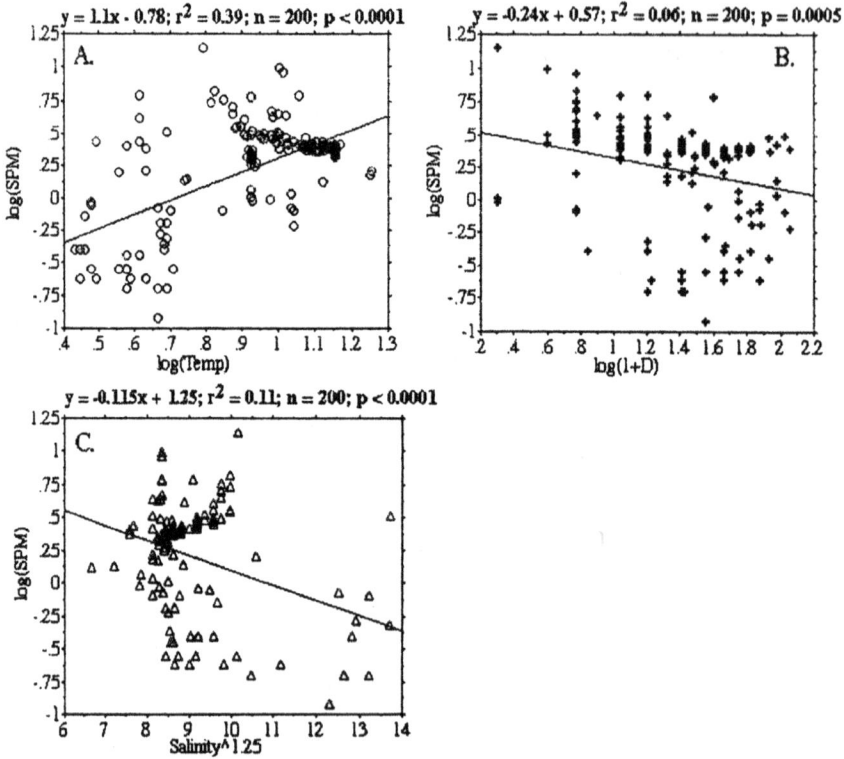

Fig. 6.40. Regressions (regression line, r^2 value, n = number of data, and p = statistical certainty) between SPM [log(SPM) on the y axis] and all the three variables, (A) temperature [log(Temp)], distance from the bottom [log(1+D)], and salinity (Salinity^1.25) characterizing the variability within given sampling sites.

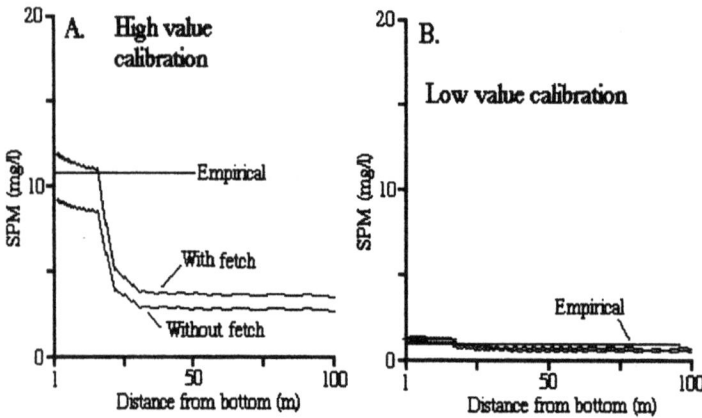

Fig. 6.41. Calibration of the dimensionless moderators for SPM in surface water under stratified conditions. This figure gives the best results using a value of 0.45 for the calibration constant.

The SPM model for the Baltic

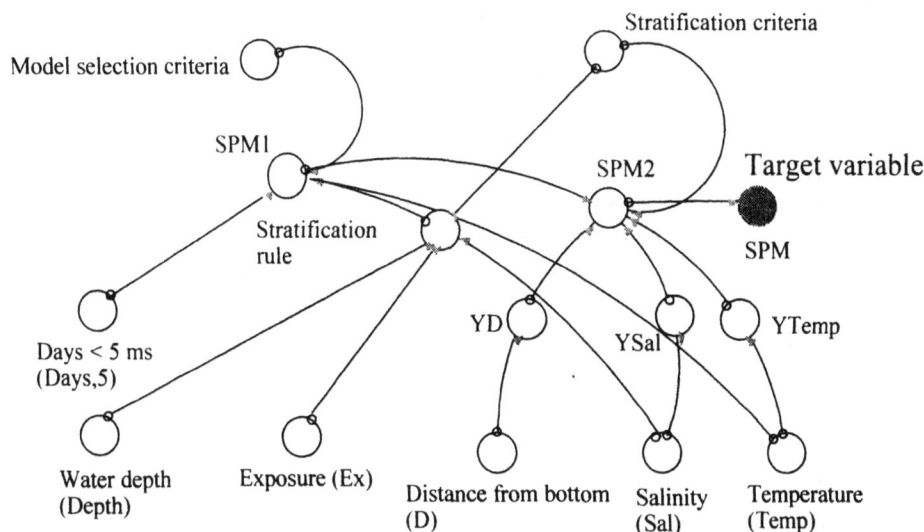

Fig. 6.42. An outline of the model for SPM.

This means that the factor for temperature influences on SPM under stratified conditions should not be given by a factor of 8 but by 0.45·8 = 3.6. This is described by the following dimensionless moderator:

$$Y_{Temp} = (1+ 1.9 \cdot (\log(2+Temp)/\log(8) - 1)) \tag{6.22}$$

where the normal temperature is set to 6°C [or rather to (2+6); log(8)]; the temperature factor is set to (2 + Temp) and not to Temp since log(Temp) is not defined if the temperature is zero; the amplitude value is set to 1.9. This means that if the actual temperature is 2.7°C, Y_{Temp} is 0.51, if the actual temperature is 17.9°C, Y_{Temp} is 1.83 and the factor is 1.83/0.51 = 3.6, as requested (see also table 6.12).

The relationship between salinity and SPM is shown in fig. 6.40C ($r^2 = 0.11$, $p < 0.0001$). It is rather weak but highly significant. One can note that if the salinity varies from 4.6 to 8.1, SPM is likely to vary with a factor of 6.4 (table 6.12), but since there are also marked correlations between salinity and temperature (table 6.10) and between salinity and distance from the bottom, also this factor is reduced by 0.45. This can then be quantified by the following dimensionless moderators:
If the salinity < 8%, the moderator is given by:

$$Y_{Sal} = (1 - 3.4 \cdot (Sal^{1.25}/8^{1.25} - 1)) \text{ else } Y_{Sal} = (1 - 0.4 \cdot (Sal^{1.25}/8^{1.25} - 1)) \quad (6.23)$$

where the normal salinity is set to 8% ($8^{1.25}$); the salinity factor is $Sal^{1.25}$; and the amplitude value in the low salinity range tested in this study is set to 3.4; the amplitude value in a high salinity range is assumed to be 0.4, but this is just a hypothesis. The main motive for this is that if one would keep the amplitude value of 3.4 in the entire salinity range (up to 30%), this would produce unrealistically low SPM values in highly saline waters. This means that if the actual salinity is 8.1%, Y_{Sal} is 0.99, if the actual salinity is 4.6%, Y_{Sal} is 2.7, and the factor 2.7/0.95 is 2.85, as requested.

The relationship between the distance from the bottom (D in m) and SPM is shown in fig. 6.40B ($r^2 = 0.06$, $p = 0.0005$). It is weak but significant. We will also reduce the factor expressing how changes in D are likely to influence SPM by 0.45, from 2.68 to 1.2 (table 6.12) for the same reasons as just given for salinity and temperature. This means that the dimensionless moderator is given by:

$$Y_D = (1 - 0.15 \cdot (\log(1+D)/\log(21) - 1)) \quad (6.24)$$

where the normal distance is set to 20 m [log(1+20)]; the factor for distance from the bottom is set to (1 + D) since D can approach zero and log(0) is not defined; and the amplitude value is set to 0.15. This gives the requested factor of 1.2 (see also table 6.12).

6.4.3.3. A statistical analysis using all data

For comparative purposes, a complementary statistical treatment using the entire dataset will be presented. Also this time, stepwise regressions of log(SPM) vs all the available variables (and not just individual data on temperature, salinity, and distance from the bottom as in table 6.11) will be used. The results are given in table 6.14 and account for both variations within and among sites/events under both mixed and stratified conditions. One can note:

- After three steps, the r^2 value is 0.58. That is, 83% of the variation that one can expect to explain ($r^2/r_r^2 = 0.58/0.70$) is explained by this model. The steps are, days with wind velocities lower than 5 m/s ($r^2 = 0.45$), fetch ($r^2 = 0.56$), and temperature ($r^2 = 0.58$). Hence, generally, these are the three most important factors for SPM variations in the Baltic proper, according to these data.

Table 6.14. Regression models for variations in suspended particulate matter (SPM); n = 200; y variable = log(SPM); F > 4.

Step	x variable	r^2	Regression
1	$(Days,5)^{1.2}$	0.45	$y=-0.13 \cdot x_1 + 0.523$
2	$Fetch^{0.5}$	0.56	$y=-0.096 \cdot x_1 - 0.0072 \cdot x_2 + 0.823$
3	$log(Temp)$	0.58	$y=-0.0819 \cdot x_1 - 0.0054 \cdot x_2 + 0.356 \cdot x_3 + 0.370$

6.4.3.4. Variations within and among sites/events

Fig. 6.43 gives some interesting plots, SPM vs the relative depth (100·distance from bottom divided by the depth of the site; the relative depth is used here to be able to compare sites with different water depths), salinity and temperature, for three sites Utö, Mysingen and Laxen under stratified conditions.

One can note:

- Under different conditions SPM can vary with a significant positive, a significant negative, and in a nonsignificant manner relative to water temperature.

- This is also the case for SPM versus the relative depth.

- However, for salinity, in these examples, there is a positive relationship between SPM and salinity in spite of the fact (fig. 6.40C) that generally SPM decreases with increasing salinity.

The main conclusion is that SPM is a most variable water variable, and the information given in fig. 6.43 stresses that many factors can influence SPM variations within and among sites. This means that one cannot draw any conclusions about general patterns in SPM variations from a few investigated sites, like the ones shown in fig. 6.43. It is important to take many samples covering a wide range of conditions. This has been the goal of this study, but evidently much more can be done. It should also be stressed that this is the first study of this kind and that it is very difficult to collect SPM data, e.g., during storm events. It is also difficult to predict the weather conditions when field trips are being planned.

The questions concerning variability "within and among" is fundamental for understanding issues related to compatible and representative values of water variables (see Håkanson and Peters, 1995). If the difference in CV_{among} and CV_{within} is large, good predictive models for water chemical variables can be obtained from models based on readily available map parameters describing variations in mean conditions among sites. Fig. 6.44 gives the CV_{within} values for the four variables, distance from the bottom (median CV = 0.62), our target variable, SPM (CV = 0.67), salinity (CV = 0.07), and water temperature (CV = 0.40). CV_{within} is logically very low for salinity from this part of the Baltic, and CV_{within} for SPM is higher than the value for water temperature.

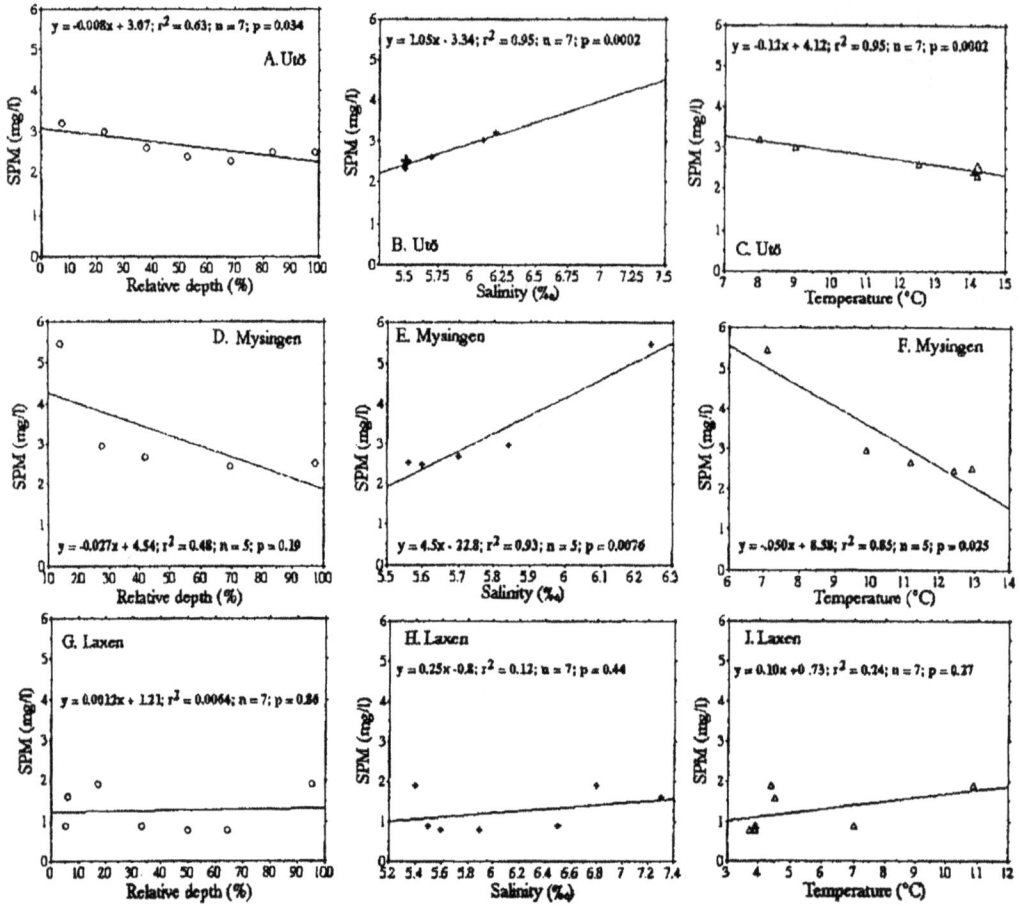

Fig. 6.43. Regressions (regression line, r^2 value, n = number of data, and p = statistical certainty) between SPM and relative depth (D_{max} = 100%), salinity and water temperature for the three stations Utö (figures A, B and C), Mysingen (figures D, E and F) and Laxen (figures (G, H and I) under stratified conditions.

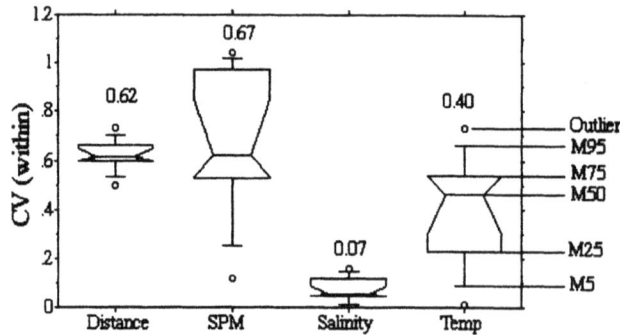

Fig. 6.44. Variations (in CV) in distance from the bottom, SPM, salinity, and temperature within the sites. The mean values are given for each parameter. The box-and-whisker plots give median values, quartiles, percentiles, and ouliers.

Corresponding information is given in fig. 6.45 for CV_{among}. Again, one can note the very high CVs for SPM and the low CVs for salinity (mean = MV and median = M50 values).

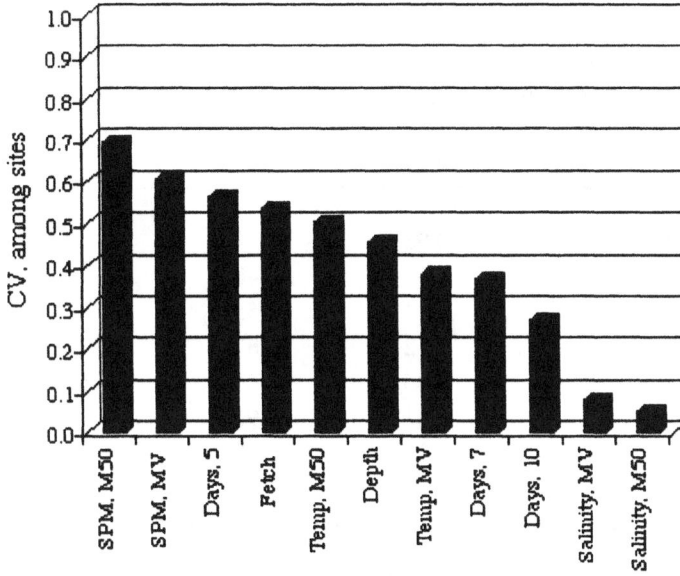

Fig. 6.45. Compilation of coefficients of variation (CV) for all studied variables for variations among the sites from all variables. SPM, M50 = median value; SPM, MV = mean value; Depth = the water depth at the sample site, Days, 5 = number of days with lower wind velocities than 5 m/s; etc.

The ratios CV_{among}/CV_{within} are given in table 6.15 for temperature, SPM, and salinity and for comparative purposes also similar information for temperature, pH, total P concentration, Secchi depth, color, hardness and conductivity for lake variables. Table 6.15B also gives the highest r^2 values for models based on map parameters. One can note that when the ratio CV_{among}/CV_{within} is high, as is it for lake conductivity, models yielding very high r^2 values have been presented, and vice versa. These principles are general and apply also to water variables from marine and brackish environments. Since the ratio CV_{among}/CV_{within} is close to 1 for SPM, one can never expect to predict SPM variations very well from models based on just map variables, one must also include, e.g., climatological variables as model variables. This conclusion can be drawn from the results given in table 6.15, and it also agrees with the results previously presented.

Table 6.15. Compilation of data on the ratio CV (among) to CV (within), the CV ratio, for (A) data on temperature, suspended particulate matter (SPM) and salinity from this study and (B) for comparative purposes corresponding data for the following lake variables, temperature, pH, total P concentration, Secchi depth, color, hardness (= Ca+Mg concentrations) and conductivity (data from Håkanson and Peters, 1995). The table also gives r^2 values for best regression model based on map variables (describing catchment area and lake morphometric characteristics).

A. Baltic data, this study		CV ratio	
Temperature	0.50/0.40	1.25	
SPM	0.69/0.67	1.03	
Salinity	0.05/0.07	0.71	
B. Lake data			r^2
Temperature		0.98	-
pH		1.55	0.40
Total P		1.91	0.55
Secchi depth		2.12	0.67
Color		2.45	0.66
Hardness		2.51	0.90
Conductivity		3.26	0.89

The model for SPM presented in the next section is based on these results.

6.4.3.5. The SPM model for open water areas

Fig. 6.42 gave a compilation of the model.

1. First, mean SPM values for given sites/events are calculated under mixed conditions using the models presented in table 6.9. If data on the fetch are available, one can use the model in table 6.9A, if not, the model in table 6.9B. These models give r^2 values of about 0.80, which is high for SPM.

2. Then, for stratified conditions, the dimensionless moderators presented in table 6.12 are used.

3. The following criteria have been applied to define stratified conditions:

   • If the difference between the surface water salinity and the deep water salinity < 0.5%, then the water is mixed.

   • If the difference between the surface water temperature and the deep water temperature < 5°C, the water is mixed.

   • Otherwise, the water is stratified.

Fig. 6.46 illustrates how the model predicts SPM values under stratified (A and B) and mixed (C, D, E, and F) conditions. For these simulations, the following default conditions have been defined (see also fig. 6.46A): The water depth of the site is 50 m, there have been 3 calm days (wind speeds lower than 5 m/s) prior to the sampling, the fetch is 1000 km, the distance from the bottom is 20 m (except in figures A and B when this distance is varied), the salinity is 6% and the temperature is 6°C.

1. One can note from figures 6.46A and 6.46D, that close to tributaries, at low salinities, one would normally expect high SPM values, but this also depends, e.g., on the distance from the bottom under stratified conditions, and the number of days with calm weather under mixed conditions.

2. At higher temperatures (fig. 6.46B and fig. 6.46E), one can expect higher SPM values, but it depends, e.g., on the stratification and the distance from the bottom.

3. If the fetch is large, SPM is likely to be low, but it depends on, e.g., the number of days with calm conditions (fig. 6.46C).

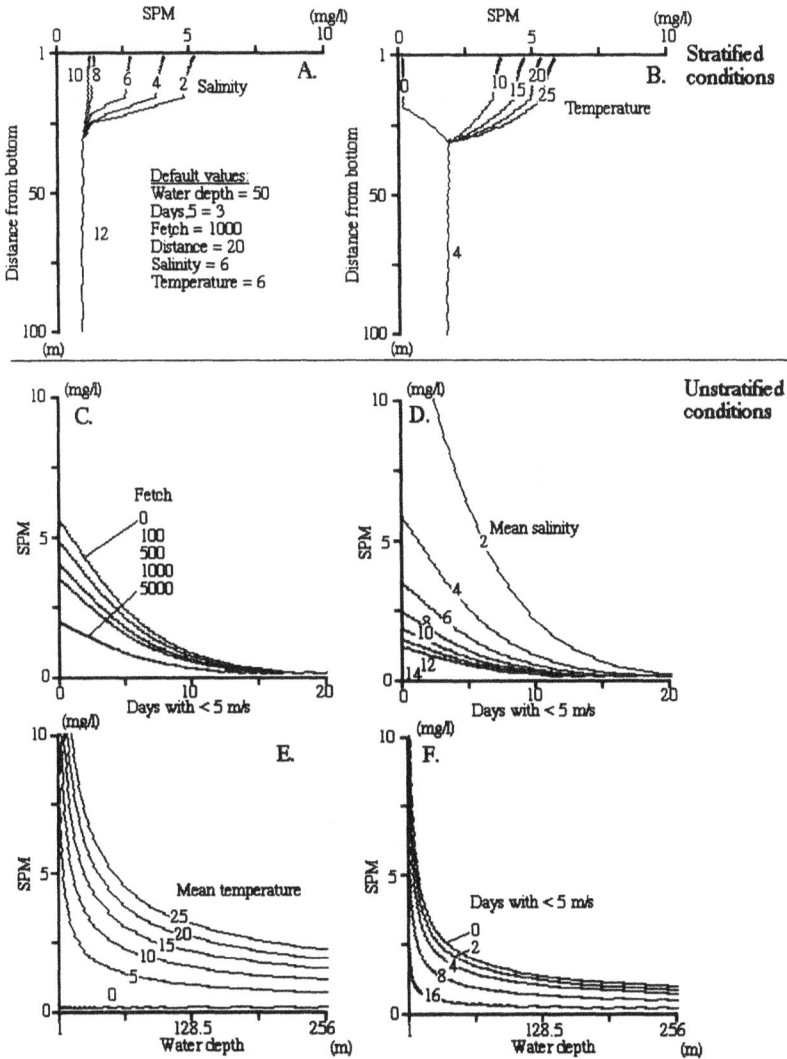

Fig. 6.46. Illustration of how variations in salinity (A) and temperature (B) under stratified conditions influence SPM values at different distances from the bottom, and how variations in the fetch (C) and mean salinity (D) under mixed conditions using the model including the fetch when the number of days with wind speeds lower than 5 m/s are also varied to influence SPM values and how variations in (E) mean temperature and (F) days with winds less than 5 m/s influence SPM values among stations with different mean depth using the model without the fetch.

6.4.4. Comments

Fig. 1.6 in chapter 1 puts these results into a more general context related to typical variations in SPM and everything related to SPM, like particulate phases of nutrients, metals, radionuclides, and organic toxins concerning the concepts "sediment focusing" and "coastal focusing." SPM (and related substances) can be expected to appear with high values in coastal areas with high river input where SPM is also added by coastal currents and a high coastal production.

The results presented here are based on statistical analyses of measured data from defined individual sites/events. Fig. 1.7 illustrated that these statistical results may be put into a more dynamic and mechanistic context where the processes influence SPM fluxes are quantified so that the measured SPM concentrations may be explain causally. The statistical analyses given in this section concern factors such as temperature, fetch, water depth, distance from the bottom, salinity, and winds (days with wind velocities lower than 5 m/s prior to the sampling). All of these factors can be related to the processes influencing SPM in coastal areas, as indicated in fig. 1.7.

This section has demonstrated that the characteristic CV for individual SPM samples in open marine sites in the Baltic Sea is 0.67, a very high value, which restricts the predictive power of any model targeting on SPM. It will also influence the predictive power of models using SPM as a model variable to predict other important variables.

Bioturbation, fish movements (Meijer et al., 1990), currents (Lemmin and Imboden, 1987), and slope processes (Håkanson and Jansson, 1983) might all influence the SPM concentration and its variation among and within sites as well as boat traffic, trawling, and dredging (Weyhenmeyer, 1998). These variables have not, however, been tested in this context. The aim here was to illustrate how much of the variations in SPM that can be statistically explained by the factors included in this study. The residual uncertainty ($R = r_r^2 - r^2$) using the r_r^2 value given in table 3.1 based on the inherent uncertainty in the individual SPM data and the empirical model given in table 6.14 is 12% ($0.12 = 0.70 - 0.58$). Other factors not accounted for in this study would have to be attributed to accounted for this residual uncertainty.

The data discussed in this section have been used to identify and rank the factors influencing the variability in SPM among and within sites and a predictive model for SPM in the open Baltic proper has been introduced. An important prerequisite for this model was that the driving variables should be easily accessed from standard monitoring programs and/or maps.

Previous knowledge regarding the SPM concentration, its variation and the factors influencing variations among and within sites was very limited and a value of 3 mg/l (from Pustelnikov, 1977) was often used for the Baltic Sea. The results discussed here represent a step forward in understanding and predicting SPM in the Baltic Sea and also in other similar systems. Evidently, it would have been preferable to have access to an even larger database, but it is very demanding (in terms of costs, manpower, ships, etc.) to collect such data.

7. Epilogue

The book has presented several databases on SPM and variables that can explain SPM variations within and among sites in lakes, rivers, and marine areas. Many statistical/empirical models and dynamical models have also been presented. This should have clarified the present state-of-the-art and also where there are major gaps in data and knowledge. Today, there is limited information on SPM and covariables from saline lakes and from many types of marine systems. Hopefully, the gaps may become smaller in the future. Then it is probable that many of the results, algorithms, and approaches discussed and presented in this book may have to be modified. One cannot falsify results from empirical models by new empirical data, but new empirical data may improve the description of the model domain and the boundary conditions when and where the model applies. Equations in dynamic model, on the other hand, may be falsified by new empirical data. Basically, there is only one avenue to increase the knowledge about how aquatic systems work, and that is by collecting new and better data. So, if this book could help the argument that more data on SPM and covariables are needed and should be integral parts of all monitoring programs, that would be a very positive result. However, one must stress that since SPM is such a "variable variable," many data are needed on SPM and since SPM can be related to allochthonous sources, autochthonous production, and internal loading, one also needs to collect data on many "variable covariables," such as tributary inflow, primary production, temperatures, and salinity.

The predictive power of dynamic ecosystem models is determined by the model structure and the equations and model constants used for the various transport processes. If there is an error in the quantification of an important flux in a model, this error has to be corrected by making at least one more compensatory error in order to calculate the target y variable well in a given system. Any model can be tuned so that it describes empirical data well in a given system. However, errors in models are often - if not always - revealed when models are blind tested against independent data from new systems. So, validations are fundamental in ecosystem modelling in disclosing deficiencies in models, and hence also in the modeller's understanding of how natural systems work. There are at least four basic criteria by which dynamic ecosystem models can be critically evaluated:

1. By the predictive power revealed by validations;

2. By the relevance of the target y variable in disclosing fundamental ecosystem structures, functional aspects of aquatic ecosystems and threshold values related to operationally applied guidelines in water management;

3. By the applicability and generality of the model, i.e., by the width of the model domain, and;

4. By the accessibility of the driving variables needed to make simulations.

Evidently, there exist very many models for lakes, rivers, and marine systems, including box models of the kind discussed in this book. At a first glance, such models may look the same, but there can also be fundamental differences between seemingly similar models because the basic structures, the equations, and the model constants may be different. To the best of the author's knowledge, no other models use the same sedimentological criteria as the models discussed in this book to define the fundamental model structures, i.e., the surface water compartment, the deep water compartment, the sediment compartment for ET areas (where there is resuspension), and the accumulation area compartment (where there is no wind/wave-induced resuspension). This also means that all key transport processes, such as sedimentation, resuspension, mixing, mineralization, and outflow, are quantified differently in this modelling approach compared with other models. All approaches to quantify these transport processes cannot be best or most relevant from a mechanistic point of view. Such a ranking of models cannot be done by arguments, only from critical validations using reliable empirical data from a wide domain of systems. The author knows of no dynamic models that provide seasonal variations for SPM - or for nutrients or water pollutants for lakes, rivers, or coastal areas - based on other structures than those discussed in this book that have been validated over such wide domains and given results even close to what has been reported here for SPM.

It is not as interesting to study how different models work as it is to study how different natural systems work. So, the latter aspect has been in focus for the comparative studies discussed in this book, and not the former.

There are a few key words that may be used to characterize the approach taken and advocated in this book:

Comparative studies. If the scientific task is to gain better understanding about how aquatic systems work, there are few more rewarding avenues than comparative studies. But comparative studies are based on data collected at individual sites, so there is no contradiction in studies based on different scale perspectives. This book is meant as a comparative study between lakes, rivers, and marine areas; this perspective is not so common. It is also a very interesting perspective. If this book could help to minimize the demarcation lines between academic subjects such as limnology, hydrology, marine ecology, environmental science, then something would have been gained.

Process-oriented mass-balances. This book has presented many statistical/empirical models, which in themselves may reveal very little about processes. However, such models are excellent tools to rank x variables influencing variations in target y variables, and in this way they can provide invaluable information in building practical and operational process-oriented mass-balance models. In such models, statistical explanation may be transformed into mechanistic explanation. Then, the aim is not to account for "everything," but to try to find and quantify the most important transport processes and omit or simplify the smaller processes. This is far easier said than done. But it can be done with the help of statistical/empirical modelling approaches, as demonstrated in this book. To find the optimal temporal and spatial scales for such dynamic process-oriented models is an important task, and also to find the most relevant modelling structure. The dynamic SPM models for lakes, rivers, and coastal areas presented in this work are all based on the same basic structure. The criteria to define the compartments, such as surface water, deep water, areas of erosion and transport, and accumulations areas, are general and from these basic building blocks, one can also define algorithms for the key processes, such as sedimentation, resuspension, mixing, mineralization, production, inflow, and outflow of SPM.

Ecosystem perspective. All models presented in this book relate to the ecosystem scale, i.e., they are basically intended for entire lakes, river stretches, and defined coastal areas and for time periods of one month. This is also a very important perspective in water management, e.g., in contexts of impact assessment, when remedial measures are discussed, and when very basic questions are asked, e.g.: What is the status on this ecosystem? What can be done to improve the conditions? Few people would be interested in the content of a sampling bottle. Most of us are interested in what this content may actually represent. That is, we are interested in a larger entity, the ecosystem. But there is no contradiction between work at this larger ecosystem scales and sampling and work at smaller scales, since the mean values characterizing ecosystem conditions and the standard deviations characterizing the variability around such mean values of necessity must emanate from sampling at individual sites.

Practical usefulness. "Everything should be as simple as possible, not simpler." This statement from Albert Einstein is also a key to obtain practical usefulness of models for aquatic ecosystems. A very important demand for all models presented in this book is that they should be practically useful, which implies that the obligatory driving variables should be easily accessed. The models should also predict well. Both the empirical models and the dynamic models for SPM discussed in this book may be driven by readily accessible data from standard maps and monitoring programs, e.g., altitude, latitude, continentality, area, mean depth, and max. depth. To calculate SPM from autochthonous production, one also needs data on chlorophyll. Meteorological data on winds and light conditions and oceanographic data on directions and speeds of currents and temperatures at individual sites have been omitted, since the models discussed in this book focus on monthly predictions at the ecosystem scale and during a period of one month, winds can blow with many speeds from many directions.

Predictive power. The ultimate model testing is not sensitivity or uncertainty tests, like those discussed in chapter 3 and exemplified many times in this book, but validations, i.e., blind tests against independent data. For dynamic models at the ecosystem scale, it is very important to critically evaluate the area- and time-compatibility of the data when modelled values are compared with empirical data. Generally, one would assume that a poor fit between modelled values and empirical data can be explained by deficiencies in the model. But empirical data are, in fact, also always uncertain. This means that it is important to control model predictions against uncertainty bands for the empirical data. This has been stressed several times in this book, and the examples given for Lake Balaton in chapter 4 should be informative on this important matter. The dynamic models for SPM in lakes, rivers, and coastal areas have been critically tested and demonstrated to give good predictive power. However, this does not mean that they will predict equally well for all systems. These models are meant to account for defined processes and factors in a general way so that characteristics SPM values can be predicted. Evidently, there may be situations which are not normal, but abnormal. Then, this modelling can provide a reference value, which would express normal conditions so that the divergency from the normal can be quantified and maybe related to the factor causing the abnormal conditions.

Pluralism. This may seem like a strange word since there is a general common base for all the empirical models and for all the dynamic models discussed in this book. But aquatic systems are very complex. This means that in order for science to progress concerning such systems, it is important to allow and stimulate different approaches, different models, and pluralism. The modelling approaches discussed in this book, and the scales and structures of these models, provide only one of many important pieces to the puzzle of SPM in aquatic ecosystems. It has been argued that "a true liberal would be willing to sacrifice his/her life to give the opponent a chance to express his/her view." That may be too much, but the statement reflects a beautiful idea that also has bearings to science. Imagine if members of different scientific schools could take that attitude! The argument here is that it would benefit science if competing and complementary approaches can be supported - not suppressed.

Table 7.1, finally, gives a few statements related to ecosystem models and modelling.

Table 7.1. The "ten commandments of aquatic ecosystem models", as modified from Håkanson and Peters (1995).

- Understanding at one scale often refers to processes at the next lower scale. This can continue down to the level of the atom and beyond. So, "understanding" is a difficult concept in ecosystem contexts, just like "holism."

- If the aim is to quantify, rank, predict, and simulate, there are few, if any, alternative approaches to models in complex ecosystems. "Verbal models," "qualitative explanations," and "logical reasoning" often deteriorate into "Environmental theology," rather than environmental science. Models are tools to analyze data in a structured manner.

- Big models are often "prescriptive" or "descriptive"; not predictive. Big models may look more objective than small, but this may be self-deception. Their complexity hides the deception as a deodorant hides a bad smell.

- Models are built and validated with empirical data. Empiricism enters at all steps from start to finish in ecosystem modelling. But empirical data, and any knowledge based on empirical data, are uncertain. Accumulated uncertainties in the models will cause accumulated uncertainties in model predictions.

- The predictive power of a model is not governed by the strength of the model's strongest part, but by the weakness of its weakest part.

- Big models are simple to build, but hard to validate. Small models are hard to build, and simple to validate.

- Small size is necessary, but not sufficient, for utility and predictive power; so useful models must be small. Small models should be based on the most fundamental processes, but that is far easier to say than it is to accomplish.

- Scientific knowledge does not lie in the model alone, nor in the empirical data alone, but in their overlap as validated, predictive models.

- The key issue is not to verify, but to falsify a model, and thereby determine its limitations.

- It is important to predict mean values, but it is equally important to predict the confidence interval around the mean and to give confidence intervals for the empirical data.

---

8. Literature References

Aalderink R.H., Lijklema L., Breukelman J., van Raaphorst W. and Brinkman A.G. 1984. Quantification of wind induced resuspension in a shallow lake. Wat. Sci. Tech., 17:903-914.

Abdel-Moati, M.A.R. 1997. Industrial dumping impact on oxygen and nitrogen fluxes in Abu Qir Bay, southeastern Mediterranean Sea. Environment International, 23(3):349-357.

Abrahamsson, O. and Håkanson, L. 1998. Modelling seasonal flow variability of European rivers. Ecol. Modelling, 114:49-58.

Abrosov, V.N. 1982. Zonal types of limnogenesis. Leningrad (in Russian).

Aertbjerg, G. (ed.), 2001. Eutrophication in Europe's coastal waters. European Environment Agency, Topic report 7/2002, Copenhagen, 86 p.

Ahl, T. 1979. Undersökningar i Kolbäcksåns vattensystem. V. Fysikalisk-kemiska undersökningar 1978. Naturvårdsverkets limnologiska undersökning (in Swedish).

Ahl, T. 1980. Undersökningar i Kolbäcksåns vattensystem. 1979 års undersökningar. Naturvårdsverket Rapport, SNV PM 1287 (in Swedish).

Ahl, T. 1981. Undersökningar i Kolbäcksåns vattensystem. XI. Fysikaliska-kemiska undersökningar 1980. Naturvårdsverket Rapport, SNV PM 1406 (in Swedish).

Ahlgren, G. 1970. Limnological studies of Lake Norrviken, an eutrophicated Swedish lake. II. Phytoplankton and its production. Schweiz. Z. Hydrol., 32:353-396.

Ahlgren, I. 1973. Limnologiska studier av sjön Norrviken. III. Avlastningens effekter. Scripta Limnologica Upsaliensia. Coll. 333.

Aiken, G.R., McKnight, D.M., Werskaw, R.L. and MacCarthy, P. 1985 (eds). Humic substances in soil, sediment and water. New York, Wiley Interscience, 692 p.

Aizaki, M., Otsuki, A., Fukushima, T., Hosomi, M. and Muraoka, K. 1981. Application of Carlson's trophic state index and other parameters. Verh. Int. Verein. Limnol., 21:675-681.

Allard, B., Borén, H. and Grimvall, A. 1991 (eds). Humic substances in aquatic and terrestrial environment. Springer, Heidelberg, 514 p.

Allen, J.R.L. 1970. Physical processes of sedimentation. London, Allen and Unwin, 248 p.

Ambio, 1976. Special issue on acid rain. Vol 5. No. 5-6.

Ambio, 1990. Special issue. Marine eutrophication, 19:102-176.

Ambio, 2000. Eutrophication and contaminants in the aquatic environment. Nr. 4-5, pp. 183-290.

Andersson, A. and Gustafsson, A. 1988. Bulk deposition of trace elements in precipitation. Swed. Univ. Agricult. Sciences, Ecohydrology 26:5.12.

Andersson, C. 2000. The influence of wind-induced resuspension on sediment accumulation rates: A study of offshore and archipelago areas in the NW Baltic proper. Master Thesis. Uppsala University.

Andreev, A., Kusakabe, M., Honda, M., Murata, A. and Saito, C. 2002. Vertical fluxes of nutrients and carbon through the halocline in the western subarctic Gyre calculated by mass balance. Deep-Sea Research II, 49:5577-5593.

Andronikova, I.N., Drabkova, V.G., Kuzmenko, K.N., Mokievskiy, K.A., Stravinskaja, E.A. and Trifonova, I.S. 1973. Production of main communities of the Red Lake and its biotic balance: Production-biological investigations of the freshwater ecosystems. Minsk, pp. 74-90 (in Russian).

Anokhina, L.E. 1999. The effect of a water body, biotope and substratum on chlorophyll "a" content and phytoperiphyton photosynthesis rate in lakes of different types: Structural-functional organization of freshwater ecosystems of different types. St. Petersburg, pp. 195-208 (in Russian).

Atkinson E. 1995. Methods for assessing sediment delivery in river systems. Hydrol. Sci., 40(2):273-280.

Baccini, P., Grieder, E., Stierli, R. and Goldberg, S. 1982. The influence of natural organic matter on the adsorption properties of mineral particles in lake water. Schweiz. Z. Hydrol., 44:99-116.

Balistrieri, L.S. and Murray, J.W. 1983. Metal-solid interactions in the marine environment: estimating apparent equilibrium binding constants. Geochim. Cosmochim. Acta, 47:1091-1098.

Balistrieri, L.S., Murray, J.W. and Paul, B. 1992. The biogeochemical cycling of trace metals in the water column of Lake Sammamish, Washington: Response to seasonally anoxic conditions. Limnol. Oceanogr., 37:510-528.

Balls, P.W. 1988. The control of trace metal concentrations in coastal seawater through partition onto particulate matter. Neth. J. Sea Res., 22:213-218.

Balls, P.W. 1989. The partition of trace metals between dissolved and particulate phases in European coastal waters: A compilation of field data and comparison with laboratory studies. Neth. J. Sea Res., 23:7-14.

Balogh, S.J., Meyer, ML. and Johnson, D.K. 1997. Mercury and suspended sediment loadings in the lower Minnesota River. Environ. Sci. Technol., 31:198-202.

Bannister, T.T. 1979. Quantitative description of steady state, nutrient-saturated algal growth, including adaptation. Limnol. Oceanogr., 24:6-96.

Beach Erosion Board, 1972. Waves in inland reservoirs. Technical Memoir 132, Beach Erosion Corps of Engineers, Washington, D. C.

Bear, J. 1979. Hydraulics of Groundwater. New York, McGraw-Hill, 567 p.

Beck, M.B. and Van Straten, G. 1983 (eds). Uncertainty, system identification and the prediction of water quality. Heidelberg, Springer, 387 p.

Beeton, A.M., Edmondson, W.T. 1972. The eutrophication problem. J. Fish. Res. Canada, 29:673-682.

Benner, R., Moran, M.A. and Hodson, R.E. 1986. Biogeochemical cycling of lignocellulosic carbon in marine and freshwater ecosystems: relative contributions of procaryotes and eucaryotes. Limnol. Oceanogr., 31:89-100.

Benoit, G. 1995. Evidence of the particle concentration effect for lead and other metals in fresh waters based on ultraclean technique analyses. Geochim. Cosmochim. Acta, 59:2677-2687.

Benoit, G., Oktay-Marshall, S.D., Cantu, A., II, Hood, E.M., Coleman, C.H., Corapcioglu, M.O. and Santschi, P.H. 1994. Partitioning of Cu, Pb, Ag, Zn, Fe, Al and Mn between filter-retained particles, colloids, and solution in six Texas estuaries. Mar. Chem., 45:307-336.

Benoit, G. and Rozan, T.F. 1999. The influence of size distribution on the particle concentration effect and trace metal partitioning in rivers. Geochim. Cosmochim. Acta, 63:113-127.

Berman, T., Yacobi, Y. and Pollingher, U. 1992. Lake Kinneret phytoplankton: stability and variability during twenty years (1970-1989). Aquat. Sci., 54:104-127.

Berman, T., Stone, L., Yacobi, Y. Z., Kaplan, B., Schlichter, M., Nishri, A., and Pollingher, U. 1995. Primary production and phytoplankton in Lake Kinneret: a long term record (1972-1993). Limnol. Oceanogr., 40, 1064-1076.

Best, E.P.H. 1982. The aquatic macrophytes of Lake Vechten. Species composition, spatial distribution and production. Studies of Lake Vechten and Tjeukemeer. The Netherlands, The Hague, pp. 59-77.

Bierman, V.J. Jr. 1980. A comparison of models developed for phosphorus management in the Great Lakes. In: Loehr, C., Martin, C.S. and Rast, W. (eds.), Phosphorus management strategies for lakes. Ann Arbor, Ann Arbor Science Publishers, pp. 235-255.

Bird, D.F. and Kalff, J. 1984. Empirical relationships between bacterial abundance and chlorophyll concentration in fresh and marine waters. Can. J. Fish. Aquat. Sci., 41:1015-1023.

Birkeland, P.W. 1974. Pedology, Weathering and Geomorphological Research. New York: Oxford University Press.

Blomqvist, S. 1992. Geochemistry of coastal Baltic sediments: processes and sampling procedures. Dr. thesis, Stockholm Univ., Sweden.

Bloesch J. 1995. Mechanisms, measurement and importance of sediment resuspension in lakes. Mar. Freshwater Res., 46:295-304.

Bloesch J. 1997. The Danube River basin – The other cradle of Europe: The Limnological dimension. Ann. Europan Acad. Sci. and Arts, 34:51-77.

Bloesch, J. and Burns, N.M. 1980. A critical review of sedimentation trap technique. Schweiz. Z. Hydrol., 42:15-55.

Bloesch, J. and Uehlinger, U. 1986. Horizontal sedimentation differences in a eutrophic Swiss lake. Limnol. Oceanogr., 31:1094-1109.

Boers, P.C.M., Cappenberg, Th.E. and van Raaphorst, W. 1993 (eds). Proceeding of the Third International Workshop on Phosphorus in Sediments. Hydrobiologia, Vol. 253, 376 p.

Bortoli, A., Dell'Andrea, E., Gerotto, M., Marchiori, M., Palonta, M. and Troncon, A. 1998. Soluble and particulate metals in the Adige river. Microchim. J., 59:19-31.

Boulion, V.V. 1985a. Some regularities of primary production in lake ecosystems: The production of populations and aquatic organisms communities and methods of its study. Sverdlovsk, pp. 19-29 (in Russian).

Boulion, V.V. 1985b. Limnological studies of Mongolia. Leningrad (in Russian).

Boulion, V.V. 1994. Regularities of the primary production in limnetic ecosystems. St. Petersburg, 222 p. (in Russian).

Boulion, V.V. 1997. General characterization of some lakes in southern Karelia differing in the acidity and humic state: The response of lake ecosystems to changes in biotic and abiotic conditions. St. Petersburg, pp. 5-28 (in Russian).

Boulion, V.V. 2001. Contribution of macrophytes and phytobenthos in primary production of lake ecosystems. VIII Congress of hydrobiological society of Russian Academy of Sciences. Kaliningrad, pp. 158-159 (in Russian).

Boyle, J.F. and Birks, H.J.B. 1999. Predicting heavy metal concentrations in the surface sediments of Norwegian headwater lakes from atmospheric deposition: An application of a simple sediment-water partitioning model. Wat. Air Soil Pollut., 114:27-51.

Boynton, W.R., Kemp, W.M. and Keefe, C.W. 1982. A comparative analysis of nutrients and other factors influencing estuarine phytoplankton production. In: Kennedy, V.S. (ed), Estuarine comparisons. London, Academic Press, pp. 69-90.

Brady, N.C. 1984. The Nature and Properties of Soils. London: Collier Macmillian Publishing, 750 p.

Brezonik, P.L. 1978. Effect of organic color and turbidity on Secchi disk transparency. J. Fish. Res. Board Canada, 35:1410-1416.

Brittain, J., Håkanson, L., Bergström, U. and Bjørnstad, H. E. 1994. The significance of hydrological and catchment processes for the transport and biological uptake of radionuclides in northern aquatic ecosystems. Proceedings of the 10th International Northern Research Basins Symposium and Workshop, Spitsbergen, Norway.

Brittain, J.E. and Brabrand, Å. 2001. Fish movement in rivers, lakes and estuaries in relation to contamination by radionuclides. Freshwater Ecology and Inland Fisheries Laboratory (LFI), Natural History Museums, University of Oslo, Norway.

Broberg, A. and Andersson, E. 1989. Circulation of caesium in limnic ecosystems (in Swedish). Inst. of Limnology, Uppsala Univ., 30 p.

Brylinsky, V. and Mann, K.H. 1973. An analysis of factors governing productivity in lakes and reservoirs. Limnol. Oceanogr., 18:1-14.

Burban, P.-Y., Lick, W. and Lick, J. 1989. The flocculation of fine-grained sediments in estuarine waters. J. Geophy. Res., 94:8223-8330.

Burban, P.-Y., Xu, Y.-J., McNeiel, J. and Lick, W. 1990. Settling speeds of flocs in fresh water and seawater. J. Geophy. Res., 95:18213-18220.

Burrough, P.A. 1986. Principles of Geographical Information Systems for Land Resources Assessment. Oxford, Clarendon Press.

Busch W.-D.N. and Sly, P.G. 1992. The development of an aquatic habitat classification system for lakes. Boca Raton, Florida, CRC Press.

Calmon, P. 2002. Estimation of consequences following an atmosheric radioactive emission in a forest ecosystem. In: Brechignac, F. and Howard, B.J. 2001 (eds), Radioactive pollutants: Impact on the environment, EDP Sciences, Les Ulis, France, pp. 403-408.

Calow, P. and Petts, G.E. 1994. (Eds). The river handbook, hydrological and ecoplogical principles. London: Blackwell Scientific Publishing, Vol. 1, 526 p. and Vol. 2, 523 p.

Canfield, D.E.JR. and Bachmann, R.E. 1981. Predication of total phosphorus concentrations, chlorophyll a, and secchi depths in natural and artificial lakes. Can. J. Fish. Aquat. Sci., 38:414-423.

Carlson, R.E. 1977. A trophic state index for lakes. Limnol. Oceanogr., 22:361-369.

Carlson, R.E. 1980. More complications in the chlorophyll - Secchi disk relationship. Limnol. Oceanogr., 25:379-382.

Carlsson, L., Persson, J. and Håkanson, L. 1999. A management model to predict seasonal variability in oxygen concentration and oxygen consumption in thermally stratified coastal waters. Ecological Modelling, 119:117-134.

Carroll, J. and Harms, I.H. 1999. Uncertainty analysis of partition coefficients in a radionuclide transport model. Wat. Res., 33:2617-2626.

Chapra, S.C. 1980. Application of the phosphorus loading concept to the Great Lakes. In: Loehr, C., Martin, C.S. and Rast, W. (eds.), Phosphorus Management Strategies for Lakes. Ann Arbor: Ann Arbor Science Publishers, pp. 135-152.

Chapra, S.C. and Reckhow, K. 1979. Expressing the phosphorus loading concept in probabalistic terms. J. Fish. Res. Bd. Can., 36:225-229.

Chapra, S.C. and Reckhow, K. 1983. Engineering approaches for lake management. Vol. 2. Mechanistic modelling. Woburn, MA, Butterworth.

Chow, V. T. 1988. Applied Hydrology. London, McGraw-Hill, Inc., 572 p.

Christman, R.F. and Gjessing, E.T. 1983 (eds). Aquatic and Terrestrial Humic Materials. New York: Ann Arbor Science, 538 p.

Cole, J.J., Likens, G.E., and Strayer, D.L. 1982. Photosynethetically produced dissolved organic carbon: An important carbon source for planktonic bacteria. Limnol. Oceanogr., 27:1080-1090.

Conan, P., Turley, C., Stutt, E., Pujo Pay, M., and Van Wambeke, F. 1999. Relationship between phytoplankton efficiency and proportion of bacterial production to primary production in the Mediterranean Sea. Aquatic Microbial Ecology, 17:131-144.

Cox, D.C. and Baybutt, P. 1981. Methods for uncertainty analysis: a comparative survey. Risk Analysis, 1:251-258.

Currie, D..J. 1990. Large-scale variability and interactions among phytoplankton, bacterioplankton, and phosphorus. Limnol. Oceanogr., 35:1437-1455.

Cuthbert, I.D. and Kalff, J. 1993. Empirical models for estimating the concentrations and exports of metals in rural rivers and streams. Wat. Air Soil Pollut., 71:205-230.

Dedkhov, A.P. and Mozzherin, V.I. 1996. Erosion and sediment yield on the Earth. International Association of Hydrological Sciences Publication, 236:29-33.

De Schmedt, F., Vuksanovic, V., Van Meerbeeck, S. and Reyns, D. 1998. A time-dependent flow model for heavy metals in the Scheldt estuary. Hydrobiologia, 366:143-155.

Dillon, P.J. and Rigler, F.H. 1974. The phosphorus-chlorophyll relationship in lakes. Limnol. Oceanogr., 19:767-773.

Dotsenko, O.H. and Raspopov, I.M. 1982. Peculiarities of macrophyte covering of north part of Gulf Big Onego. Limnological investigations of Gulf Big Onego in Lake Onega. Leningrad, pp. 114-117 (in Russian).

Douglas, R.W., Rippey, B. and Gibson, C.E. 2003. Estimation of the in-situ settling velocity of particles in lakes using a time series sediment trap. Freshwater Biology, 48:512-518.

Draper, N. R. and Smith, H. 1966. Applied Regression Analysis. New York: John Wiley and Sons, 352 p.

Dubko, N.V. 1985. The labile and stable organic matter: Ecological system of Naroch lakes. Minsk, pp. 233-237 (in Russian).

Dunne, T. and Leopold, L.B. 1978. Water in environmental planning. San Francisco, Freeman and Co., 818 p.

Duursma, E.K. and Bewers, J.M. 1986. Application of Kds in marine geochemistry and environmental assessement. In: T.H. Sibely and C. Myttenae (Editors), Application of Distribution Coefficients to Radiological Assessement Models. Essex, UK: Elsevier Applied Science Publishing Ltd., Essex, pp. 139-165.

Dyer, J.L. 1972. Estuarine hydrography and sedimentation. Cambridge: Cambridge University Press, 230 p.

Eckhéll, J., Jonsson, P., Meili, M and Carman, R. 2000. Storm influence on the accumulation and lamination of sediments in deep areas of the northwestern Baltic proper. Ambio 29:238-245.

Edwards, A.C., Creasy, J. and Cresser, M.S. 1984. The conditioons and frequency of sampling for elucidation of transport mechanisms and element budgets in upland drainage basins. pp. 187-202. In: Eriksson, E. (ed.). Hydrochemical Balances of Freshwater Systems. IAHS-AISH Publication No 150.

Efford, I.E. 1972. An interim review of the Marion Lake Project. Productivity problems of freshwaters. Warszawa-Krakow: Polish Science Publishers, pp. 89-109.

Einstein, H.A. 1950. The bed-load function for sediment transportation in open-channel flowes. U.S. Dept. Agric. Soil. Cons. Serv., T.B., No. 1026.

Ekzertsev, V.A. 1958. Production of coastal-water plants in Ivankovskoe reservoir. Bulletin of Institute of Reservoir Biology, 1:19-21 (in Russian).

Ekzertsev, V.A. and Dovbnja, I.V. 1973. About annual production of hydrophilic plants in littoral of Gorkovskoe reservoir. Turnover of matter and energy in lakes and reservoirs. Listvenichnoe on the Baikal, pp. 139-141 (in Russian).

Eliassen, A. and Saltbones, J. 1983. Modelling of long-range transport of sulphur over Europe: A two-year model run and some model experiments. Atm. Env., 17:1457-1473.

Elster, H.J. 1958. Das limnologische Seetypensystem. Ruckblick and Ausblick. Verh. Int. Verein. Limnol., 13:101-120.

Erel, Y. and Stolper, E.M. 1993. Modeling of rare-earth element partitioning between particles and solution in aquatic environments. Geochimica Cosmochimica Acta, 57:513-518.

Eriksson, E. 1974. Water, the carrier of chemicals (in Swedish). Forskning och Framsteg, 5:41-45.

Eriksson, E. 1985. Principles and Applications of Hydrochemistry. London, Chapman and Hall, 187 p.

Evans M.S., Arts M.T. and Robarts R.D. 1996. Algal productivity, algal biomass, and zooplankton biomass in a phosphorus-rich, saline lake: deviations from regression model predictions. Can. J. Fish. Aquat. Sci., 53:1048-1060.

Floderus, S. 1989. The effect of sediment resuspension on nitrogen cycling in the Kattegatt: variability in organic matter transport. Thesis, Uppsala University, UNGI Report, 71.

Floderus, S. and Håkanson, L. 1989. Resuspension, ephermal mud blankets and nitrogen cycling in Laholmsbukten, south east Kattegat. Hydrobiologia, 176/177:61-75.

Förstner, U. and Salomons, W. 1981. Trace metal analysis in polluted sediments. Delft Hydraulics Lab., Publ. No. 248, pp. 1-13.

Foster, I.D.L. 1978. A multivariate model of storm-period solute behavior. J. Hydrol., 39:339-353.

Foster, I.D.L., Baban, S.M.J., Wade, S.D., Charlesworth, S., Buckland, P.J. and Wagstaff, K. 1996. Sediment-associated phosphorus transport in the Warwickshire River Avon, UK. IAHS Publ. (Proc. Exeter Symposium), 236:303-312.

Foster, I.D.L., Baban, S.M.J., Charlesworth, S.M. Jackson, R., Wade, S., Buckland, P.J., Wagstaff, K. and Harrison, S. 1997. Nutrient concentrations and planktonic biomass (chlorophyll a) behavior in the catchment of the river Avon, Warwickshire IAHS Publ. (Proc. Rabat Symposium), 243:167-176.

Foster, I.D.L., Wade, S., Sheasby, J., Baban, S., Charlesworth, S., Jackson, R. and McIlroy, D. 1998. Eutrophication in controlled waters in Lower Severn Area. Environment Agency, Severn-Trent Region (Final Report), Vol. 1, 94 pp., Vol. 2, 163 pp., Tewkesbury, UK.

Fritz, B., Massabuau, J.C. and Ambroise, B. 1984. Physicochemical characteristics of surface waters and hydrological behavior of a small granitic basin (Visges massif, France: Annual and daily variations. pp. 249-261. In: Eriksson, E. (ed.). Hydrochemical Balances of Freshwater Systems. IAHS-AISH Publication No 150.

Gak, D.Z., Gurvich, V.V., Korelyakova, I.L., Kostikova, L.E., Konstantinova, N.A., Olivari, G.A., Primachenko, A.D., Tseeb, Ya.Ya., Vladimirova, K.S. and Zimbalewskaya L.N. 1972. Productivity of aquatic organism communities of different trophic levels in Kiev Reservoir. Productivity problems of freshwaters. Warszawa-Krakow: Polish Scientific Publishers, pp. 447-455.

Gaudie, A. (ed.) 1981. Geomorphological Techniques. London: Allen and Unwin, 395 p.

Gaudie, A. (ed.) 1990. Geomorphological Techniques 2nd Edition, London: Unwin Hyman Ltd, 570 p.

Gilbert, R.O. 1987. Statistical Methods for Environmental Pollution Monitoring. New York, Van Nostrand Reinold Co.

Golterman, H.L., Sly, P.G. and Thomas, R.L. 1983. Study of the relationship between water quality sand sediment transport. Unesco, Tech. Paper No. 26, 231 p.

Gorbunova, Z.A., Gordeeva, K.N., Gritsevskaja, G.L., Dmitrienko, Yu.S. and Ryzhkov, L.P. 1973. Biological productivity ja Chedenjarvi Lake. Production-biological investigations of freshwater ecosystems. Minsk, pp. 44-54 (in Russian).

Gorham, E., Dean, W.E. and Sanger, J.E., 1983. The chemical composition of lakes in the north-central United States. Limnol. Oceanogr., 28:287-301.

Gorham, E., Underwood, J.K., Martin, F.B. and Ogden III, J.B. 1986. Natural and antropogenic causes of lake acidification in Nova Scotia. Nature, 324:451-453.

Gray, J.R., Glysson, G.D., Turcios, L.M. and Schwarz, G.E. 2000. Comparability of suspended sediment concentrations and total suspended solids data. U.S. Geol. Survey, Report 00-4191, Reston, Virginia, 14 p.

Grip, H. 1982. Water chemistry and runoff in forest streams at Kloten. Thesis, Uppsala Univ., UNGI Report 58, 144 p.

Grip, H. and Rodhe, A. 1985. Vattnets väg från regn till bäck (The flow of water from rain to river; in Swedish). Forskningsrådens Förlagstjänst, Karlshamn, 156 p.

Gustafsson, Ö. and Gschwend, P.M. 1997. Aquatic colloid: Concepts, definitions, and current challenges. Limnol. Oceanogr., 42:519-528.

Gyllenhammar, A. 2004. Predictive modelling of aquatic ecosystems at different scales using mass-balances and GIS. Uppsala Univ., Dr. thesis.

Håkanson, L. 1977. The influence of wind, fetch, and water depth on the distribution of sediments in Lake Vänern, Sweden. Can. J. Earth Sci., 14:397-412.

Håkanson, L. 1978. Erken - morfometri. Scripta Limnologica Upsaliensia, Scriptum 468, 23 p.

Håkanson, L. 1991. Physical Geography of the Baltic. The Baltic University. Session 1. Uppsala University.

Håkanson, L. 1995. Fiskodling och miljöeffekter i sjöar - nya resultat motiverar nya bedömningsunderlag. Vatten 51, Bloms Boktryckeri, Lund, 103 sid.

Håkanson, L. 1999. Water pollution: methods and criteria to rank, model and remediate chemical threats to aquatic ecosystems. Leiden: Backhuys Publishers, 299 p.

Håkanson, L. 2000. Modelling radiocesium in lakes and coastal areas: new approaches for ecosystem modellers. A textbook with Internet support. Dordrecht, Kluwer Academic Publishers, 215 p.

Håkanson, L. 2003a. Liming as a method to remediate lakes contaminated by radiostrontium. J. Env. Radioactivity, 65:47-75.

Håkanson, L. 2003b. Quantifying burial, the transport of matter from the lake biosphere to the geosphere. Internat. Rev. Hydrobiol., 88:539-560.

Håkanson, L. 2004a. Break-through in predictive modelling opens new possibilities for aquatic ecology and management – a review. Hydrobiologia, 518:135-157.

Håkanson, L. 2004b. Lakes – form and function. Caldwell, New Jersey, The Blackburn Press, 201p.

Håkanson, L. 2004c. A new generic submodel for radionuclide fixation in large catchments from continuous and single pulse fallouts, as used in a river model. J. Env. Radioactivity, 77:247-273.

Håkanson, L. 2004d. Modelling the transport of radionuclides from land to water. J. Env. Radioactivity, 73:267-287.

Håkanson, L. 2004e. Internal loading: A new solution to an old problem in aquatic sciences. . Lakes & Reservoirs: Research & Management, 9:3-23.

Håkanson, L., Blenckner, T. and Malmaeus, J.M. 2004c. New, general methods to define the depth separating surface water from deep water, outflow and internal loading for mass-balance models for lakes. Ecol. Modelling, 175:339-352.

Håkanson, L. and Boulion, V. 2002. The Lake Foodweb - modelling predation and abiotic/biotic interactions. Leiden, Backhuys Publishers, 344 p.

Håkanson, L. and Boulion, V.V. 2003. A model to predict how individual factors influence Secchi depth variations among and within lakes. Internat. Rev. Hydrobiol., 88:212-232.

Håkanson, L. and Eckhéll, J. 2004. Suspended particulate matter (SPM) in the Baltic: empirical data and models. Manuscript, Dept. of Earth Science, Uppsala University

Håkanson, L., Ervik, A., Mäkinen, T. and Möller, B, 1988. Basic concepts concerning assessments of environmental effects of marine fish farms. Nordic Council of Ministers, NORD88:90, Copenhagen, 103 p.

Håkanson, L., Floderus, S. and Wallin, M. 1989. Sediment trap assemblages - a methodological description. Hydrobiologia, 176/177:481-490.

Håkanson, L., Gyllenhammar, A. and Brolin, A. 2004a. A dynamic model to predict sedimentation and suspended particulate matter in coastal areas. Ecol. Modelling, 175:353-384.

Håkanson, L. and Jansson, M. 1983. Principles of Lake Sedimentology. Caldwell, The Blackburn Press, 316 p.

Håkanson, L. and Karlsson, M. 2004. A dynamic model to predict phosphorus fluxes, concentrations and eutropications effects in Baltic coastal areas. In: Karlsson, M., 2004. Predictive Modelling – a Tool for Aquatic Environmental Management. Thesis, Department of Earth Sciences, Uppsala Univiversity, 108 p.

Håkanson, L., Kulinski, I. and Kvarnäs, H. 1984. Water dynamics and bottom dynamics in coastal areas (in Swedish, Vattendynamik och bottendynamik i kustzonen). SNV PM 1905, Solna, 228 p.

Håkanson, L., Kvarnäs, H. and Karlsson, B. 1986. Coastal morphometry as regulator of water exchange: a Swedish example. Estuarine, Coastal & Shelf Science, 23:1-15.

Håkanson, L. and Lindström, M. 1997. Frequency distributions and transformations of lake variables, catchment area and morphometric parameters in predictive regression models for small glacial lakes. Ecol. Modelling, 99:171-201.

Håkanson, L., Malmaeus, J.M., Bodemar, U. and Gerhardt, V. 2003a. Coefficients of variation for chlorophyll, green algae, diatoms, cryptophytes and blue-greens in rivers as a basis for predictive modelling and aquatic management. Ecol. Modelling, 169:179-196.

Håkanson, L., Mikrenska, M., Petrov, K., and Foster. I.D.L. 2004b. Suspended particulate matter (SPM) in rivers – empirical data and models. Ecol. Modelling (in press).

Håkanson, L., Ostapenia, A. and Boulion, V.V. 2003b. A mass-balance model for phosphorus accounting for biouptake and retention in biota. Freshw. Biol., 48:928-950.

Håkanson, L., Parparov, A. and Hambright, K.D. 2000a. Modelling the impact of water level fluctuations on water quality (suspended particulate matter) in Lake Kinneret, Israel. Ecol. Modelling, 128:101-125.

Håkanson, L., Parparov, A., Ostapenia, A. and Boulion, V.V. 2000b. Development of a system of water quality as a tool for management. Final report to INTAS, Uppsala Univ., Dept. of Earth Sci., 2000-11-07, 19 p.

Håkanson, L. and Peters, R.H. 1995. Predictive limnology: Methods for predictive modelling. Amsterdam, SPB Academic Publishing, 464 p.

Håkanson, L. och Rosenberg, R. 1985. Praktisk kustekologi. SNV PM 1987, Solna, 110 sid.

Håkanson, L., and Sazykina, T. 2001. A blind test of the MOIRA lake model for radiocesium for Lake Uruskul, Russia, contaminated by fallout from the Kyshtym accident in 1957. J. Env. Radioactivity, 54:327-344.

Håkanson, L., Sazykina, T.G., and Kryshev, I.I. 2002. A general approach to transform a lake model for one radionuclide (radiocesium) to another (radiostrontium) and critical model tests using data for four Ural lakes contaminated by the fallout from the Kyshtym accident in 1957. J. Env. Radioactivity, 60:319-350.

Hamby, D.M. 1995. A comparison of sensitivity analysis techniques. Health Physics, 68:195-204.

Harden-Jones, F. R. 1968. Fish Migration. London: Edward Arnold Publishers, 325 pp.

Harper, D.M. and Stewart, W.D.P. 1987. The effects of land use upon water chemistry, particulary nutrient enrichment in shallow lowland lakes: comparative studies in three lochs in Scotland. Hydrobiologia, 148:211-229.

Hawley, N., Robbins, J.A. and Eadie, B.J. 1986. The partition of [7]beryllium in fresh water. Geochim. Cosmochim. Acta, 50:1127-1131.

Hayes, M.H.B., MacCarthy, P., Malcom, R.L. and Swift, R.S. 1989 (eds). Humic substances II. In search of structure. New York, Wiley Interscience, 764 p.

Hecky, R.E. and Kilham, P. 1988. Nutrient limitation of phytoplancton in freshwater and marine environments: A review of recent evidence on the effects of enrichment. Limnol. Oceanogr., 33:796-822.

Heiskanen, A.-S. and Tallberg, P. 1999. Sedimentation and particulate nutrient dynamics along a coastal gradient from a fjord-like bay to the open sea. Hydrobiologia, 393:127-140.

HELCOM, 1998. Baltic Sea Environment commission – Helsinki commission, 1998. Fourth periodic assessment of the state of the marine environment of the Baltic Sea, 1994-1998.

Hellström, T. 1991. The effect of resuspension on algal production in a shallow lake. Hydrobiologia, 213:183-190.

Hillel, D. 1982. Introduction to Soil Physics. New York, Academic Press, 364 p.

Hinton, T. G., 1993. Sensitivity analysis of ecosys-87: an emphasis on the ingestion pathway as a function of radionuclide and type of disposition. Health Physics, 66:513-531.

Hjulström, F. 1935. Studies on the morphological activity of rivers as illustrated by the River Fyris. Bull. Geol. Inst., Uppsala, 25:221-527.

Honeyman, B.D. and Santschi, P.H. 1988. Metals in aquatic systems. Environ. Sci. Technol., 22:862-871.

Howarth, R.W., 1988. Nutrient limitation of net primary production in marine ecosystems. Ann. Rev. Ecol., 19:89-110.

Howarth, R.W. and Cole, J.J. 1985. Molybdenum availability, nitrogen limitation, and phypoplankton growth in natural waters. Science, 229:653-655.

Huet, M. 1949. Outline of the relations between slope and fish populations in running waters. Schweiz. Zeit. Hydrol., 11:332-351.

Hurley, J.P, Cowell, S.E., Shafer, M.M. and Hughes, P.E. 1998a. Partitioning and transport of total and methyl mercury in the lower Fox river, Wisconsin. Environ. Sci. Technol., 32:1424-1432.

Hurley, J.P. Cowell, S.E., Shafer, M.M. and Hughes, P.E. 1998b. Tributary loading of mercury to lake Michigan: importance of seasonal events and phase partitioning. Sci. Tot. Environ., 213:129-137.

Hutchinson, G.E. 1957. A Treatise on Limnology. Vol. 2. Geography, physics and chemistry. New York: John Wiley and Sons.

Hutchinson, G.E. 1967. A Treatise on Limnology. II. Introduction to lake biology and the limnoplankton. New York: John Wiley and Sons, 1115 p.

IAEA, 1998. International Atomic Energy Agency. Modelling of radiocesium in lakes. Tec. Doc. (in print), Vienna.

IAEA, 2000. Modelling of the transfer of radiocaesium from deposition to lake ecosystems. Report of the VAMP Aquatic Working Group. International Atomic Energy Agency, Vienna, IAEA-TECDOC-1143, 343 p.

Illies, J. 1961. Attempt to a general biocoenotic classification of running waters. Int. Rev. Ges. Hydrobiol., 46:517-523.

Istvánovics, V., Pettersson, K., Rodrigo, M.A., Pierson, D.C. Padisak, J. and Colum, W., 1993. The colonial cyanobacteria *Gloeotrichia echinulata has* a unique phosphorus uptake and life strategy. J. Plankton Res., 15:531-552.

Istvánovics V. and Somlyódy L. 2001. Factors influencing lake recovery from eutrophication: the case of Basin 1 of Lake Balaton. Wat. Res., 35(3):729-735.

Jackson, T.A. and Hecky, R.E. 1980. Depression of primary productivity by humic matter in lake and reservoir waters of the boreal forest zone. Can. J. Fish. Aquat. Sci., 37:2301-2317.

Jackson, D.A., Harvey, H.H. and Somers, K.M. 1990. Ratios in aquatic sciences: Statistical shortcomings with mean depth and the morpoedaphic index. Can. J. Fish. Aquat. Sci., 47:1788-1795.

James, W.F. and Barko, J.W. 1993. Sediment resuspension, redeposition, and focusing in a small dimictic reservoir. Can. J. Fish. Aquat. Sci., 50:1023-1028.

Jansson, M. 1982. Land erosion by water in different climates. UNGI Report 57, Uppsala University, 151 p.

Janus, L.L. and Vollenweider, R.A. 1981. The OECD cooperative report on eutrophication: Canadian contribution. Summary Rep. Sci. Ser. No. 132.

Johansson H. and Håkanson L. 2004. Models to predict the particulate and dissolved fraction of phosphorus in lakes. Uppsala Univ., Dept. of Earth Sci., Manuscript.

Johansson, T., Håkanson, L., Borum, K. and Persson, J. 1998. Direct flows of phosphorus and suspended matter from a fish farm to the wild fish in Lake Southern Bullaren, Sweden. Aquacult. Engineering, 17:111-137.

Johansson, H., Lindström, M. and Håkanson, L. 2001. On the modelling of particulate and dissolved distributions of substances in aquatic ecosystems: sedimentological and ecological interactions. Ecol. Modelling, 137:225-240.

Jonasson, S.A., Lång, L.-O. and Swedberg, S.E. 1985. Factors controlling pH and alkalinity. A study of acidic well-water in south-eastern Sweden (in Swedish). SNV Report 3021, 84 p.

Jonsson, A, 1997. Whole lake metabolism of allochthonous organic material and the limiting nutrient concept in lake Örträsket, a large humic lake in northern Sweden. Dissertation, Dept. of Physical Geography, Umeå University, Sweden. ISBN 91-7191-382-3.

Jonsson, P. 1992. Large-scale changes of contaminants in Baltic Sea sediments during the twentieth century. Thesis, Uppsala University, Sweden.

Jørgensen, S.E. and Johnsen, J. 1989. Principles of environmental science and technology (2nd edition). Studies in Environmental Science, 33. Amsterdam: Elsevier, 628 p.

Kajak, Z., Hillbricht-Ilkowska A. and Pieczynska, E. 1972. The Production Processes in Several Polish Lakes: Productivity Problems of Freshwaters. Warszawa-Krakow, Polish Scientific Publishers, pp. 129-147.

Kalff J. 2002. Limnology. New Jersey: Prentice Hall, 592 p.

Kenney, B.C. 1982. Beware of spurious self-correlations! Wat. Resour. Res., 18:1041-1048.

Khusainova, N.Z., Mitrofanov, V.P., Mamilova, R.K. and Sharapova, L.I. 1973. Biological productivity of the Lake Karakul: Production-biological investigations of the freshwater ecosystems. Minsk, pp. 32-43 (in Russian).

Kirk, J.T.O. 1983. Light and photosynthesis in aquatic ecosystems. Cambridge: Cambridge University Press.

Kirkby, H.J. (ed.), 1978. Hillslope Hydrology. Chichester: John Wiley and Sons, 389 p.

Kljukina, E.A. and Freindling, A.V. 1983. Distribution and production of macrophytes in small water bodies of Middle Karelia. Hydrobiological Journal, 19:40-45 (in Russian).

Knoechel, R. and Campbell, C.E. 1988. Physical, chemical, watershed and plankton characteristics of lakes on the Avalon Peninsula, Newfoundland, Canada: a multivariate analysis of interrelationships. Verh. Internat. Verein. Limnol., 23:282-296.

Knowlton, M.F. and Jones, J.R. 1993. Testing models of chlorophyll and transparency for midwest lakes and reservoirs. Lake Reserv. Manage., 8:13-16.

Kochanova, E.I. 1976. Macrophytes and its production in lakes of Kharbey system. Productivity of lakes of Bolshezemelskaia tundra. Leningrad, pp. 79-89 (in Russian).

Koelmans, A.A. and Lijklema, L. 1992. Sorption of 1,2,3,4-tetrachlorobenzene and cadmium to sediments and suspended solids in Lake Volkerak/Zoom. Wat. Res., 26:327-337.

Koelmans, A.A. and Radovanovic., H. 1998. Prediction of trace metal distribution coefficient (KD) for aerobic sediments. Wat. Sci. Technol., 37:71-78.

Konitzer, K. and Meili, M. 1997. Redistribution of sedimentary Cs-137 in small Swedish lakes after the Chernobyl fallout 1986. In: Desmet, G. (et al., editors), Freshwater and Estuarine Radioecology, pp. 167-172. Amsterdam: Elsevier.

Konoplev, A., Bulgakov, A., Hilton, J., Comans, R. and Popov, V. 1997. Long-term kinetics of radiocesium fixation by soils. In: Desmet, G. (et al., editors), Freshwater and Estuarine Radioecology, pp. 173-182. Amsterdam, Elsevier.

Krambeck, H.-J. 1995. Application and abuse of statistical methods in mathematical modelling in limnology. Ecol. Model., 78:7-15.

Kranck, K. 1973. Flocculation of suspended sediment in the sea. Nature, 246:348-350.

Kranck, K. 1979. Particle matter grain-size characteristics and flocculation in a partially mixed estuary. Sedimentology, 28:107-114.

Kristensen, P., Søndergaard, M. and Jeppesen, E. 1992. Resuspension in a shallow eutrophic lake. Hydrobiologia, 228:101-109.

Kronvang, B., Laubel, A. and Grant, R. 1997. Suspended sediment and particulate phosphorus transport and delivery pathways in an arable catchment, Gelbæk stream, Denmark. Hydrol. Process., 11:627-642.

Kuznetsov , S.I. 1970. Microflora of lakes and its geochemical activity. Leningrad (in Russian).

Lalonde, S. and Downing, A. 1991. Epiphyton biomass is related to lake trophic status, depth, and macrophyte architecture. Can. J. Fish. Aquat. Sci. 48:2285-2291.

Lam, D.C.L. and Bobba, A.G. 1985. Modelling watershed runoff and basin acidification. In: Johansson (ed.), Hydrological and hydrogeochemical mechanisms and model approaches to the acidification of ecological systems, NHP-report No. 10, Internat. Hydrol. Program (IHP) Workshop, Uppsala, Sweden, pp. 205-237.

Lemmin, U and Imboden, D. M. 1987. Dynamics of bottom currents in a small lake. Limnology and Oceanography, 32:62-75.

Leveque, C., Carmouze, J.P., Dejoux, C., Durand, J.R., Gras, R., Iltis, A., Lemoalle, J., Loubens, G., Lauzanne, L. and Saint-Jean, L. 1972. Recherches sur les biomasses et la productivity du Lac Tchad. Productivity Problems of Freshwaters. Warszawa-Krakow: Polish Scientific Publishers, pp. 165-181.

Li, Y.-H., Burkhardt, L., Buchholtz, M. B., O'Hara, P. and Santschi, P.H. 1984. Partition of radiotracers between suspended particles and seawater. Geochim. Cosmochim. Acta, 48:2011-2019.

Lick, W., Lick, J. and Ziegler, C.K. 1992. Flocculation and its effect on the vertical transport of fine-grained sediments. Hydrobiologia, 235/236:1-16.

Lindström, M. 2000. Distribution of particulate and dissolved mercury in surface water of Swedish forest lakes: An empirically based predictive model. Manuscript, Institute of Earth Sciences, Uppsala University, Uppsala, Sweden.

Lindström, M., Håkanson, L., Abrahamsson, O. and Johansson, H. 1999. An empirical model for prediction of lake water suspended matter. Ecol. Modelling, 121:185-198.

Loeb. S.L., Reuter, J.E. and Goldman, C.R. 1983. Littoral zone production of oligotrophic lakes. The contributions of phytoplankton and periphyton. Periphyton of freshwater ecosystems. The Hague: Dr. W. Junk Publishers, pp. 161-167.

Luettich R.A., Harleman D.R.F. and Somlyódy L. 1990. Dynamic behavior of suspended sediment concentrations in a shallow lake perturbed by episodic wind events. Limnol. Oceanogr., 35(5):1050-1067.

Lundin, L.C. 1999. Water in Society. Sustainable water management in the Baltic Sea basin, The Baltic Univ. Progr., Uppsala Univ., 244 p.

Lundin, L.C. 2000a. The Waterscape. Sustainable water management in the Baltic Sea basin, The Baltic Univ. Progr., Uppsala Univ., 207 p.

Lundin, L.C. 2000b. River Basin Management. Sustainable water management in the Baltic Sea basin, The Baltic Univ. Progr., Uppsala Univ., 244 p.

Madruga, M.J. and Cremers, A. 1997. On the differential binding mechanisms of radiostrontium and radiocesium in sediments. In: Desmet, G. (et al., editors), Freshwater and Estuarine Radioecology, Amsterdam: Elsevier Publishers.

Makarevich, T.A. 1985. Epiphyton. Ecolgical system of Naroch lakes. Minsk, pp. 99-112 (in Russian).

Mann, K.H. 1982. Ecology of coastal waters: A systems approach. Blackwell Scientific Publications, 322 p.

Malmaeus, J.M. and Håkanson, L. 2003. A dynamic model to predict suspended particulate matter in lakes. Ecol. Modelling, 167:247-262.

Manukas, I.L. 1973. Biological productivity of the Mjatjalai lakes. Production-biological investigations of the freshwater ecosystems. Minsk, pp. 54-60 (in Russian).

Matteucci, G. and Frascari, F. 1997. Fluxes of suspended materials in the north Adriatic Sea (Po prodelta area). Water, Air and Soil Pollution, 99:557-572.

McCave, I.N. 1981. Location of coastal accumulations of fine sediments around the southern North Sea. Rapp. P. v. Reun. Int. Explor. Mer., 181:15-27.

McDowall, R. M. 1988. Diadromy in Fishes. Portland, Oregon: Timber Press, 308 p.

McMahon, T.A., Finlayson, B.L., Haines, A., and Srikanthan, R. 1987. Runoff variability: a Global perspective. International Association of Hydrological Sciences Publication, 168:3-11.

Meeuwig, J.J., Kauppila, P and Pitkänen, H. 2000. Predicting coastal eutrophication in the Baltic: a limnological approach. Can, J. Fish. Aquat. Sci., 57:844-855.

Meijer, M. L., Dehaan, M. W., Breukelaar, A. W. and Buiteveld, H. 1990. Is reduction of benthivorous fish an important cause of high transparency following biomanipulation in shallow lakes? Hydrobiologia, 200-201:303-315.

Menshutkin, V.V., 1971. Mathematical modelling of populations and communities of aquatic animals. Leningrad (in Russian).

Merilehto, K., Kenttämies, K. and Kämäri, J. 1988. Surface water acidification in the ECE region. Nordic Council of Ministers, NORD 1988:89, 156 p.

Milius, A. 1982. Indicators of eutrophication and indices of trophy for small Estonian lakes. Proceedings of the Estonian Academy of Sciences, 31:302-309.

Monitor, 1981. Acidification of land and water (in Swedish). SNV, Solna, 175 p.

Monitor, 1991. Acidification and liming of Swedish freshwaters. Swedish Environmental Protection Agency, Solna, 144 p.

Monte, L. 1997. A collective model for predicting the long-term behavior of radionuclides in rivers. The Science of the Total Environment, 201:17-29.

Morel, F.M.M. and Gschwend, P.M. 1987. The role of colloids in the partitioning of solutes in natural waters. In: W. Stumm (Ed.), Aquatic Surface Chemistry. New York: Wiley-Interscience, pp. 405-422.

Mosteller, F. and Tukey, J. W. 1977. Data Analysis and Regression: a second course in statistics. Reading, MA, Addison-Wesley Publ., 588 p.

Muir Wood, A.M. 1969. Coastal Hydraulics. London: Macmillan, 187 p.

Muller, F.L.L., Tranter, M. and Balls, P.W. 1994. Distribution and transport of chemical constituents in the Clyde estuary. Estuar. Coast. Shelf Sci., 39:105-126.

Nazarova, E.I. and Shishkin, B.A. 1981. Elements of organic matter balance in Lake Arakhley. Biological productivity of Lake Arakhley (Trans-Baikal region). Novosibirsk, pp. 41-46 (in Russian).

Neronova, G.A. and Karasev, G.L. 1977. Composition, biomass, and production of water plants in Eravno-Kharga lakes in connection with acclimatization of herbivorous fishes. Proceedings of Baikal Department of Siberian Fish-project Institute, 1:94-105 (in Russian).

Newman, M.C. 1993. Regression analysis of log-transformed data: statistical bias and its correction. Env. Tox. & Chem., 12:1129-1133.

Newton, R.M., Weintraub, J. and April, R. 1987. The relationship between surface water chemistry and geology in the North Branch of the Moose River. Biogeochem., 3:21-35.

Nikulina, V.N. 1979. Periphyton of north coast of lake Issyk Kul: Experimental and field investigations of biological basis of lake productivity. Leningrad, p.p. 50-57 (in Russian).

Nilsson, Å. 1992. Statistical modelling of regional variations in lake water chemistry and mercury distribution. Thesis, Umeå University, Sweden.

Nilsson, Å. and Håkanson, L. 1992. Relationship between drainage area characteristics and lake water characteristics. Env. Geol. and Water Sci., 19:75-81.

Nilsson, J. 2002. Variation pattern for suspended particulate matter in the Baltic Sea: empirical data and model. M.Sc thesis, Uppsala University, Sweden, 29 p.

Nixon, S.W. 1990. Marine eutrophication: a growing international problem. Ambio, 3:101.

Nixon, S.W., and Pilson, 1983. Nitrogen in estuarine and coastal marine ecosystems. In: Carpenter, E.J. and Capone, D.G. (eds.), Nitrogen in the Marine Environment. New York, Academic Press, pp. 565-648.

Nõges, P., Tuvikene, L., Nõges, T. and Kisand, A. 1999. Primary production, sedimentation and resuspension in large shallow Lake Võrtsjärv. Aquatic Sciences, 61:161-182.

Nordvarg, L. 2001. Predictive models and eutrophication effects of fish farms. Dr. Thesis, Uppsala University, Sweden.

Northcote, T. G. 1978. Migratory strategies and production in freshwater fishes. In: Gerking, S.D. (Ed.). Ecology of Freshwater Fish production. Oxford: Blackwell Scientfic Publications, pp. 326-359.

Nürnberg, G.K. and Shaw, M. 1998. Productivity of clear and humic lakes: nutrients, phytoplankton, bacteria. Hydrobiologia, 382:97-112.

Nyström, U. 1985. Transit time distribution of water in two small forested catchments. Ecol. Bull., Stockholm, 37:98-100.

O'Connor, D.J. and Connelly, J.P. 1980. The effect of concentration of adsorbing solids on the partition coefficient. Wat. Res., 14:1517-1526.

OECD, 1982. Eutrophication of waters. Monitoring, assessment and control. Paris, OECD, 154 p.

Ohle, W. 1956. Bioactivity, production, and energy utilization of lakes. Limnol. Oceanogr., 1:139-149.

Ohle, W. 1958. Typologische Kennzeichnung der Gewasser auf Grund ihrer Bioaktivitat. Verh. Int. Verein. Limnol., 13:196-211.

Ostapenia, A.P. 1985. Ratio between the components of seston. Ecological system of Naroch lakes. Minsk, pp. 232-233 (in Russian).

Ostapenia, A.P. 1987. Ratio between particulate and dissolved organic matter in waters of different types: Production-hydrobiological investigations of water ecosystems. Leningrad, pp. 109-115 (in Russian).

Ostapenia, A.P. 1989. Seston and detritus as structural and functional components of water ecosystems. Thesis for a Doctor's degree. Kiev (in Russian).

Ostapenia, A.P., Pavljutin A.P. and Zhukova T.V. 1985. Conditions and factors determining quality of waters and trophic status of lakes. Ecological system of Naroch lakes. Minsk, pp. 263-269 (in Russian).

Ottosson, F. and Abrahamsson, O. 1998. Presentation and analysis of a model simulating epilimnetic and hypolimnetic temperatures in lakes. Ecol. Modelling, 110:223-253.

Overbeck, J. 1972. Distribution pattern of phytoplankton and bacteria, microbial decomposition of organic matter and bacterial production in eutrophic, stratified lake. Productivity problems of freshwater. Warszawa-Krakow: Polish Scientific Publishers, 227-237.

Overrein, L.N., Seip, H.M and Tollan, A. 1980. Acid precipitation - effect on forest and fish.. Final report on the SNSF-project 1972-1980, Oslo-Ås, 175 p.

Pearson, K. 1897. On a form of spurious correlation which may arise when indices are used in the measurements of organs. Proc. R. Soc., London, 60:489-502.

Pearson, T.H. and Rosenberg, R. 1976. A comparative study on the effects on the marine environment of wastes from cellulose industries in Scotland and Sweden. Ambio, 5:77-79.

Persson, G. and Jansson, M. 1988. Phosphorus in freshwater ecosystems. Hydrobiologia, Vol. 170, 340 p.

Persson, J., Håkanson, L. and Pilesjö, P. 1994. Prediction of surface water turnover time in coastal waters using digital bathymetric information. Environmentrics, 5:433-449.

Persson, J. and Håkanson, L. 1995. Prediction of bottom dynamic conditions in coastal waters. Mar. and Freshw. Res., 46:359-371.

Persson, J. and Håkanson, L. 1996. A simple empirical model to predict deepwater turnover time in coastal waters. Can. J. Fish. Aq. Sci., 53:1236-1245.

Persson, J. and Jonsson, P. 2000. Historical development of laminated sediments: An approach to detect soft sediment ecosystem changes. Mar. Pollut. Bull., 40:122-134.

Peters, R.H. 1981. Phosphorus availability in Lake Memphremagog and its tributaries. Limnol. Oceanogr., 26:1150-1161.

Peters, R.H. 1986. The role of prediction in limnology. Limnol. Oceanogr., 31:1143-1159.

Peterson, R.C. Jr. 1991. The contradictory biological behavior of humic substances in aquatic environment. pp. 369-390. In: Allard, B., Borén, H. and Grimvall, A. 1991 (eds). Humic Substances in Aquatic and Terrestrial Environment. Heidelberg, Springer, 514 p.

Pettersson, K. 1985. Vattenöversikt. Broströmmens vattensystem 1984. Inst. of Limnology, Uppsala Univ., Report 1985, B:8.

Pettersson, K. and Istvanovics, V. 1988. Sediment phosphorus in Lake Balaton - forms and mobility. Arch. Hydrobiol. Beih. Ergebn. Liomnol., 30:25-41. Pfaffenberger, R.C. and Patterson, J.H. 1987. Statistical methods. Illinois: Irwin, 1246 p.

Pierson, D.C., Pettersson, K., Istvánovics, V. 1992. Temporal changes in biomass specific photosynthesis during the summer: regulation by environmental factors and the importance of phytoplankton succession. Hydrobiologia, 243-244/Dev. Hydrobiol., 79:119-135.

Pilesjö, P., Persson, J. and Håkanson, L. 1991. Digital bathymetric information for calculations of morphometrical parameters and surface water retention time for coastal areas (in Swedish). National Swedish Environmental Protection Agency (SNV) Report no. 3916, Solna, Sweden.

Pohl, C. and Hennings, U. 1999. The effect of redox processes on the partitioning of Cd, Pb, Cu, and Mn between dissolved and particulate phases in the Baltic Sea. Mar. Chem., 65:41-53.

Pollingher, U., Berman, T., Chaplain, B., and Scharf, D. 1988. Lake Kinneret phytoplankton: Response to N and P enrichments in experiments and in nature. Hydrobiologia, 166:65-75.

Pokrovskaja, T.N., Mironova N.Ja. and Shilkrot G.S. 1983. Macrophyte lakes and their eutrophication. Moscow, Nauka, 153 p. (in Russian).

Postma, H. 1967. Sediment transport and sedimentation in the estuarine environment. In: Lauff, G.H. (ed.) Estuaries. Washington, DC: American Association for the Advancement of Science, pp. 158-179.

Postma, H. 1982. Sediment transport and sedimentation. In: Olausson, E. and Cato, I. (eds), 1982, Chemistry and Biogeochemistry of Estuaries, Chichester: Wiley & Sons, pp. 153-186.

Prairie, Y. 1996. Evaluating the predictive power of regression models. Can. J. Fish. Aquat. Sci., 53:490-492.

Preisendorfer, R.W. 1986. Secchi disk science: Visual optics of natural waters. Limnol. Oceanogr., 31:909-926.

Priymachenko, A.D. 1983. Ecological peculiarities of phytoplankton photosynthesis and its role in ecosystems of Dnieper reservoirs. Hydrobiology, 19:57-66 (in Russian).

Pustelnikov, O.S. 1977. Geochemical features of suspended matter in connection with recent processes in the Baltic Sea. Ambio, 5:157-162.

Quémerais, B., Cossa, D., Rondeau, B., Pham, T.T. and Fortin, B. 1998. Mercury distribution in relation to iron and manganese in the waters of St. Lawrence river. Sci. Tot. Environ., 213:193-201.

Rasmussen, J.B. and Kalff, J. 1987. Empirical models for zoobenthic biomass in lakes. Can. J. Fish. Aquat. Sci., 44:990-1001.

Raspopov, I.M. 1973. Phytomass and production of macrophytes in Lake Onega. Microbiology and primary production of Lake Onega. Leningrad, pp. 123-142 (in Russian).

Raspopov, I.M. 1978. Higher water vegetation in lakes. Hydrobiology of lakes Vozhe and Lacha. Leningrad, pp. 12-27 (in Russian).

Redfield, A.C. 1958. The biological control of chemical factors in the environment. Am. Sci., 46:205-222.

Reed, J.L. 1921. On the correlation between any two functions and its application to the general case of spurious correlation. J. Wash. Acad. Sci., 11:449-455.

Remane, A. 1934. Die Brackwasserfauna. Verh. Dt. Zool. Ges., 36:34-74.

Riise, G., Björnstad, H.E., Oughton, D.H. and Salbu, 1990. A study on radionuclide associations with soil components using sequential extraction procedure. J. Radioanal. Nucl. Chem., 142:531-538.

Riley E.T. and Prepas E.E. 1985. Comparison of the phosphorus-chlorophyll relationships in mixed and stratified lakes. Can. J. Fish. Aquat. Sci., 42:831-835.

Rodhe, A. 1987. The origin of streamwater traced by oxygen-18. Thesis, Uppsala University, Series A, No. 41, 260 p.

Rodhe, W. 1958. Primarproduktion und Seetypen. Verh. Int. Verein. Limnol., 13:121-141.

Rose, K.A., McLean, R.I. and Summers, J.K. 1989. Development and Monte Carlo analysis of an oyster bioaccumulation model applied to biomonitoroing. Ecol. Mod., 45:111-132.

Rosenberg, R. 1985. Eutrophication: the future marine coastal nuisance? Mar. Pollut. Bull., 16:227-231.

Ryther, J.H. and Dunstan, W.M. 1971. Nitrogen, phosphorus, and eutrophication in the coastal marine environment. Science, 171:1008-1013.

Salomons, W. and Förstner, U. 1984. Metals in the Hydrocycle. Heidelberg, Springer, 349 p.

Sandberg, J., Elmgren, R. and Wulff, F. 2000. Carbon flows in Baltic Sea food webs: a re-evaluation using a mass balance approach. J. Mar. Systems, 25:249-260.

Santschi, P.H. and Honeyman, B.D. 1991. Radioisotopes as tracers for the interactions between trace elements, colloids and particles in natural waters. In: J.P. Vernet (Editor), Heavy metals in the Environment. Amsterdam: Elsevier, pp. 229-246.

Scheffer, M. 1990. Multiplicity of stable states in freshwater systems. Hydrobiologia, 200/201:475-486.

Scheffer, M., Brock, W and Westley, F. 2000. Socioeconomic mechanisms preventing optimum use of ecosystem services: An interdisciplinary theoretical analysis. Ecosystems, 3:451-4571.

Schindler, D.W. 1977. Evolution of phosphorus limitation in lakes. Science, 195:260-262.

Schindler, D.W. 1978. Factors regulating phytoplankton production and standing crop in the world's freshwaters. Limnol. Oceanogr., 23:478-486.

Schofield, N.J. and Ruprecht, J.K. 1989. Regional analysis of stream salinisation in southwest Western Australia. J. Hydrol., 112:19-39.

Seibold, E. and Berger, W.H. 1982. The sea floor. Heidelberg, Springer-Verlag, 288 p.

Serruya, C. (Editor). 1978. Lake Kinneret. The Hague: Dr. W. Junk Publishers.

Shafer, M.M., Overdier, J.T., Phillips, H., Webb, D., Sullivan, J.R. and Armstrong, D. 1999. Trace metal levels and partitioning in Wisconsin rivers. Wat. Air Soil Pollut., 110:273-311.

Shapiro, J. 1957. Chemical and biological studies on the yellow organic acids of lake water. Limnol. Oceanogr., 2:161-179.

Shiller, A.M. and Boyle, E.A. 1991. Trace elements in the Mississipi River delta outflow region: behavior at high discharge. Geochim. Cosmochim. Acta, 55:3241-3251.

Skidmore, A.K. 1990. Terrain position as mapped from a gridded digital elevation model. Int. Journal of Geographical Information Systems, 4:33-49.

Slotton, D.G. and Reuter, J.E. 1995. Heavy metals in intact and resuspended sediments of a California reservoir, with emphasis on potential bioavailability of copper and zinc. Mar. Freshwat. Res., 46:257-265.

Sly, P.G. 1978. Sedimentary Processes in Lakes. In: Lerman, A. (ed.), Lakes: Chemistry, Geology, Physics. Berlin, Springer, pp. 65-89.

Smith, I.R. and Sinclair, I.J. 1972. Deep water waves in lakes. Freshwater Biol. 2:387-399.

Smith, V.H. 1979. Nutrient dependence of primary productivity in lakes. Limnol. Oceanogr., 24:1051-1064.

Solo-Gabriele, H.M. and Perkins, F.E. 1997. Metal transport within a small urbanised watershed. J. Irrig. Drain. Enging., 123:114-122.

Somlyódy L. and Koncsos L. 1991. Influence of sediment resuspension on light conditions and algal growth in Lake Balaton. Ecol. Model. 57, 173-192.

Somlyódy L. and van Straten G. 1981. (eds.) Modelling and managing shallow lake eutrophication, with application to Lake Balaton. Berlin, Springer, 386 p.

Spence, D.H.N. 1982. The zonation of plants in freshwater lakes. Adv. Ecol. Res., 12:37-125.

Stanley, D.J. and Swift, D.J. (eds), 1976. Marine Sediment Transport and Environmental Management. New York: John Wiley and Sons, 602 p.

Stokes, G.G. 1851. Collected papers. Vol. III. Cambridge Trans. Vol. IX (see, e.g., Lamb, H., 1945. Hydrodynamics. New York: Dover Publ., 450 p.)

Strahler, A.N. 1963. The Earth Sciences. New York: Harper & Row, Publ., 681 p.

Stumm, W. and Morgan, J.J. 1981. Aquatic Chemistry. London: Wiley Interscience, 780 p.

Sung, W. 1995. Some observations on surface partitioning of Cd, Cu, and Zn in Estuaries. Environ. Sci. Technol., 32:620-625.

Sverdrup, H. 1985. Calcite dissolution kinetics and lake neutralization. Thesis, Lund University, Sweden.

Sverdrup, H. and Warfvinge, P. 1990. The role of weathering and forestry in determining the acidity of lakes in Sweden. Water, Air and Soil Poll., 52:71-78.

Talbot, R.W. and Andren, A.W. 1984. Seasonal variations of ^{210}Pb, and [210]Po concentrations in an oligotrophic lake. Geochim. Cosmochim. Acta, 48:2053-2063.

Taylor, J.K. 1990. Statistical Techniques for Data Analysis. Chelsea, Lewis, 200 p.

Tessier, A., Carignan, R., Dubreuil, B. and Rapin, F. 1989. Partitioning of zinc between the water column and the oxic sediments in lakes. Geochim. Cosmochim. Acta, 53:1511-1522.

Thienemann, A. 1928. Die Binnengewasser. Bd 4. Der Sauerstoff im eutrophen and oligotrophen See. Stuttgart.

Thierfelder, T. 1999. The role of catchment hydrology in the characterization of water quality in glacial/boreal lakes. J. Hydrol., 216:1.16.

Thomas, R.L. 1972. The distribution of mercury in the sediments of Lake Ontario. Can. J. Earth Sci., 9:636-651.

Thomas, R.L., Kemp, A.L.W., Lewis, C.F.M. 1972. Distribution, composition and characteristics of the surficial sediments of Lake Ontario. J. Sed. Pet., 42:66-84.

Thomas, R.L., Jacquet, J.M., Kemp, A.L.W., Lewis, C.F.M. 1976. Surficial sediments in Lake Erie. J. Fish. Res. Bd Can., 33:385-403.

Tiwari, J.L. and Hobbie, J.E. 1976. Random differential equations as models of ecosystems: Monte Carlo simulation approach. Math. Biosci., 28:25-44.

Törnblom, E. and Rydin, E. 1998. Bacterial and phosphorus dynamics in profundal Lake Erken sediments following the deposition of diatoms: a laboratory study. Hydrobiologia, 364:55.63.

Turner, A. 1996. Trace-metal partitioning in estuaries: importance of salinity and particle concentration. Mar. Chem., 54:27-39

Turner, A. 1999. Diagnosis of chemical reactivity and pollution sources from particulate trace metal distribution in estuaries. Estuar. Coast. Shelf Sci., 48:177-191.

Turner, A., Millward, G. E., Bale, A. J. and Morris, A. W. 1993. Application of the Kd concept to the study of trace metal removal and desorption during estuarine mixing. Estuar., Coast. and Shelf Sci., 36:1-13.

Turner, A., Hyde, T.L. and Rawling, M.C. 1999. Transport and retention of hydrophobic organic micropollutants in estuaries: Implications of the particle concentration effect. Estuar. Coast. Shelf Sci., 49:733-746.

Turner, J.V. 1984. Hydrogeochemical evolution and variability of groundwater in experimental basins of S.W. Western Australia. pp. 41-54. In:Eriksson, E. (ed.). Hydrochemical Balances of Freshwater Systems. IAHS-AISH Publication No. 150.

Ugolini, F.C. 1986. Processes and rates of weathering in cold and polar desert environments, pp. 193-235. In: Colman, S.M. and Dethier, D.P. (eds). Rates of Chemical Weathering of Rocks and Minerals. New York: Academic Press.

van der Weijden, C.H., Ten Haven, H.L., Boer, H.A., Hopstaken, C.F.A.M. and Vried, S.P. 1984. Geochemical studies in the drainage basin of the Rio Vouga (Portugal). pp. 263-276. In: Eriksson, E. (ed.). Hydrochemical Balances of Freshwater Systems. IAHS-AISH Publication No 150.

Vannote, R.-L., Minshall, G.W., Cummins, K.W., Sedell, J.R., and Cushing, C.E. 1980. The river continuum concept. Can. J. Fish. Aq. Sci., 37:103-137.

Velimorov, B. 1991. Detritus and the concept of non-predatory loss. Arch. Hydrobiol., 121:1-20.

Verstraeten G. and Poesen J. 2001. Factors controlling sediment yield from small intensively cultivated catchments in a temperate humid climate. Geomorphology, 40:123-144.

Vlasova, T.A., Baranovskaja, V.K., Getsen, M.V., Popova, E.I. and Sidorov, G.P. 1973. Biological productivity of the Kharbei lakes of Bolshezemelskaia tundra. Production-biological investigations of the freshwater ecosystems. Minsk, pp. 147-163 (in Russian).

Voipio, A. (ed.), 1981. The Baltic Sea. Amsterdam: Elsevier Oceanographic Series, 418 p.

Vollenweider, R.A. 1958. Sichttiefe und Production. Verh. Int. Ver. Limnol., 13:142-143.

Vollenweider, R.A. 1960. Beitrage zür Kenntnis optischer Eigenschaften der Gewässer und Primärproduktion. Mem. Ist. Ital. Idrobiol., 12:201-244.

Vollenweider, R.A. 1968. The scientific basis of lake eutrophication, with particular reference to phosphorus and nitrogen as eutrophication factors. Tech. Rep. DAS/DSI/68.27, OECD, Paris, 159 pp.

Vollenweider R.A. 1969. A Manual on Methods for Measuring Primary Production in Aquatic Environments. IBP, Oxford, Handbook, 12, 213 p.

Vollenweider, R.A. 1976. Advances in defining critical loading levels for phosphorus in lake eutrophication. Mem. Ist. Ital. Idrobiol., 33:53-83.

Vollenweider, R.A. 1990. Eutrophication: conventional and non-conventional considerations on selected topics. In: de Bernardi, R., Giussani, G. and Barbanti, L. (eds), 1990. Scientific Perspectives in Theoretical and Applied Limnology. Memorie dell'Istituto Italiano di Idrobiologia Dott. Marco de Marchi, Vol. 47, Pallanza, 378 p.

Vollenweider, R.A. and Kerekes , J. 1980. OECD cooperative programme on monitoring of inland waters (eutrophication control). Synthesis Rep. (from Peters, 1986).

Vorobev, G.A. 1977. Landscape types of lake overgrowing. Natural conditions and resources of North European part of the Soviet Union. Vologda, pp. 48-60 (in Russian).

Walker, W.W. 1979. Use of hypolimnetic oxygen depletion rates as trophic state indicator for lakes. Water Resources Research, 15:1463-1470.

Wallin, M. and Håkanson, L. 1991. Morphometry and sedimentation as regulating factors for nutrient recycling in shallow coastal waters. Hydrobiologia, 75:33-46.

Wallin, M., Håkanson, L. and Persson, J. 1992. Load models for nutrients in coastal areas, especially from fish farms (in Swedish with English summary). Nordiska ministerrådet, 1992:502, Copenhagen, 207 p.

Walling D.E. 1983. The sediment delivery problem. Journal of Hydrology, 65:209-237.

Walling, D.E. 2000. Linking land use, erosion and sediment yields in river basins. Hydrobiologia, 410:223-240.

Walling, D.E. and Amos, C.M. 1999. Source, storage and mobilisation of fine sediment in a chalk stream system. Hydrological Processes, 13:323-40.

Ward, J.W. 1989. The four-dimensional nature of lotic ecosystems. J. North Am. Bent. Soc., 8:2-8

Ward, J.W. 1998. Riverine landscapes: Biodiversity patterns, disturbance regimes and aquatuc conservation. Biol. Conserv., 83:269-278.

Ward, J,V. and Stanford, J.A. 1995. The serial discontinuity concept: extending the model to floodplain rivers. Regulated Rivers, 10:159-168.

Warfvinge, P. 1988. Modeling acidification mitigation in watersheds. Thesis, Lund Univ., Sweden.

Warren, L.A. and Zimmerman, A.P. 1994. The influence of temperature and NaCl on cadmium, copper and zinc partitioning among suspended particulate and dissolved phases in an urban river. Water Res., 28:1921-1931.

Watras, C.J., Morrison, K.A., Host, J.S. and Bloom, N.S. 1995a. Concentration of mercury species in relationship to other site specific factors in the surface waters of northern Wisconsin lakes. Limnol. Oceanogr., 40:556-565.

Watras, C.J., Morrison, K.A. and Bloom, N.S. 1995b. Chemical correlates of Hg and methyl-Hg in northern Wisconsin lake waters under ice-cover. Wat. Air Soil Pollut., 84:253-267.

Watras, C.J., Back, R.C., Halvorsen, S., Hudson, R.J.M., Morrison, K.A. and Wente, S.P., 1998. Bioaccumulation of mercury in pelagic freshwater food webs. Sci. Tot. Environ., 219:183-208.

Weber, W.J.JR., McGinley, P.M. and Katz, L.E. 1991. Sorption phenomena in subsurface systems: Concepts, models and effects on contaminant fate and transport. Wat. Res., 25:499-528.

Weyhenmeyer, G.A. 1996. The significance of lake resuspension in lakes. Uppsala dissertations from the faculty of science and technology 225. Sweden.

Wershaw, R.L., Burcar, P.J. and Goldberg, M.C. 1969. Interaction of pesticides with natural organic material. Env. Sci. Tech., 3:271-273.

Wetzel, R.G. 1964. A comparative study of the primary productivity of higher aquatic plants, periphyton, and phytoplankton in a large, shallow lake. Int. Rev. Ges. Hydrobiol., 49:1-61.

Wetzel, R.G. 1983. Attached algal-substrata interactions: fact or myth, and when and how? Periphyton of freshwater ecosystems. The Hague: Dr. W. Junk Publishers, pp. 207-215.

Wetzel R.G. 2001. Limnology. London: Academic Press, 1006 p.

Whicker, F.W. and Schultz, V. 1982. Radioecology: Nuclear energy and the environment. Volume 1, Boca Raton, FL, CRC Press, 228 p.

Winberg, G.G. 1960. Primary production of water bodies. Minsk (in Russian).

Winberg, G.G. 1970. Main features of ecological system of the Lake Drivjaty. Biological productivity of eutrophic lake. Minsk, pp. 185-195 (in Russian).

Winberg, G.G. 1977. Comparison of phytoplankton and zooplankton biomasses. Hydrobiological Journal, 13:14-23 (in Russian).

Winberg, G.G. 1980. General characteristics of freshwater ecosystems based on Soviet IBP studies. The functioning of freshwater ecosystems. Cambridge, pp. 481-492.

Winberg, G.G. 1985a. Main features of production process in the Naroch lakes. Ecological system of Naroch lakes. Minsk, pp. 269-284 (in Russian).

Winberg, G.G. 1985b. Some results of practical use of production-hydrobiological methods. The production of populations and aquatic organisms communities and methods of its study. Sverdlovsk, pp. 19-29 (in Russian).

Winberg, G.G. 1986. Principles of composition of biotic balance. Investigation of the relationship between food base and fish productivity. Leningrad, pp. 188-194 (in Russian).

Winberg, G.G., Alimov, A.F., Umnov, A.A. and Norenko, D.S. 1986. Productivity and rational use of lakes of Eravno-Kharga system. Investigation of the relationship between food basis and fish productivity. Leningrad, pp. 212-219.

Winberg, G.G., Babitsky, V.A., Gavrilov, S.I., Gladky, G.V., Zakharenkov., I.S., Kovalevskaya, R.Z., Mikheeva, T.M., Nevyadomskaya P.S., Ostapenya A.P., Petrovich, P.G., Potaenko, J.S. and Yakushko, O.F. 1972. Biological Productivity of Different Types of Lakes. Productivity Problems of Freshwaters. Warszawa-Krakow: Polish Scientific Publishers, pp. 383-404.

Worley, B.A. 1987. Deterministic uncertainty analysis. Oak Ridge National Laboratory Report ORNL-6428, Oak Ridge, U.S.A., 53 p.

Wotton, R. J. 1990. Ecology of Teleost Fishes. New York: Chapman and Hall, 404 pp.

Wulff, F., Rahm, L. and Larsson, P. (eds), 2001. A systems analysis of the Baltic Sea. Berlin: Springer-Verlag, Ecological studies 148, 455 p.

Yan, L., Stallard, R.F., Key, R.M. and Crerar, D.A. 1991. Trace metals and dissolved organic carbon in estuaries and offshore waters of New Jersey. Geochim. Cosmochim. Acta, 55:3647-3656.

You, C-F., Lee, T. and Li, Y-H. 1989. The partition of Be between soil and water. Chem. Geol., 77:105-118.

Young, L.B. and Harvey, H.H. 1992. The relative importance of manganese and iron oxides and organic matter in the sorption of trace metals by surficial lake sediments. Geochim. Cosmochim. Acta, 56:1175-1186.

Zakharenkova, G.F. 1970. Water plants of Lake Drivjaty. Proceedings of All-Union Hydrobiological society, 15:71-80 (in Russian).

Zhou, J.L., Fileman, T.W., Evans, S., Donkin, P., Readman, J.W., Mantoura, R.F.C. and Rowland, S. 1999. The partition of fluoranthene and pyrene between suspended particles and dissolved phase in the Humber Estuary: a study of the controlling factors. Sci. Tot. Environ., 244:305-321.

Zolotareva, L.N. 1981. Higher water plants in Lake Arakhley: Biological productivity of Lake Arakhley (Trans-Baikal region). Novosibirsk, pp. 31-41 (in Russian).

9. Appendices

9.1. Databases for SPM

9.1.1. Lakes

A dataset for lakes from several investigations has been compiled by Lindström et al. (1999) and used in this work:

1. 14 lakes from a study 1978-1980 of the River Kolbäcksåns water system (Ahl, 1979, 1980, 1981); Väsman, Ö. Hillen, N. Hillen, Leran, Haggen, N. Barken, S. Barken, Noren, St. Aspen, Åmänningen, Magsjön, Östersjön, Freden, and Virsbosjön.

2. Lake Erken in Uppland county, Sweden, studied by, e.g., Weyhenmeyer (1997); the data used here emanate from her studies.

3. Lake Örträsket from northern Sweden (data from Jonsson, 1997).

4. Lakes Flatsjön and Siggeforasjön, Sweden (unpublished data from Marcus Sundbom, Dept. of Limnology, Uppsala University, Sweden).

5. Data from the VAMP project; Lake Iso Valkjärvi, Finland, Lake Bracciano, Italy, Lake Øvre Heimdalsvatn, Norway, Lake IJsselmeer, Holland, Lake Hillesjön, Sweden, Lake Devoke Water, UK, and Lake Esthwaite Water, UK (data from IAEA, 2000).

6. Data Lake Zürich, Switzerland (from Jim Smith, Institute of Freshwater Ecology, UK).

Mean (= characteristic) values are used for the target variable, SPM (mg/l), and for all the selected covariables, such as total phosphorous concentration, TP (µg/l) reflecting allochthonous production, pH reflecting autochthonous influences, mean depth, D_m (m), maximum depth, D_{max} (m), area of the lake, Area (m^2), theoretical water retention time, T (yr), mean annual water discharge (Q = V/T, where V is the lake volume, m^3. The dynamic ratio, DR, is defined as $\sqrt{(Area)}/D_m$; Area in km^2 and D_m in m. The various form and size parameters reflect mechanisms related to differences among lakes in, e.g., resuspension. The dataset is presented in table 9.1.

9.1.2. The river database

This database has been compiled and described by Håkanson et al. (2004b). It consists of three parts. Further information about the data, number of samples, sample sites, methods of analyses, etc. are provided in the literature references given below:

1. The European database (table 9.2; the data emanate mainly from the United Nations Environmental Programme, GEMS/Water). Data exist from 89 stations in Belgium, Finland, France, Germany, Hungary, the Netherlands, Poland, Portugal, Russia, Switzerland, and UK on:
   - Mean water discharge (Q in m^3/s).
   - SPM; values in mg/l.
   - Latitude (Lat, °N), longitude (Long, °E), continentality (Cont, i.e., distance from the ocean in km), and altitude (Alt, m.a.s.l.).

Table 9.1. Data from the lake database (from Lindström et al., 1999). SPM in mg/l, Area in km^2, mean depth in m, max. depth in m, total phosphorus in mg/l, theoretical lake water retention time in yr, mean tributary water discharge (Q) in $m^3/yr \cdot 10^6$, dynamic ratio, DR = $\sqrt{Area}/D_m$, and form factor, Vd = $3 \cdot D_m/D_{max}$.

Lake	SPM	Area	D_m	D_{max}	pH	TP	T	Q	DR	V_d
Flatsjön	2.52	0.61	2.5	4	8	20	0.94	1.6	0.312	1.875
Ijsselmeer	40.00	1147	4.3	10	8.5	60	0.41	12.0	7.876	1.290
Iso Valkjärvi	0.50	0.042	3.1	8	5.1	11	3	43.4	0.066	1.163
Bracciano	0.50	57	89.5	150	8.5	5	137	37.27	0.084	1.790
Øvre Heimdalsvatn	0.30	0.78	4.7	20	6.8	10	0.17	21.6	0.188	0.705
Hillesjön	5.00	1.6	1.7	3.1	7.3	20	0.36	7.6	0.744	1.645
Devoke Water	0.70	0.34	4	15	6.5	10	0.24	5.7	0.146	0.800
Esthwaite Water	3.55	1.0	6.4	15	8	25	0.19	33.7	0.156	1.280
Siggeforasön	2.10	0.73	4.2	11	7.2	10	0.78	3.9	0.203	1.145
Zürich	0.93	67.2	49	143	7	15	1	3.3	0.167	1.028
Väsman	1.34	38.6	10.6	53	6.69	14.9	1.08	379	0.586	0.600
Övre Hillen	1.58	5.0	10.6	40	6.72	21.2	0.14	379	0.211	0.795
Nedre Hillen	2.44	2.7	3.7	10	6.85	30.8	0.042	238	0.444	1.110
Leran	3.04	2.8	3.6	10	6.8	25.4	0.042	240	0.465	1.080
Haggen	1.51	7.3	10.1	29	6.51	14.2	1.175	62.7	0.268	1.045
Norra Barken	1.94	19.5	10.1	36.5	6.82	20.7	0.375	525	0.437	0.830
Södra Barken	2.44	11.0	5.7	24	6.81	24.3	0.108	581	0.582	0.713
Noren	4.00	1.3	3	5	6.94	33	0.077	50.69	0.380	1.800
Stora Aspen	3.81	5.9	7	25	6.88	31.9	0.27	153	0.347	0.840
Åmänningen	2.72	21.9	6.2	30	6.97	20.2	0.19	715	0.755	0.620
Magsjön	4.62	0.6	2.1	7.5	6.92	27.4	0.0017	741	0.369	0.840
Östersjön	4.67	1.3	3	7	7.05	38.8	0.005	780	0.380	1.286
Freden	5.88	3.3	5.5	15	7.04	40.8	0.025	726	0.330	1.100
Virsbosjön	3.76	49.0	4.1	12	6.85	27.5	0.025	8.0	1.707	1.025
Erken	2.45	23.7	9	20.65	7.95	25.9	6.6	32.3	0.541	1.308
Örträsket	1.41	7.3	23	64	6.5	21	0.247	680	0.117	1.078

2. The UK database (table 9.3 and Foster et al., 1996, 1997, 1998). This database provides a range of data for 79 monitoring sites in UK rivers. Sampling was undertaken between December 1994 and September 1996. Except where continuous flow monitoring data were available, river discharge was measured at each site on the day of sampling. Two water samples were collected at each site, one of which was filtered (0.45 μm) immediately in the field. The filtered and unfiltered samples were subsequently analyzed for a range of properties (e.g., SPM, BOD, COD, Kjeldahl-N [unfiltered], total phosphorus [filtered and unfiltered], ammonia [filtered], TON [filtered], orthophosphate [filtered and unfiltered] and Si and chlorophyll-a]. Sampling frequencies ranged from weekly to monthly and sampling.

3. The Swedish database for 95 catchments (see Håkanson and Peters, 1995). Data are complied in table 9.4. This database is used to address questions related to how catchment factors influence the variability among sites in Secchi depth (and hence also SPM). The lakes may be regarded as integrators of river transport of suspended matter. The idea is to identify the most important features of the catchment areas regulating variability among sites in SPM/Secchi depth and rank the relative role of these features. The Swedish database includes data on:

    o   Mean water discharge (Q, m^3/s).

    o   Mean monthly Secchi depth (m).

    o   Comprehensive water chemistry (table 9.4).

    o   Comprehensive catchment area descriptions, including drainage area zonation (DAZ; Håkanson and Peters, 1995) of many features (soil types, bedrocks, vegetation, etc.), and data of catchment characteristics close to the lake ("near area," as defined by the DAZ method).

The European database has mainly been used for a statistical treatment of correlations between SPM and Q, Lat, Long, Cont and Alt, which could influence the variability in mean SPM values among the river sites. The UK database is used to address the problem of temporal variability within river sites (weekly data are available for measurement series covering several years), and hence also the representativeness of the samples, and the predictive power of the models for SPM (see chapter 3).

Table 9.2. Data from the European database.

No.	Station	Station name	Country	Q (m³/s)	SPM (mg/l)	Lat (°N)	Long (°E)	Cont (km)	Alt (m)
1	51001	ESCAUT RIVER AT BLEHARIES	Belgium	24	21.00	50.3	3.25	18.00	25
2	51007	WARNETON - LYS RIVER	Belgium		23.50	50.45	2.57	54.00	
3	51008	LEERS/NORD - ESPIERRE RIVER	Belgium		201.50	50.41	3.16	58.50	
4	51009	DOEL - SCHELDT RIVER	Belgium		64.00	51.19	4.14	90.00	
5	51011	ERQUELINNES - SAMBRE RIVER	Belgium		27.00	50.18	4.07	117.00	
6	51012	HEER/AGIMONT - MEUSE RIVER	Belgium		18.00	50.1	4.5	151.00	
7	51013	LANAYE/TERNAAIEN - MEUSE RIVER	Belgium		14.50	50.47	5.41	157.50	
8	51014	MARTELANGE -SURE RIVER	Belgium		12.00	49.51	5.43	247.50	
9	51015	ZELZATE - GHENT/TERNEUZEN	Belgium		12.00	51.12	3.48	31.50	
10	65001	TORNIONJOKI RIVER STN 14100	Finland	366	5.00	66.43	23.54	112.50	78
11	65002	KYMIJOKI STN 5610	Finland	83	4.67	60.3	26.55	0.00	4
12	65003	KALKKINEN STN 4800	Finland	209	1.00	61.17	25.36	85.50	78
13	12031		France		34.00	48.51	2.21	137.20	
14	12032	SEINE RIVER - MELUN	France		14.50	48.32	2.33	168.70	
15	12033	SEINE RIVER - MONTEREAU	France		22.00	48.23	2.59	213.70	
16	12034	SEINE RIVER - MERY-SUR-SEINE	France		14.50	48.31	3.54	254.20	
17	12041	LOIRE RIVER - ORLEANS	France		35.00	47.53	1.55	209.20	
18	12042	LOIRE RIVER - INGRANDES	France		31.00	47.25	0.55	218.20	
19	12043	LOIRE RIVER-ROANNE	France		9.00	46.07	4.04	274.50	
20	12044	LOIRE RIVER - VEAUCHE	France		7.67	45.4	4.12	240.70	
21	12051	GARONNE RIVER - COUTHURES	France		37.67	44.31	0.04	105.70	
22	12052	GARONNE RIVER - VALENCE D'AGEN	France		31.00	44.05	0.05	112.50	
23	12053	GARONNE RIVER - TOULOUSE	France		16.33	43.41	1.22	148.50	
24	12061	RHONE RIVER - ST. VALLIER	France		20.67	45.11	4.49	189.00	
25	12062	SAONE RIVER - LYON	France		18.67	45.46	4.48	270.00	
26	12063	RHONE RIVER - LYON	France		16.50	45.48	4.51	272.20	
27	12064	RHONE RIVER - COLLONGES	France		19.67	46.06	5.52	285.70	
28	12065	SAONE RIVER - AUXONNE	France		17.67	47.1	5.21	405.00	
29	12162	RHONE RIVER - LYON	France		24.00	45.35	4.48	229.50	
30	135011	GEESTHACHT - ELBE RIVER	Germany		21.00	53.26	10.22	72.00	
31	66001	TISZA RIVER AT SZOLNOK	Hungary	589	28.50	47.02	20.02	480.00	81
32	46001	RHINE RIVER AT GERMAN FRONTIER	Netherlands	2200	33.67	51.51	6.06	60.70	11
33	46002	BOVEN MERWEDE (ARM OF RHINE)	Netherlands		37.50	51.49	4.58	45.00	1
34	46003	LEX (ARM OF RHINE)	Netherlands		22.00	51.59	5.04	38.20	2
35	46004	IJSSEL (ARM OF RHINE)	Netherlands		27.67	52.33	5.55	18.00	0
36	46005	MAAS RIVER AT BELGIAN FRONTIER	Netherlands	250	21.00	52.46	5.42	9.00	44
37	21001	VISTULA RIVER - KRAKOW	Poland	90	29.00	50.01	19.48	472.50	
38	21002	VISTULA RIVER -WARSZAWA	Poland	450	35.67	52.14	21.02	277.50	
39	21003	VISTULA RIVER - KIEZMARK	Poland	1010	22.00	54.08	18.45	45.00	
40	21004	ODRA RIVER - CHALUPKI	Poland	43	25.00	49.56	18.19	502.50	
41	21005	ODRA RIVER - WROCLAW	Poland	133	23.33	51.07	17.03	375.00	
42	21006	ODRA RIVER - KRAJNIK	Poland	563	23.33	52.46	14.19	277.50	
43	73001	RIVER TEJO AT SANTAREM	Portugal	450	13.50	39.13	8.4	157.50	6
44	26002	SELENGA RIVER	Russia	781	26.00	52	106.4	1680.00	463
45	26003	BELAYA RIVER	Russia	745	113.33	54.58	55.58	1440.00	83
46	26004	TOM RIVER	Russia	1080	23.00	56.34	84.54	1660.00	71
47	26005	IRTYSH RIVER	Russia	905	51.33	55.12	73.12	1480.00	69
48	26008	VOLGA RIVER	Russia	7250	21.67	46.45	47.48		24
49	26009	AMUR RIVER	Russia	1020	27.00	52.22	140.3	60.00	2
50	26010	YENISEI RIVER	Russia	1810	2.67	67.26	86.3	780.00	2
51	26011	OB RIVER	Russia	1250	9.67	66.37	66.33	200.00	0.4
52	26012	PECHORA RIVER	Russia	4380	7.67	67.35	52.1	100.00	0.8
53	26013	SEVERNAYA DVINA RIVER	Russia	2580	10.33	64.08	41.55	0.00	2
54	26014	NEVA RIVER	Russia	2510	4.50	59.5	30.3	0.00	0.9
55	26015	DON RIVER	Russia	683	124.67	47.2	40.3	0.00	3

No.	Station	Station name	Country	Q (m³/s)	SPM (mg/l)	Lat (°N)	Long (°E)	Cont (km)	Alt (m)
56	26016	KUBAN RIVER	Russia	320	45.33	45.1	38.13	0.00	7
57	26017	KOLYMA RIVER	Russia	3440	30.33	68.43	158.5	120.00	2
58	26018	LENA RIVER	Russia	1350	26.00	70.4	127.2	220.00	4
59	26019	NINELEN RIVER	Russia	111	30.67	52.33	136.3	340.00	95
60	26020	DZHIDA RIVER	Russia	34	6.33	50.2	103.5		902
61	26027	MEZEN RIVER.	Russia	80	6.67	63.37	49.27		105
62	26028	LENA RIVER	Russia	14900	23.00	72.2	126.4	0.00	0
63	26103	BELAYA RIVER	Russia	745	119.33	54.58	55.58		83
64	26104	TOM RIVER.	Russia	1080	23.33	56.34	84.54		71
65	26105	IRTYSH RIVER	Russia	905	44.33	55.12	73.12		69
66	26110	YENISEI RIVER	Russia	1810	4.67	67.26	86.3		2
67	26111	OB RIVER	Russia	1250	10.33	66.37	66.33		0.4
68	26113	SEVERNAYA DVINA RIVER	Russia	2580	12.00	64.08	41.55		2
69	26203	BELAYA RIVER	Russia	745	91.33	54.58	55.58		83
70	26204	TOM RIVER	Russia	1080	19.33	56.34	84.54		71
71	26205	IRTYSH RIVER	Russia	905	29.67	55.12	73.12		69
72	26210	YENISEI RIVER	Russia	1810	2.33	67.26	86.3		2
73	26211	OB RIVER	Russia	1250	9.33	66.37	66.33		0.4
74	26213	SEVERNAYA DVINA RIVER	Russia	2580	10.33	64.08	41.55		22
75	200001	RHINE - DIEPOLDSAU	Switzerland	230	127.00	47.23	9.38	283.50	410
76	200002	RHINE - REKINGEN	Switzerland	440	17.50	47.34	8.19	339.70	323
77	200003	RHINE - BASLE	Switzerland	1067	16.00	47.37	7.34	355.50	243
78	200004	AARE - BRUGG	Switzerland	314	16.00	47.29	8.11	353.20	332
79	200005	RHONE - PORTE DU SCEX	Switzerland	172	117.00	46.21	6.53	263.20	377
80	200006	RHONE - CHANCY	Switzerland	327	26.00	46.09	5.58	288.00	347
81	27001	RIVER THAMES	UK	78	12.67	51.25	0.19	40.00	5
82	27002	RIVER AVON	UK	18	21.33	51.25	2.29	0.00	16
83	27003	RIVER EXE	UK	16	13.67	50.48	3.3	40.00	26
84	27004	RIVER TRENT	UK	82	21.00	52.56	1.08	49.50	16
85	27005	RIVER DEE	UK	30	8.67	53.08	2.52	40.50	4
86	27006	RIVER LEVEN	UK	39	3.33	55.58	4.34	45.00	4
87	27007	RIVER MERSEY	UK	21	15.33	53.23	2.34	40.50	6
88	27008	RIVER TWEED ABOVE GALAFOOT	UK	33	5.67	55.36	2.46	45.00	92
89	27009	RIVER CARRON	UK	8	1.33	57.26	5.26	36.00	4

Table 9.3. The UK database. This table gives statistical information (n = number of data, MV = mean value, SD = standard deviation, CV = coefficient of variation) for 10 of 79 stations for many of the variables included in the database. DO = dissolved oxygen, Cond = conductivity, Redox = redox potential (= Eh), SPM = suspended particulate matter, BOD = biological oxygen demand, COD = chemical oxygen demand, Chl = chlorophyll-a, TP = total phosphorus, PP = particulate phosphorus, PartP = particulate phosphorus, DP = dissolved phosphorus, and PF = particulate fraction of phosphorus.

Site name		Q m³/s	DO %	Temp °C	pH	Cond MSM	Redox MV	SPM mg/l	BOD mg/l	COD mg/l	Chl µg/l	TP mg/l	PP mg/l	PartP mg/g	DP mg/l	PF
Avon (Evesham)	n	86	91	93	77	79	77	93	93	93	56	93	93	93	93	93
	MV	13.2	94.9	12.5	8.2	1.0	198.3	24.3	2.6	34.1	42.2	1.9	0.2	15.2	1.6	0.1
	SD	15.1	24.4	5.4	0.3	0.4	40.5	62.3	1.4	27.8	53.5	0.9	0.4	35.0	1.0	0.2
	CV	1.14	0.26	0.44	0.04	0.42	0.20	2.56	0.52	0.82	1.27	0.47	2.25	2.30	0.61	1.54
Stratford STW (Milcote)	n															
	MV	46	45	45	37	39	37	46	46	45	0	46	46	46	46	46
	SD	0.2	87.6	14.6	7.7	1.1	178.5	21.8	16.8	82.0		5.5	0.7	36.0	4.0	0.1
	CV	0.0	19.5	4.5	0.2	0.4	31.7	6.2	6.2	28.8		1.9	1.2	72.2	2.3	0.2
		0.26	0.22	0.31	0.03	0.42	0.18	0.28	0.37	0.35		0.35	1.79	2.01	0.58	1.43
Avon (Stratford)	n	43	46	47	38	40	38	47	47	47	27	47	47	47	47	47
	MV	8.9	95.2	12.3	7.9	1.0	207.3	15.8	2.8	38.3	26.5	2.3	0.2	42.9	2.0	0.1
	SD	11.9	17.3	5.6	0.2	0.4	33.6	27.0	1.2	24.8	29.6	1.1	0.5	163.1	1.1	0.2
	CV	1.34	0.18	0.46	0.03	0.45	0.16	1.71	0.44	0.65	1.12	0.46	2.48	3.80	0.54	1.85
Avon (Barford)	n	86	90	92	76	78	76	93	93	93	56	93	93	93	93	93
	MV	6.9	88.3	12.6	7.8	1.0	213.2	13.8	2.9	41.6	25.3	2.6	0.2	32.5	2.3	0.1
	SD	9.0	22.0	5.1	0.2	0.5	40.7	22.5	1.0	35.2	30.0	1.2	0.5	97.5	1.2	0.1
	CV	1.30	0.25	0.41	0.03	0.46	0.19	1.63	0.34	0.85	1.18	0.44	2.57	3.00	0.52	1.63
Warwick STW (Longbridge)	n	91	90	91	74	75	74	91	91	91	0	91	91	91	91	91
	MV	0.4	72.6	13.7	7.3	1.0	183.6	5.8	4.8	49.6		4.1	0.3	97.8	3.4	0.1
	SD	0.1	20.6	3.9	0.3	0.5	53.1	4.6	3.1	27.5		1.7	0.6	223.1	1.6	0.2
	CV	0.29	0.28	0.29	0.04	0.44	0.29	0.80	0.64	0.55		0.41	1.75	2.28	0.46	2.01
Avon (Stare Bridge)	n	87	88	90	74	76	74	93	93	93	56	93	93	93	93	93
	MV	1.8	106.4	11.9	8.2	0.8	168.7	9.8	1.8	38.7	14.8	1.0	0.1	17.6	0.8	0.1
	SD	2.3	26.2	5.1	0.4	0.3	29.1	15.0	0.6	59.7	12.4	0.6	0.2	50.6	0.6	0.1
	CV	1.27	0.25	0.43	0.05	0.41	0.17	1.53	0.36	1.54	0.84	0.59	3.00	2.88	0.76	1.64
Avon (Lawford)	n	83	83	84	69	71	69	89	89	89	53	89	89	89	89	89
	MV	1.4	95.4	12.4	7.9	0.8	197.4	16.9	2.2	38.3	12.4	1.5	0.1	13.3	1.2	0.1
	SD	1.9	21.4	5.5	0.3	0.4	33.5	31.2	0.8	28.4	8.9	1.1	0.4	23.4	1.1	0.2
	CV	1.30	0.22	0.44	0.04	0.42	0.17	1.85	0.38	0.74	0.71	0.79	3.07	1.76	0.97	1.32
Rugby STW (Newbold)	n	82	84	86	70	72	70	88	88	88	0	88	88	88	88	88
	MV	0.3	72.7	15.0	7.4	0.9	214.0	6.5	4.0	51.6		2.8	0.3	69.1	2.2	0.1
	SD	0.1	22.8	4.4	0.2	0.4	36.9	5.9	3.0	30.2		1.5	0.7	179.1	1.5	0.1
	CV	0.32	0.31	0.30	0.03	0.47	0.17	0.90	0.75	0.59		0.56	2.56	2.59	0.72	1.36
Avon (Clifton)	n	82	86	88	72	74	72	88	88	88	52	88	88	88	88	88
	MV	0.6	86.1	10.9	7.9	0.6	206.6	11.6	1.5	25.6	11.1	0.2	0.0	13.9	0.1	0.3
	SD	1.2	21.8	4.9	0.3	0.3	39.6	27.6	0.7	19.7	12.6	0.1	0.1	40.6	0.1	0.3
	CV	1.81	0.25	0.44	0.04	0.45	0.19	2.37	0.50	0.77	1.14	0.60	1.57	2.91	1.13	1.07
Stanford Reservoir	n	0	38	40	34	35	34	40	40	40	23	40	40	40	40	40
	MV		104.4	11.9	8.7	0.4	174.7	14.1	3.3	29.9	60.3	0.1	0.0	3.5	0.0	0.3
	SD		24.2	5.9	0.6	0.1	36.0	17.1	5.8	14.4	188.9	0.1	0.1	4.9	0.1	0.4
	CV		0.23	0.49	0.06	0.34	0.21	1.22	1.74	0.48	3.13	0.53	1.23	1.40	5.75	1.06

Table 9.4. Basic data from the 95 Swedish catchments and lakes (from Håkanson and Peters, 1995).

	Mean	Min.	Max.	n		Mean	Min.	Max.	n
A. Drainage area (no DAZ correction; abbreviated %)					**D. Lake morphometric parameters**				
A_{DA} (km^2)	15.8	0.36	81	95	length, L_{max} (km)	1.82	0.22	7.80	95
RDA	41.0	8	194	95	max. depth, D_{max} (m)	12.4	1.7	52.0	95
Lake%	2.3	0	13.1	95	lo (km)	7.01	0.98	35.30	95
Forest%	80.0	60.9	100	95	total area (km^2)	0.79	0.02	5.53	95
Mire%	14.2	0	35.3	95	area (km^2)	0.77	0.02	5.26	95
Open land (OL)%	3.4	0	20.8	95	area/A_{DA}	0.073	0.002	0.391	95
Rock%	9.1	0	77.0	95	volume (km^3)	0.0041	0.00005	0.0495	95
Till%	67.7	3.2	96.0	95	mean width, B_m (km)	0.35	0.06	1.05	95
Coarse sed (CS)%	5.1	0	40.3	95	mean depth, D_m (m)	4.18	0.90	10.10	95
Fine sed (FS)%	1.6	0	28.4	95	relative depth, D_{rel}	1.71	0.18	7.39	95
Basic%	2.3	0	50.0	95	shore irregularity, F	2.43	1.40	8.20	95
Intermediate rock%	24.1	0	100	95	forn factor, Vd	1.18	0.50	2.20	95
Acid%	73.6	0	100	95	dynamic ratio, DR	0.24	0.04	1.13	95
					ET (% of area)	26	7	90	95
B. Drainage area (with DAZ correction; abbreviated Z)					discharge, Q (m^3/s)	0.164	0.010	0.850	95
LakeZ	2.3	0	14.6	95	retention time, T (yr)	0.873	0.020	4.790	95
ForestZ	80.0	61.7	100	95					
MireZ	14.0	0	35.7	95	**E. Water chemical variables (mean annual values)**				
OLZ	3.6	0	23.1	95	Secch depth (m)	2.5	1.1	6.3	88
RockZ	9.1	0	76.9	95					
TillZ	67.5	3.8	96.0	95	**Abbreviations**				
CSZ	5.3	0	39.6	95	A_{DA}=Drainage area				
FSZ	1.9	0	32.4	95	RNA=Relief near area = dH/$\sqrt{ADA}$				
BasicZ	2.3	0	48.2	95	RDA=Relief drainage area				
Interm.Z	23.8	0	100	95	Lake%=Lake% of ADA (etc.)				
AcidZ	73.9	0	100	95	ForestN=Forest in % of near area (etc.)				
					%=Entire drainage area, no DAZ correction				
C. Near area parameters (abbreviated N)					Z=DAZ corrected values				
RNA	43.8	4	177	95	N=values from the near area (from DAZ method)				
ForestN	78.5	30.0	100	95	ET=Bottom areas of fine sediment erosion and				
LakeN	1.8	0	29.8	95	transportation				
MireN	13.4	0	45.5	95	lo=Shoreline length				
OLN	6.2	0	70.0	95					
RockN	7.9	0	76.9	95					
TillN	65.6	7.0	100	95					
CSN	5.7	0	59.0	95					
FSN	5.6	0	77.1	95					
BasicN	1.6	0	54.3	95					
Interm.N	24.4	0	100	95					
AcidN	74.0	0	100	95					

9.1.3. Marine areas

This database consists of two parts, one for SPM and covariables in Baltic coastal areas (from Wallin et al., 1992 and Håkanson et al., 2004a); the other for SPM and covariables from open marine areas (also in the Baltic; from Håkanson and Eckhell, 2004).

The 17 coastal areas are located in the Baltic Sea. Five of the areas are in the St. Anna archipelago off the Swedish east coast, 7 areas are located in the Blekinge archipelago, in the south of Sweden, and the remaining 5 areas in the Åboland archipelago, SW of Finland, fig 9.1. The Baltic Sea is brackish with a salinity ranging from 5 to 10% in a north-south gradient. Compared to other seas the Baltic Sea, with its mean depth of only 56 m, is shallow. It is almost entirely surrounded by land and the tidal variation is small (< 20 cm; see Voipio, 1981). The Åboland archipelago, Finland, is the largest archipelago in the Baltic Sea. It reaches from Åland to the Finnish main land. The average depth is low but tectonic faults as deep as 50 to 60 m do exist. The St Anna archipelago has many islands. The bays are deep and long and several have thresholds towards the sea. The Blekinge archipelago in S. Sweden is narrow and the water circulation is generally good (Persson et al., 1994). Table 6.2 gave a compilation of coastal data.

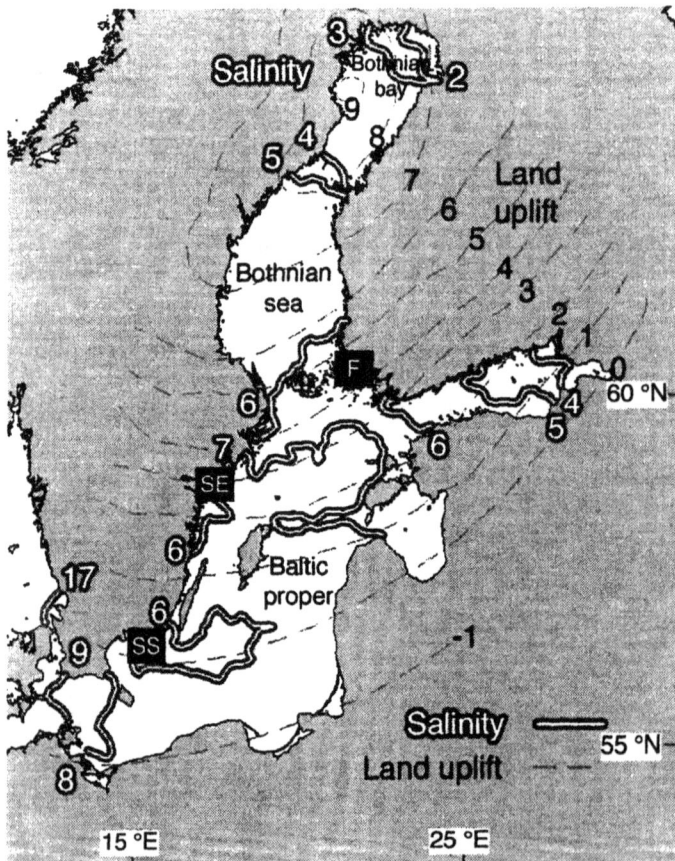

Fig. 9.1 Geographical overview of the Baltic Sea and location of the studied coastal areas. SS = S. Sweden, SE = Swedish east coast, F = Finland. Land uplift in mm/yr. Salinity in %. The scale is given by the latitudes and longitudes.

This work will also use a database from an open marine area in the Baltic Sea (tables 9.5 and fig. 9.2; data from Håkanson and Eckhell, 2004). Samplings from the research vessel M/S Sunbeam were carried out at five sites from three areas at the Swedish site of the Baltic Sea (see

Nilsson, 2002, for a more detailed description of the sampling sites). In September, 2001, samples were taken at the two sites, Utö and Mysingen. The Mysingen site, with a water depth of about 35 m, is situated in the outer Stockholm archipelago. Samples were taken at various depths (1, 10, 20, 25, 30, and 35 m) at six different occasions during three days. The Utö site is situated far out in the archipelago and has a water depth of about 65 m. At this site, samples were taken every 10 m at nine different occasions during four days. The sampling started one day earlier than in Mysingen, but stopped at the same time. The intention was to carry out sampling during a storm event, i.e., before, during and after a storm, to study the storm effects on the SPM concentration.

Sampling was carried out in 1999 at the sites Gälnan, Arholma, and Laxen. The Gälnan site, with a water depth of about 30 m, is situated in the Stockholm archipelago. The samples were taken once in August and once in October at various water depths. The Arholma site, with a water depth of about 59 m, is situated in an open position to all wind directions except from the west. Samples were taken at various depths during two days in October. The Laxen site is situated offshore, between the Swedish mainland and the island of Gotland. Samples were taken during two days in June. The water depths at the Laxen site were: 60, 90, 110, 110, and 80 m.

The water samples were filtered and the dry weight measured. Temperature and salinity profiles were also taken for all 17 sites/sampling events. The data are used to study the variations in mean SPM values among the sites/events (see section 6.4).

Wind data from three different wind stations close to the sites were obtained from the Swedish Meteorological and Hydrological Institute (SMHI). Every third hour, the average wind speed was calculated from a 10-minute period preceding the observation. Data from the Landsort wind station were used for the two sites Utö and Mysingen, data from the Svenska Högarna wind station for the Gälnan and Arholma sites, and data from Gotska Sandön wind station for Laxen. These data are used to calculate the number of days with wind speeds lower than 10, 7, 5, and 3 m/s for the 17 sites/events to get a measure of the number of calm days before the sampling (table 9.5). Since there was no period with wind speeds lower than 3 m/s, these data have been omitted from the following analyses.

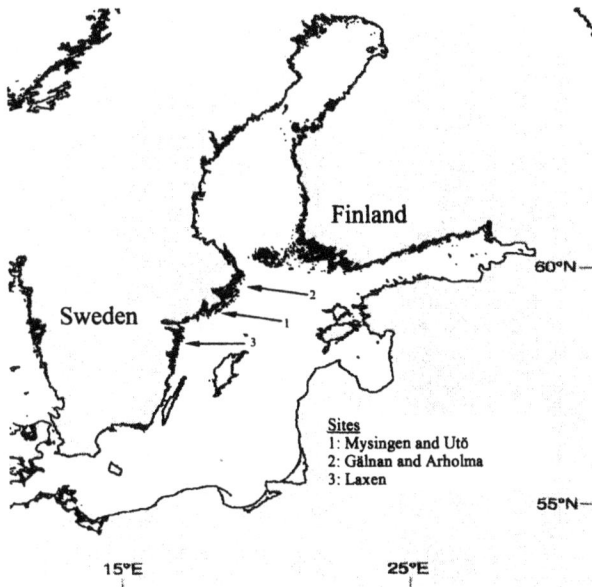

Fig. 9.2. Location map illustrating the geographical positions of the investigated sites in the open Baltic Sea.

Table 9.5. Information about the sampling sites from the open water areas in the Baltic Sea.

	Date	Depth	Number of samples
Utö	9/9-01 - 12/9-01	65	58
Mysingen	10/9-01 - 12/9-01	35	31
Gälnan	15/8-99, 20/10-99	30	12
Arholma	15/8-99, 20/10-99	59	30
Laxen	4/6-99, 7/6-99	30 - 115	69
		Sum:	200

9.2. The LakeWeb modelling approach

The details of the LakeWeb model have been presented by Håkanson and Boulion (2002) will not be repeated here, where the aim is to present the basic and specific principles of this foodweb model.

9.2.1. Outline of the model

The LakeWeb model is a general dynamic model quantifying key aquatic foodweb interactions and biotic/abiotic feedbacks for entire systems and key functional groups, which are present in most types of aquatic systems (lakes, rivers, and marine areas). The functional groups are: (1) predatory fish, (2) prey fish, (3) zoobenthos, (4) predatory zooplankton, (5) herbivorous zooplankton, (6) phytoplankton, (7) bacterioplankton, (8) benthic algae, and (9) macrophytes. The functional group "predatory" fish does the work of eating "prey fish," which does the job of consuming two functional groups of zooplankton (herbivorous and predatory) and zoobenthos, etc. LakeWeb model uses ordinary differential equations (compartment modelling for defined lakes or marine areas) and gives weekly variations on production and biomasses of the nine functional groups.

Basically, the LakeWeb model consists of two parts: (1) the foodweb submodel, which is driven by chlorophyll data and calculates production (kg ww per week) and biomass (kg ww) of the functional groups (fig. 9.3), and (2) the mass-balance model (LakeMab) for nutrients (P and/or N), which is based on transport processes (fig. 9.4), which appear in most system and for most substances (nutrients, organics, radionuclides, etc.): 1. Sedimentation, i.e., the transport from water to sediments; 2. Resuspension, i.e., the transport of matter from sediments back to water; 3. Diffusion, i.e., the transport of dissolved substances from sediments back to water; 4. Mineralization, i.e., the bacterial decomposition of matter; 5. Mixing, i.e., the upward and downward transport of dissolved and suspended matter across the thermocline or the wave base. Some systems are constantly mixed and do not stratify, but many systems develop a thermocline in the summer if they are deep enough; 6. Bioturbation is the mixing of the deposited materials from the movement of the bottom fauna (from their eating, digging and foraging activities); 7. Compaction, i.e., the vertical change in sediment porosity due to the weight of overlying sediments; 8. Burial, i.e., the transport from surficial sediments to deeper sediments; and 9. Biouptake and retention in biota. This is one of the linkages between the mass-balance model for nutrients (LakeMab) and the foodweb model within the LakeWeb model. The mass-balance model for phosphorus has been presented by Håkanson et al. (2003b).

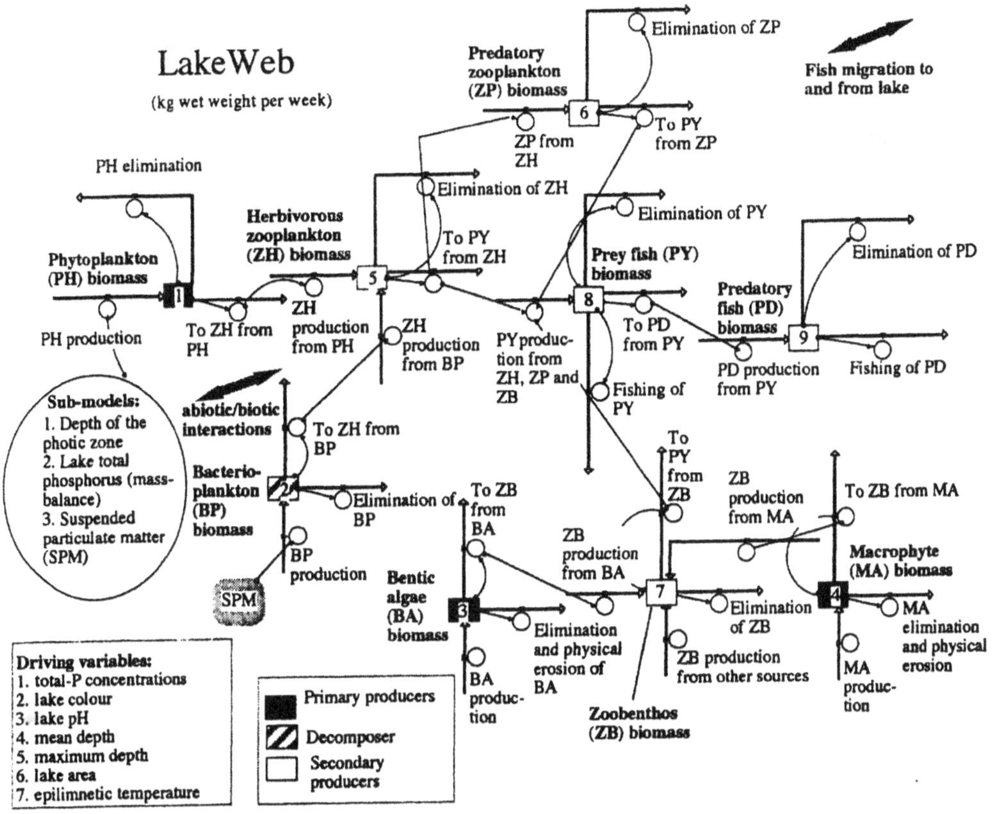

Fig. 9.3. An outline of LakeWeb, a model to quantify all important foodweb interactions (production, grazing, etc.), including biotic/abiotic feedbacks in a general manner.

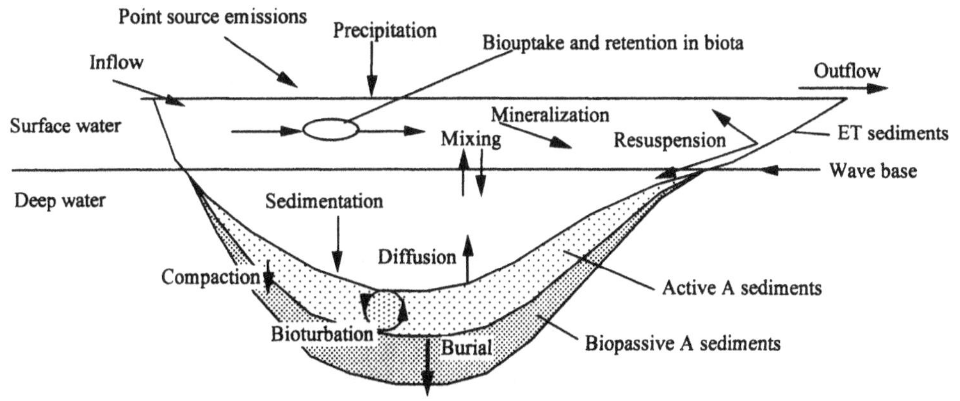

Fig. 9.4. Illustration of general and fundamental transport processes to, within and from aquatic systems used in the mass-balance model for phosphorus in the LakeWeb model. The "ET areas" shown in the figure are the erosion and transportation areas with resuspension, and the areas with "active A sediments" are the biologically active sediment areas where fine sediments are continuously being deposited (accumulation areas).

Fig. 9.5 gives an example of the predictive power of the LakeWeb model along a trophic state gradient for herbivorous zooplankton (fig. A), predatory zooplankton (fig. B), prey fish (fig. C) and phytoplankton (fig. D). Note that this is not a calibration or a tuning, it is a comparison between dynamically modelled values and values given by empirical models, which provide normal reference values (often called norms) (table 2.4). To produce the results in fig. 9.5, there were no changes in the model or the driving variables except for the given five steps in TP concentrations (10, 30, 100, 300, and 1000 µg/l). Similar tests have been carried out with the same positive results for all nine key functional groups of organisms included in the LakeWeb model. Tests like those shown in fig. 9.5 have also been carried out with good results along humic level gradients, temperature gradients, and lake size gradients.

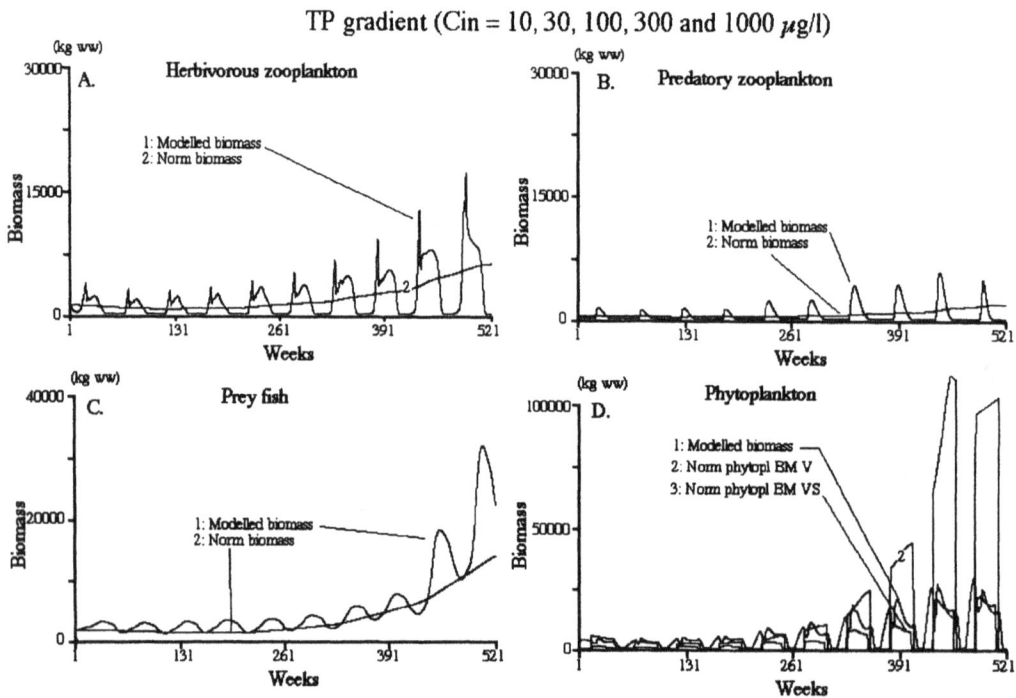

Fig. 9.5. Critical model testing (sensitivity analyses) for the following four variables (for a lake with an area = 1 km², mean depth = 10 m, and a pH = 7) along a trophic state gradient. The driving variable is the tributary TP concentration, which has been varied in 2-year steps from 10 to 1000 µg/l, while all else is constant. Note that the empirical models (called norms) often do not provide any seasonal patterns, only mean values for the growing season.
A. Gives predicted biomass of herbivorous zooplankton (curve 1) and the corresponding empirical regression line (= the norm; curve 2).
B. Gives modelled biomass of predatory zooplankton (curve 1) and the empirical norm (curve 2).
C. Gives biomass of prey fish (curve 1) and the corresponding norm (curve 2).
D. Gives predicted biomass of phytoplankton (curve 1) and the two empirical norms, the upper, maximum curve (2) as calculated for the entire lake volume, and curve 3 (the mean value) as calculated for the volume of the photic zone.

Several practical scenarios describe how the model can be applied to address important management issues, like consequences of biomanipulations (extensive fishing), changes in land-use, deposition of acid rain, global temperature changes and eutrophication and remedial measures to reduce toxins in fish (Håkanson and Boulion, 2002). LakeWeb is a powerful tool to simulate remedial measures and to get realistic expectations of positive and negative consequences of such measures.

Mass-balance modelling of nutrients is a central paradigm in aquatic ecology and management (Vollenweider, 1968; OECD, 1982) and by including this mass-balance model for nutrients in the foodweb model, it is also possible to calculate the uptake and retention of nutrients in biota. Fig. 1.10 gave an example of how the mass-balance model within the LakeWeb model predicts TP concentrations in Lake Miastro, Belarus.

Even though figures 9.5 and 1.10 indicate that the LakeWeb model can predict well, there will always be situations which it will not describe well. This is, in fact, one important motive for the model. It is meant to capture normal interactions among the most important factors influencing a given y variable so that all other phenomena can be interpreted in relation to these factors.

9.2.2. Basic mathematical structures of the primary production unit

This section first gives the basic mathematical structure of each primary production unit in the model. Fig. 9.6 illustrates the principle set-up of each primary compartment (e.g., phytoplankton, benthic algae, or macrophytes) and the obligatory driving variables.

$$BM_{PU}(t) = BM_{PU}(t - dt) + (IPR_{PU} - CON_{PUSU} - EL_{PU})\cdot dt \qquad (9.1)$$

Where BM_{PU} is the biomass (BM) of the primary unit (PU; in kg ww).

Since this is a primary production unit, the production is directly related to abiotic limiting factors, such as the concentration of nutrients, water clarity, temperature, etc. This is accounted for in the LakeWeb model in various ways, generally by dimensionless moderator techniques. That is:

$$IPR_{PU} = PR_{PU}\cdot Y_X\cdot Y_{temp} \qquad (9.2)$$

where

IPR_{PU} = the initial production of the primary unit; generally in kg ww per week;

PR_{PU} = the mass production rate of the primary unit; for phytoplankton, this is calculated from chlorophyll (kg ww per week);

Y_X or Y_{temp} = dimensionless moderators expressing how changes in environmental conditions, like nutrient concentrations (X), water clarity or water temperature influence primary production. Y_X has the following general definition:

$$Y_X = (1 + amp\cdot(X_{act}/X_{norm} - 1)) \qquad (9.3)$$

Where the amplitude value (amp) quantifies how changes in actual values (X_{act}) relative to a normal (= reference) value (= X_{norm}) will influence the production; if $X_{act} = X_{norm}$, then $Y_X = 1$.

Biomass SU

BMsu

Metabolic efficiency ratio SU for PU

Initial production of SU from PU

Turnover time SU
(Tsu)

Normal consumption
rate SU (NCRsu)

Normal biomass of SU
(NBMsu)

Consumption
rate PU to SU

Consumption
of PU by SU

**Production of
PU (PRpu)**

Biomass

Normal biomass
of PU
(NBMpu)

BMpu

Total phosphorus
(TP)

Volume (Vol)

Production rate
(Rpu)

Elimination
(EL)

Initial
production
of PU

Turnover time
of PU (Tpu)

Area
(A)

Mean depth
(Dm)

Environmental
factor X (e.g.,
pH)

Dimensionless
moderator for X
(Yx)

Dimensionless
moderator for
temp, Ytemp

Water
temperature

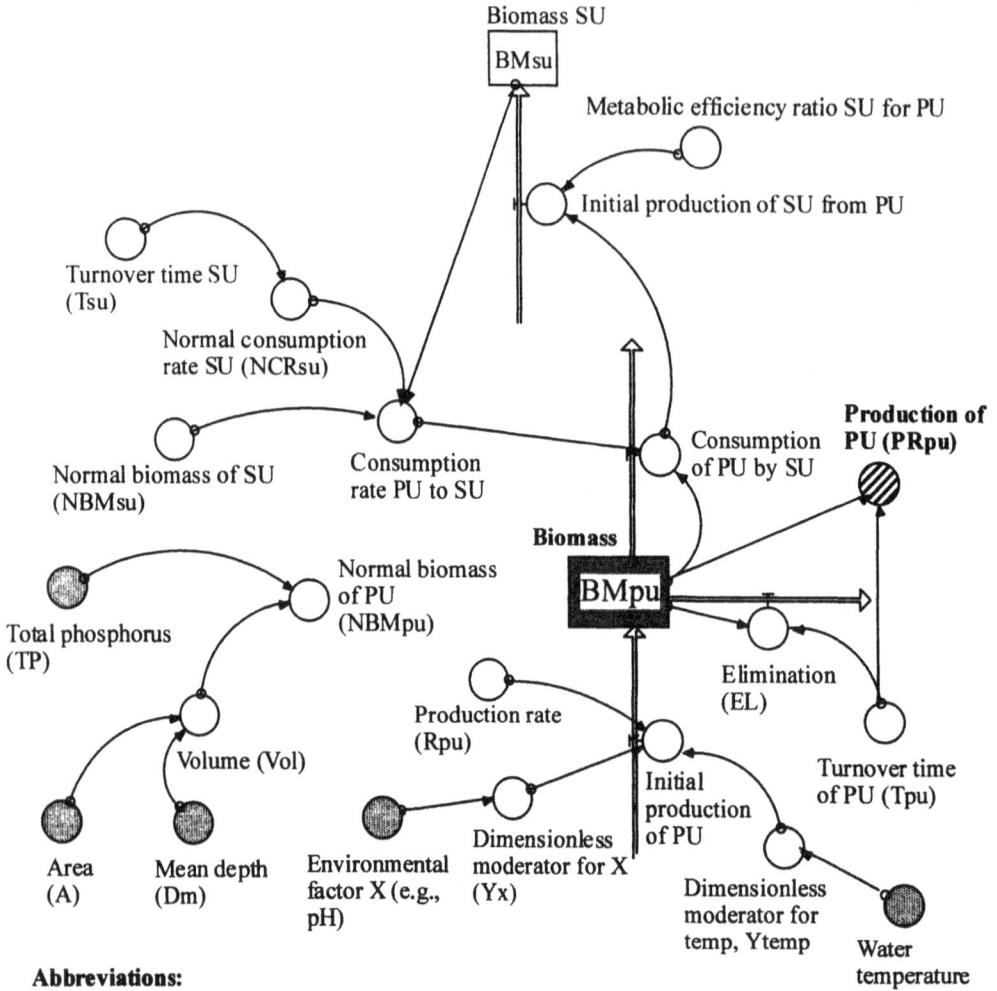

Abbreviations:
BM = Biomass
CON = Consumption (= outflow)
CR = Actual consumption rate
EL = Elimination
IPR = Initial production (= inflow)
MER = Metabolic efficiency ratio
NCR = Normal consumption rate
PR = Production
PU = Primary unit (e.g., phytoplankton)
R = Rate
SU = Secondary unit (e.g., herbivorous zooplankton)
T = Turnover time
Y = Dimensionless moderator

Set-up using:
• 1 primary production unit,
the target unit in this set-up
• 1 secondary unit

⬤ = Driving variables

Fig. 9.6. Generalized set-up of the mathematical structure of each primary production unit in the LakeWeb model. The figure also gives general abbreviations.

Table 9.6. Characteristic turnover times for key functional groups of organisms (T = BM/PR; where BM = biomass in kg ww PR = biomass production in kg ww/day). Based mainly on data from Winberg (1985a). From Håkanson and Boulion (2002).

Group	Turnover time (days)
Phytoplankton	3.2
Bacteria	2.8
Benthic algae	4.0
Herbivorous zooplankton	6.0
Predatory zooplankton	11.0
Prey fish	300
Predatory fish	450
Zoobenthos	128
Macrophytes	300

The loss of biomass is given by two processes, (1) elimination (EL), which is related to the mean characteristic turnover time of the functional group (T_{PU} in weeks) and (2) the consumption (= CON, i.e., the predation or grazing) by a secondary production unit (SU; e.g., zooplankton, see next section). Elimination (EL_{PU}) is generally given by:

$$EL_{PU} = BM_{PU} \cdot 1.386/T_{PU} \tag{9.4}$$

Where 1.386 is the halflife constant (-ln(0.5)/0.5 = (0.693/0.5; see Håkanson and Peters, 1995) and T_{PU} is the mean, characteristic turnover time of the organisms in the given compartment. Table 9.6 gives a compilation of characteristic turnover times for the nine functional groups of organisms included in the LakeWeb model. Evidently, the turnover times among the single species constituting a functional group can vary considerably. The turnover time (T) of a given functional group of organisms is defined in the traditional way as T = BM/PR, where BM = the biomass of the organism in kg ww and PR = the mass production in kg ww/week. Note that the initial production (IPR; see eq. 9.2) is higher than the actual production (PR = BM/T) because IPR gives production without losses from grazing and elimination.

An important feature of the LakeWeb model concerns the technique to calculate predation and feedbacks. The predation by a secondary unit on the biomass of a primary unit is given by:

$$CON_{PUSU} = BM_{PU} \cdot CR_{SU} \tag{9.5}$$

Where CON_{PUSU} is the flow of biomass per time unit (kg ww/week) out of the primary unit from consumption by the secondary unit. The actual consumption rate (it is not called a rate constant because it is not a constant), CR_{PU} (1/week) is defined by:

$$CR_{SU} = (NCR_{SU} + NCR_{SU} \cdot (BM_{SU}/NBM_{SU} - 1)) \tag{9.6}$$

where

NCR_{SU} = the normal consumption rate of the given secondary production unit (1/week);

NBM_{SU} = the normal (= reference) biomass of the secondary unit (kg ww/week); e.g., calculated from table 2.4.

Basically, the consumption rate for any given functional group is related to three factors:

1. The ratio between the actual biomass (BM) and the normal biomass (NBM) of the predator. The higher this ratio, the higher the predation pressure on the given prey.

2. The number of first order food choices (NR; fig. 9.7 for illustration). The structure of the LakeWeb model involves several simplifications and there are always either one or two first order food choices. This means that NR is 1 or 2.

3. The inverse of the turnover time (T_{SU}) of the predator. Animals with quick turnover times create a greater predation pressure on their prey than animals with long turnover times (and a higher value of the actual consumption rate, CR_{SU}). So, predatory fish will eat a relatively small fraction of the total available biomass of its prey per time unit. Small herbivorous zooplankton, on the other hand, are likely to consume a larger percentage of their prey (such as phytoplankton and bacterioplankton) per unit of time.

That is, for animals in the secondary unit:

$$NCR_{SU} = NR_{SU}/T_{SU} \tag{9.7}$$

A consumption rate of 0.2 means that 20% of the biomass of the primary unit is consumed by the animals in the secondary unit per week.

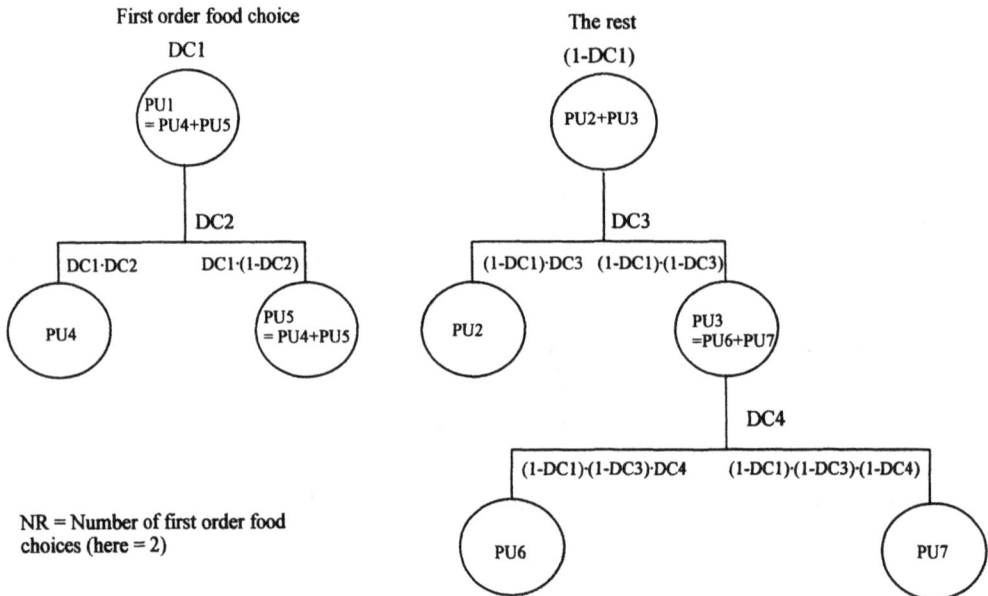

Fig. 9.7. Schematic outline of a food choice panel for a secondary unit with two first order food choices, four second order and two third order food choices. DC = Distribution coefficient (dimensionless).

9.2.3. Basic mathematical structures of the secondary production unit

Fig. 9.8 gives an illustration of the modelling approach and abbreviations. The basic equation is given by:

$$BM_{SU}(t) = BM_{SU}(t - dt) + (IPR_{SUPU1} + IPR_{SUPU2} - CON_{SUPU1} - CON_{SUPU2} - EL_{SU}) \cdot dt$$
$$(9.8)$$

Where BM_{SU} is the biomass (BM) of the secondary production unit (SU; in kg ww).

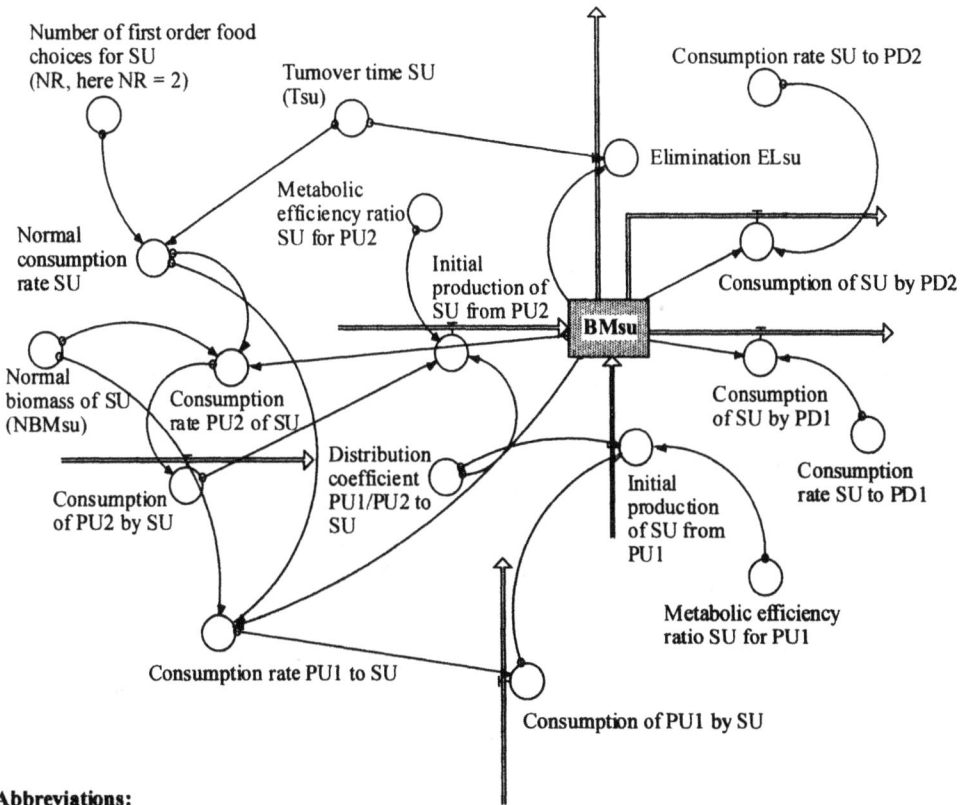

Abbreviations:
BM = Biomass
EL = Elimination
NR = Number of first order food choices
PU1 and PU2 = Primary unit 1 and 2
PD1 and PD2 = Predator unit 1 and 2 (feeding on the secondary unit)
SU = Secondary unit (e.g., herbivorous zooplankton)
T = Turnover time

for fish also:
MIGin = migration into the lake
MIGout = migration from the lake

Set-up using:
• 2 primary production units,
• 2 predatory units and
• 1 secondary unit, the target unit in this set-up

Fig. 9.8. Set-up of each secondary unit in the LakeWeb model. The figure also gives general abbreviations.

In the set-up shown in fig. 9.8, the initial secondary production (IPR) is related to two fluxes, from primary units 1 and 2 (PU_1 and PU_2). For each flux, IPR is given by:

$$IPR_{SUPU1} = DC_1 \cdot CON_{PU1SU} \cdot MER_{PU1SU} \qquad (9.9)$$

Where DC_1 is the first order distribution coefficient. The LakeWeb model uses a simple general system to assign weights on food choices and adjust the consumption rates for the number of food choices. There are, for example, three food choices for zoobenthos (benthic algae, macrophytes, and "sediments"); prey fish have a menu of three food alternatives (predatory zooplankton, herbivorous zooplankton, and zoobenthos); and predatory zooplankton only eats herbivorous zooplankton. The secondary unit in fig. 9.9 has two first order food choices, so NR = 2. If there are more than two food choices, they are first differentiated by a distribution coefficient (DC_1) into two first order food choices and then by a second distribution coefficient (DC_2) into second order food choices, etc.

CON_{PU1SU} is the consumption of biomass from the compartment PU_1 from grazing by the animals in compartment SU (kg ww/week). The actual consumption rate (CR_{SUPU1}) is calculated in the same manner as already discussed for the primary production unit.

The metabolic efficiency ratio (dimensionless), MER_{PU1SU}, gives the fraction of the food that actually increases the biomass of the secondary unit (the consumer). Table 9.7 gives a compilation of characteristic MER values used for all groups of organisms in the LakeWeb model. Note that the MER value for zoobenthos eating sediments is low (25% of the MER value for zoobenthos feeding on benthic algae).

Table 9.7. Metabolic efficiency ratios for key functional groups (MER = PR/CON, dimensionless). The MER value is basically calculated from the mass-balance equation, CON = PR + RES + FAE, where CON = consumption, PR = production, RES = respiration, FAE = unassimilated food (feces) and T = turnover time (= BM/PR, days; BM = biomass). The actual consumption rate, CR, quantifies reductions in biomass of prey organisms per unit of time. The values used for PR, RES, FAE, and MER are mainly based on data from Winberg (1985a). From Håkanson and Boulion (2002).

	CON	PR	RES	FAE	MER	T	CR	Consumes	Consumed by
Zooherb	100	24	36	40	0.24	6.0	0.17	Phytopl., Bacteriopl. Zoopred	Prey fish
Zoopred	100	32	48	20	0.32	11.0	0.091	Zooherb	Prey fish
Zoobenthos	100	15	35	50	0.15	65	0.015	Macrophytes, Benthic algae	Prey fish
Prey fish	100	16	64	20	0.16	300	0.016	Zoobenthos, Zooherb, Zoopred	Pred. fish
Predatory fish	100	25	55	20	0.25	450	0.0013-0.02	Prey fish	Pred. fish

The initial production of the secondary unit, SU, from consumption of the second primary unit (IPR_{SUPU2}) is handled in the same manner using the same consumption rate but ($1-DC_1$) instead of DC_1. Then, this initial production is given by:

$$IPR_{SUPU2} = (1-DC_1) \cdot CON_{PU2SU} \cdot MER_{PU2SU} \qquad (9.10)$$

Each consumption flow is given as a function of, (1) the normal consumption rates (here NCR_{SUPD1} or NCR_{SUPD2}), (2) the normal biomasses of the two predatory units (NBM_{PD1} and NBM_{PD2}), and the actual biomasses of the two predatory units (BM_{PD1} and BM_{PD2}), as these are calculated by the LakeWeb model in the same manner as already discussed for the primary production unit (eqs 9.5 and 9.6).

Note that the model quantifies changes in the actual consumption rate of the prey unit related to changes in the biomass of the consumer: More animals in the secondary unit (the higher BM_{SU}) means a higher actual consumption rate, CR_{SU}. If the actual biomass of the secondary unit is equal to the normal biomass of the secondary unit, then $BM_{SU}/NBM_{SU} = 1$, and $CR_{SU} = NCR_{SU}$. If the actual biomass of the secondary unit, BM_{SU}, is twice the normal biomass of the secondary unit, then $CR_{SU} = 2 \cdot NCR_{SU}$. So, the model gives a linear increase in consumption with increases in biomass of the secondary unit.

Elimination (EL_{SU} = the loss of biomass in kg ww/week from the secondary compartment) is given as:

$$EL_{SU} = BM_{SU} \cdot 1.386/T_{SU} \qquad (9.11)$$

where 1.386 is the halflife constant.

Fish migrate between habitats for feeding and spawning (Harden-Jones, 1968; Northcote, 1978; McDowall, 1988; Wotton, 1990; Brittain and Brabrand, 2001). This is accounted for in the LakeWeb model in a simplified manner. One should expect a net inflow of fish via up and downstream systems or via communicating subbasins if the actual fish biomass (BM_F) is lower than the normal fish biomass (NBM_F), since fish are likely to migrate to and stay in better habitats, and vice versa (Busch and Sly, 1992). The migration in lakes is a function of the tributary water discharge (Q_{mv}); if the system is large and divided into communicating subbasins, the water transport between the basins may be used in the same manner as tributary water discharge.

Prey fish is the most complex group of all in the LakeWeb model. Prey fish feed on three other groups, zoobenthos, herbivorous zooplankton, and predatory zooplankton. Neither prey fish nor predatory fish are, however, "permitted" to display cannibalism in the LakeWeb model, i.e., feeding on their own group (like roe and/or young-of-the-year). Cannibalism exists in natural aquatic systems among fish (see Menshutkin, 1971), but to gain simplicity, the LakeWeb model calculates net production of fish. Using this set-up for prey fish, calibrations have indicated that unrealistically low production values would generally be obtained for prey fish (with 3 food choices) if one sets the normal consumption rate for prey fish to $3/T_{PY}$. A more realistic value seems to lie between the NCR values used for predatory zooplankton and predatory fish. The main reason for this is related to the structuring of the prey fish compartment, i.e., that all types of prey fish are compiled into one functional group.

Macrophytes can appear with relatively high biomasses in aquatic systems and they can play several important roles, e.g., to bind nutrients (this is handled in the LakeWeb model by the mass-balance model for phosphorus), as a substrate for zoobenthos (this is handled in LakeWeb in the submodel for zoobenthos), and by providing shelter for fish and thereby increasing fish production. It is important to note that the latter is not related to fish feeding on macrophytes (although single species, like carp, do), but indirectly, by providing more protected living conditions for the small fish. Also this process is included in the LakeWeb model.

To conclude, fundamental concepts in the LakeWeb model are: (1) Consumption rates - "how much of the prey biomass is consumed per time unit by the predator?" (2) Metabolic efficiency ratios for each compartment - "how much of the food consumed will increase the biomass of the consumer?" (3) Turnover or retention rates for each compartment - "how long is the mean, characteristic lifespan of the group?" (4) Food choices - "if there is a food choice, how much is consumed of each food type?" (5) Migration rates for fish - "how much of the fish will leave and enter the system per unit of time?" The LakeWeb model is based on a general production unit that may be mechanistically understood and repeated for different functional groups of organisms.

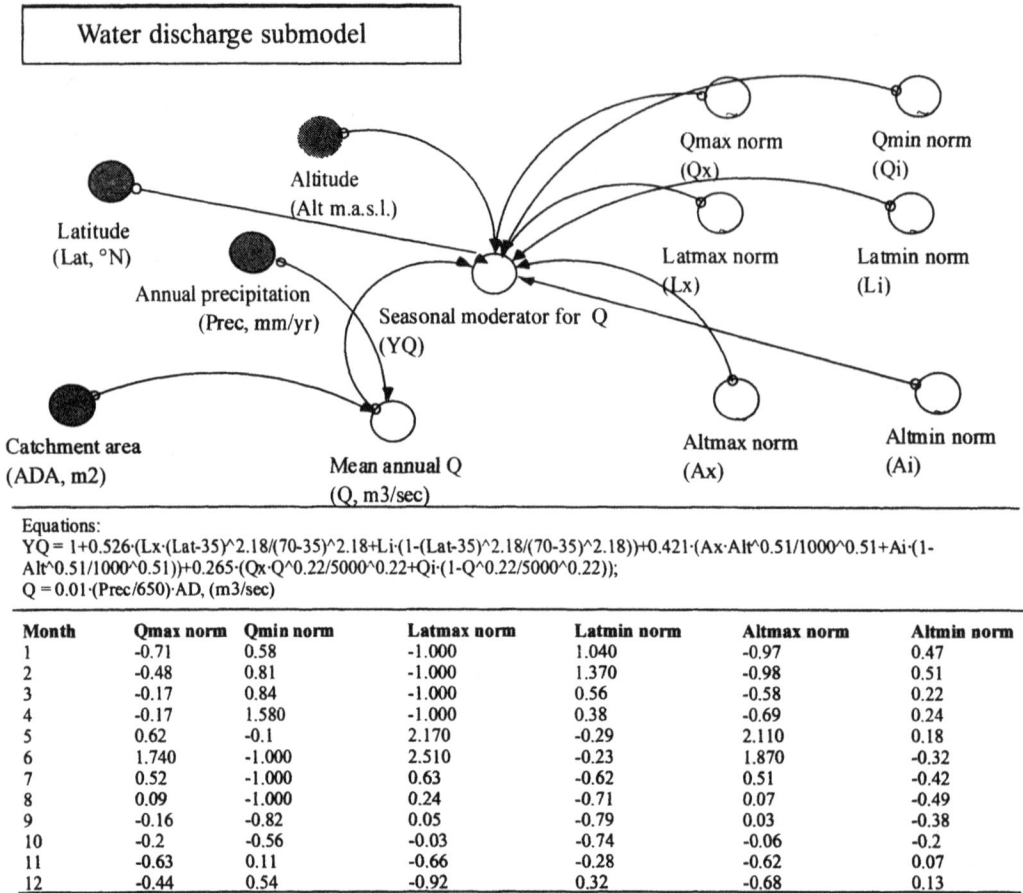

Equations:
$YQ = 1+0.526 \cdot (Lx \cdot (Lat-35)^{2.18}/(70-35)^{2.18}+Li \cdot (1-(Lat-35)^{2.18}/(70-35)^{2.18}))+0.421 \cdot (Ax \cdot Alt^{0.51}/1000^{0.51}+Ai \cdot (1-Alt^{0.51}/1000^{0.51}))+0.265 \cdot (Qx \cdot Q^{0.22}/5000^{0.22}+Qi \cdot (1-Q^{0.22}/5000^{0.22}));$
$Q = 0.01 \cdot (Prec/650) \cdot AD, (m3/sec)$

Month	Qmax norm	Qmin norm	Latmax norm	Latmin norm	Altmax norm	Altmin norm
1	-0.71	0.58	-1.000	1.040	-0.97	0.47
2	-0.48	0.81	-1.000	1.370	-0.98	0.51
3	-0.17	0.84	-1.000	0.56	-0.58	0.22
4	-0.17	1.580	-1.000	0.38	-0.69	0.24
5	0.62	-0.1	2.170	-0.29	2.110	0.18
6	1.740	-1.000	2.510	-0.23	1.870	-0.32
7	0.52	-1.000	0.63	-0.62	0.51	-0.42
8	0.09	-1.000	0.24	-0.71	0.07	-0.49
9	-0.16	-0.82	0.05	-0.79	0.03	-0.38
10	-0.2	-0.56	-0.03	-0.74	-0.06	-0.2
11	-0.63	0.11	-0.66	-0.28	-0.62	0.07
12	-0.44	0.54	-0.92	0.32	-0.68	0.13

Fig. 9.9. The submodel for tributary water discharge (compiled from Abrahamsson and Håkanson, 1998).

9.3. Water discharge predicted from map parameters

River discharge depends on many more or less stochastic processes and has a high degree of variability between years for a given river. This means that it is difficult to give a reliable prediction of the water discharge (Q) for a specific river site at a given time. A standard procedure is then to measure the river discharge for a long period of time (decades) and give a statistical estimate of a probability that Q is going to be within a certain range at a certain time. That method is appropriate for many purposes, providing that a sufficiently long and reliable set of empirical data is available (Chow, 1988). However, if empirical data on Q are not available, which is certainly the case for a very large number of rivers, other methods are necessary, e.g., statistical/empirical methods to predict Q from, e.g., soil type distributions, vegetation types, etc. Such models can be very precise and valuable, but they often require field data and site specific catchment data for the calibration.

The model discussed here (fig. 9.9) has been presented by Abrahamsson and Håkanson (1998) to meet specific demands in ecosystem modelling rather than in hydrology. The first requirement is that this model must be based on readily available driving variables, preferably from standard maps. There are many uncertainties in ecosystem models, but all uncertainties are not of equal importance for the predictive success of the model. There will always be uncertainties concerning the proper value for Q. The model presented here is meant to yield predictions of Q, which can be accepted in ecosystem models where the focus is on the predictive power for the concentration of SPM, or toxins in water, or in ecosystem modelling when river discharge and/or lake water retention time are used, i.e., when the target variables to be predicted are biological variables (like fish biomasses) and/or chemical variables (like lake pH and phosphorus concentration). In such contexts, the inevitable uncertainties in the predicted values of Q associated with this simple submodel for Q can be accepted. This is the main reason why this model is based on readily available map parameters, such as latitude, altitude, and precipitation.

To calibrate and validate this Q model, an extensive data set from more than 200 European rivers were used. The discharges of the chosen rivers were not affected by regulation for hydropower or irrigation purposes since that produce unnatural seasonal flow patterns. The time series for the monthly data were at least six years long, and some as long as 80 years. The data sets were divided in two parts of equal size; one for the calibration and the other for the validation. From fig. 9.9, one can note that the only obligatory driving variables for this Q model are, altitude, latitude, mean annual precipitation, and catchment area.

To simulate the monthly variations in Q, six seasonal variability norms are utilized for Europe (fig. 9.9). A seasonal variability norm is used to add a seasonal pattern to an annual value (Håkanson and Peters, 1995). Two of these norms should represent the typical seasonal flow pattern in the most southern and northern parts of Europe, respectively. Two other norms should describe the effect of altitude on monthly variability in Q and two should represent the typical flow pattern of rivers with very small and very large mean annual discharges. Depending on the location and the mean annual discharge of the specific river, the six seasonal variability norms are weighted together and a site-specific seasonal variability norm for Europe is calculated. In the model calibrations, it became obvious that the addition of longitude did not significantly increase the degree of explanation of the model and longitude was, therefore, excluded.

To quantitatively account for how latitude (Lat), altitude (Alt), and mean annual discharge (Q_{mv}) for a specific river influence the seasonal (monthly) variability, different weighing factors ranging from zero to one were developed. For example, the weight factor for latitude should be zero for a river in the southernmost part of Europe (35°N) and one in the northernmost part (70°N). To account for the fact that the relations must not have to be linear, each weighting factor was given an exponent. The exponents are used to control in which range the changes in the parameters are most critical. This is, of course, still a very simple approach to simulate the influence of different parameters. This approach gave the equation for the seasonal moderator for Q (Y_Q) given in fig. 9.9. The values for the six norms are also given in fig. 9.9.

A smoothing function (fig. 9.10) is used to average out seasonal variability given by the seasonal variability norm, which is defined to yield extreme values for Q.

The equation that specifies this calculation is a smoothing or averaging function, which is based on the five, easily accessible factors given above. The SMTH function uses a first-order exponential equation (fig. 9.10) to smooth the input (here the seasonal variability norm for Q). The SMTH function works in the same way as the one- and two-sided running mean values. It may be written as:

$$\text{SMTH} = \text{SMTH(input, averaging function, initial value)} \qquad (9.12)$$

This function smooths the seasonal variability norm for Q by applying a specified averaging function, which operates over a specified time interval, to an input (here the seasonal variability norm), given an initial value for that input. The initial value is simply the mean value of the seasonal variability norm for Q (namely 1). However, since the initial results depend on this initial value, this choice is not trivial.

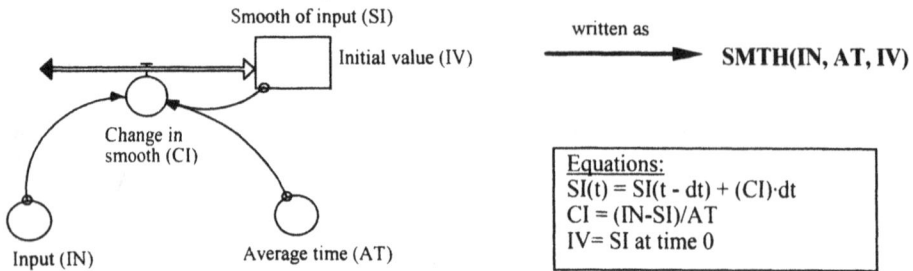

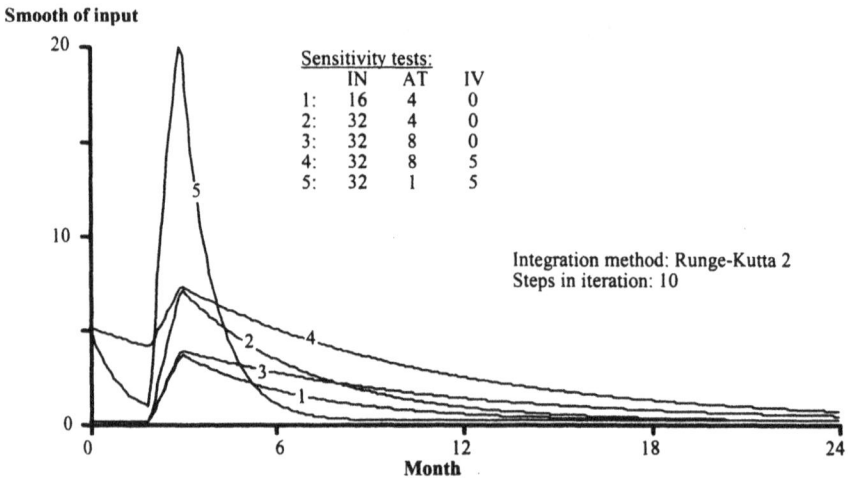

Fig. 9.10. Illustration of the smoothing function.

In spite of the fact that river discharge is a variable with great temporal and spatial variability, this approach has proven to yield good predictions of monthly average Q. The best results of the validation were achieved for the rivers with a mean annual discharge in the range 1 - 500 m³/s. More uncertain predictions are obtained for the smallest and largest rivers. The range is, however, large enough to include most European rivers (88 of 114 in the calibration and 90 of 119 in the validation). Given that limitation, the model certainly has a wide range of applicability.

Note that to apply this Q model for other parts of the world, e.g., for China and/or South America, one must recalibrate the norm and the weight factors for latitude, altitude, and water discharge using empirical data on water discharge from as many rivers as possible.

Fig. 9.11 exemplifies how differences in altitude and latitude influence the seasonal moderator for Q. The default conditions are given by a lake with a catchment area of 10 km², a mean annual precipitation of 650 mm/yr, at an altitude of 75 m.a.s.l. and a latitude of 60°N. Curve 1 in fig. 9.11 gives the characteristic seasonal variations in Y_Q. If a similar lake is situated at an altitude of 1000 m.a.s.l., it is likely that the precipitation is more evenly distributed over the year, and the seasonal variability in Q is smaller. On the other hand, if the lake is placed at latitude 40°N, there is a more pronounced seasonal variability pattern in Y_Q. Note that this is based on extensive calibrations and validations based on empirical Q series from many European catchments areas.

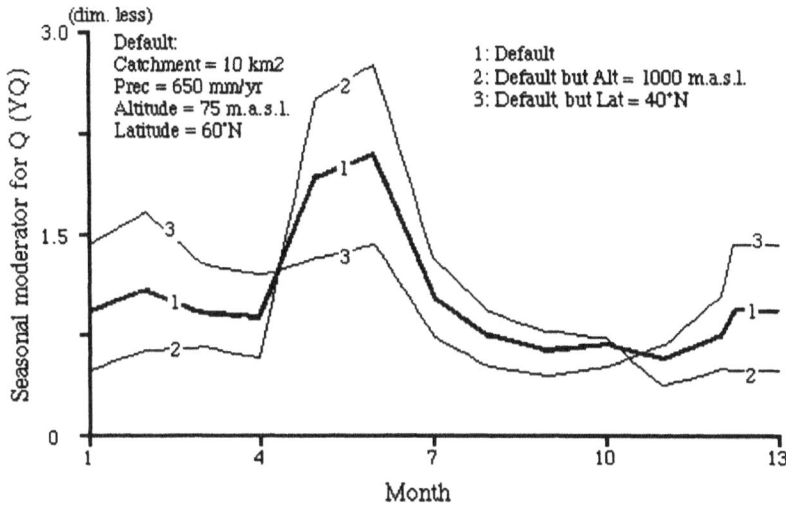

Fig. 9.11. Predicted seasonal patterns in tributary water discharge (Q), as expressed by the dimensionless moderator, Y_Q under default conditions (curve 1), for a catchment at altitude 1000 m.a.s.l. (curve 2) and for a catchment at latitude 40°N (curve 3).

9.4. The submodel for water temperatures

The temperature submodel is illustrated in fig. 9.12. It has been presented by Ottosson and Abrahamsson (1998). It is well known that water (and air) temperature is governed by many complicated climatological relationships. This approach assumes that the following factors are important:

- Altitude (Alt, in m above sea level). The higher the altitude, the lower the lake temperature and the greater the seasonal variability in temperature, if everything else is constant.

- Latitude (Lat). The higher the latitude, the lower the temperature and the greater the seasonal variability in temperature, if all else is constant. Latitude is given in °N.

- Continentality, or distance from ocean (in km). The idea is to have a relevant, simple measure describing the degree of continental influence on lake temperature: The farther away from the ocean, the more continental, the colder and the greater the seasonal variability in temperature, if all else is constant. The distance is given in km.

- Lake volume (V in m^3). The larger the volume, the smaller the seasonal variability in temperature.

Temperature submodel

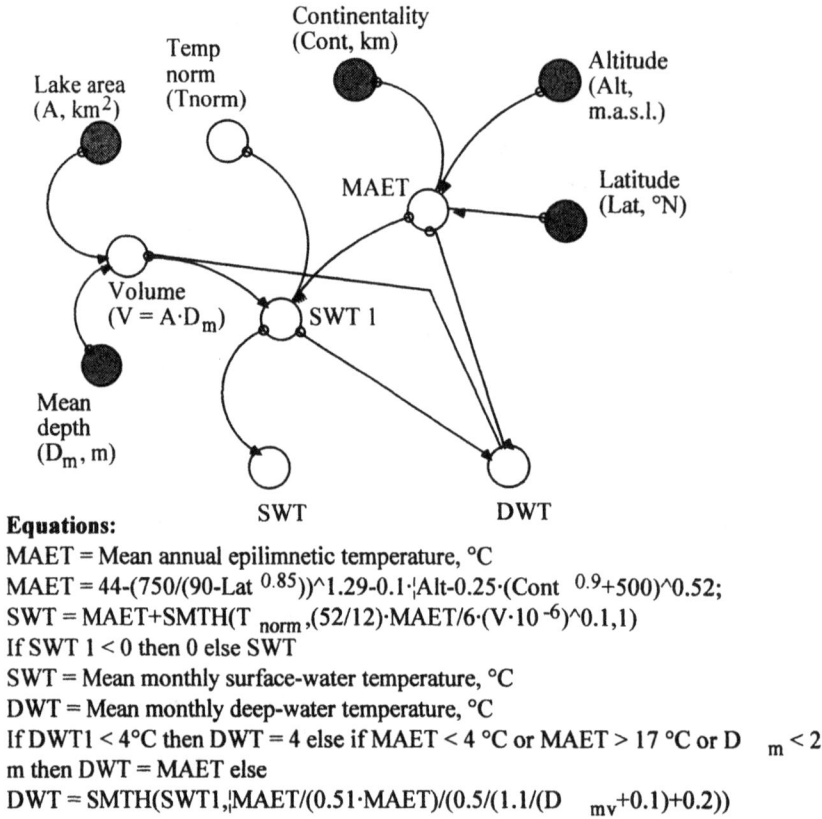

Fig. 9.12. The submodel for lake temperatures (compiled from Ottosson and Abrahamsson, 1998).

Equations:

MAET = Mean annual epilimnetic temperature, °C

MAET = $44-(750/(90-Lat^{0.85}))^{\wedge}1.29-0.1\cdot|Alt-0.25\cdot(Cont^{0.9}+500)^{\wedge}0.52$;

SWT = $MAET+SMTH(T_{norm},(52/12)\cdot MAET/6\cdot(V\cdot10^{-6})^{\wedge}0.1,1)$

If SWT 1 < 0 then 0 else SWT

SWT = Mean monthly surface-water temperature, °C

DWT = Mean monthly deep-water temperature, °C

If DWT1 < 4°C then DWT = 4 else if MAET < 4 °C or MAET > 17 °C or $D_m < 2$

m then DWT = MAET else

DWT = $SMTH(SWT1,|MAET/(0.51\cdot MAET)/(0.5/(1.1/(D_{mv}+0.1)+0.2))$

Month:	1	2	3	4	5	6	7	8	9	10	11	12
Temp norm:	-8.0	-2.0	0	2.0	8.0	20.0	8.0	2.0	0	-2.0	-8.0	-20.0

In this approach, both seasonal variability of surface and deep water temperature can be predicted from these four map parameters. This approach uses:

1. A seasonal variability norm for surface water temperature. The monthly data constituting this norm are given in fig. 9.12. The norm is constructed to illustrate a standardized case of extreme seasonal variability in surface water temperature. Note that this norm is constructed to yield realistic values of seasonal variations when the norm is being smoothed. This norm has several features. The monthly values are not dimensionless but given in °C. The dimensionless moderator will be defined by the ratio between predicted monthly mean temperatures and the predicted mean annual temperature. The range between the lowest value (-20°C for December) and the highest value (+20°C for June) should be high. The main point again is evidently not that this particular norm should give the most realistic description of mean monthly temperature in an extreme lake or river, but rather that this seasonal variability norm for temperature and an appropriate smoothing function should give realistic predictions of monthly temperature for all systems for which the norm is meant to apply.

2.   A smoothing function based on the four map parameters is used to level out the norm. A first-order exponential smooth (SMTH) of the input (the norm for temperature) is used (fig. 9.10 for information on smoothing functions).

The mean annual surface water temperature, SWT (in °C), is utilized as an averaging function. It is calculated from the following expression:

$$MAET = 44 - ((750/(90-Lat))^{0.85})^{1.29} - 0.1 \cdot Alt^{0.5} - 0.25 \cdot (Cont^{0.9}+500)^{0.52} \quad (9.13)$$

Where Lat = Latitude (°N); Alt = Altitude (m.a.s.l.); and Cont = distance (in km) from the lake to the ocean. These empirical constants have been obtained after calibrations using data for a wide range of lakes (see Ottosson and Abrahamsson, 1998).

MAET should be high close to the equator (Lat = 0). The value decreases with the square root of altitude (all else being constant). MAET is rather independent of continental cooling near the oceans (given by $(Cont^{0.9}+500)^{0.52}$).

The mean monthly surface water temperature (SWT) of a given lake may then be estimated from the following expression:

$$SWT = MAET + SMTH(T_{norm}, (MAET/6) \cdot ((Vol \cdot 10^{-6})^{0.1}), 1) \qquad (9.14)$$

where T_{norm} = the norm for surface temperature (fig. 9.12); Vol = volume (km³).

The input is the T_{norm}. The averaging function is MAET/6. The higher the MAET, the smoother the curve for monthly surface temperature. The factor of 6 is an empirical constant used to get a relevant smooth, since high values of MAET, like 25 (°C), would produce almost a straight line, and 25/6 = 4.2, gives a more realistic smooth. The initial value is 1. This means that one first calculates MAET, then, for the summer period, positive values are added to MAET, and during the winter, negative values are added to MAET. When the lake water is perennially frozen, i.e., if MAET < -10 then, SWT is set to 0. The size of the lake (Vol) is also used in the averaging function to smooth the norm (but Vol will not influence the mean annual water temperature) in such a way that lakes with a large volume will get a smaller monthly variability in surface temperature than small lakes. This is given by: $Vol^{0.1}$. The exponent 0.1 is an empirical constant applied to obtain a realistic smooth.

The values of deep water temperatures (°C) are characteristic for dimictic lakes, which appear from approximately latitudes 30°N to 60°N and altitudes 0 to 4000 (see Wetzel, 2001). A limit should also be set for very shallow lakes at such latitudes and altitudes. Lakes larger than 1 km² with a mean depth less than 1-2 m may not be dimictic but polymictic. For lakes with mean depth greater than 2 m, this approach assumes that the relationship between surface and bottom temperatures depends on the mean depth: The smaller the mean depth, the smaller the difference between surface and bottom temperatures. The curve for the mean monthly deep water temperature (DWT) should be markedly smoother than the curve for the mean monthly surface temperature (SWT). If the predicted mean annual water temperatures (MAET) are higher than 17 (°C), the lake would probably be warm monomictic; if MAET is below about 4 (°C), the lake would be cold monomictic. Between these limits, the deep water temperatures are predicted from:

$$DWT = SMTH(SWT, MAET^{0.5}, 0.51 \cdot SWT)/(0.5/(1.1/(D_m+0.1)+0.2))) \quad (9.15)$$

Where the function to be smoothed (the input) is the predicted mean monthly surface temperature (SWT). The seasonal variability for the mean monthly deep water temperature (DWT) should be smaller than for SWT. This is given by the averaging value of $MAET^{0.5}$.

Fig. 9.13 shows predictions of SWT and DWT with this model, which is meant to be generally applicable for European lakes.

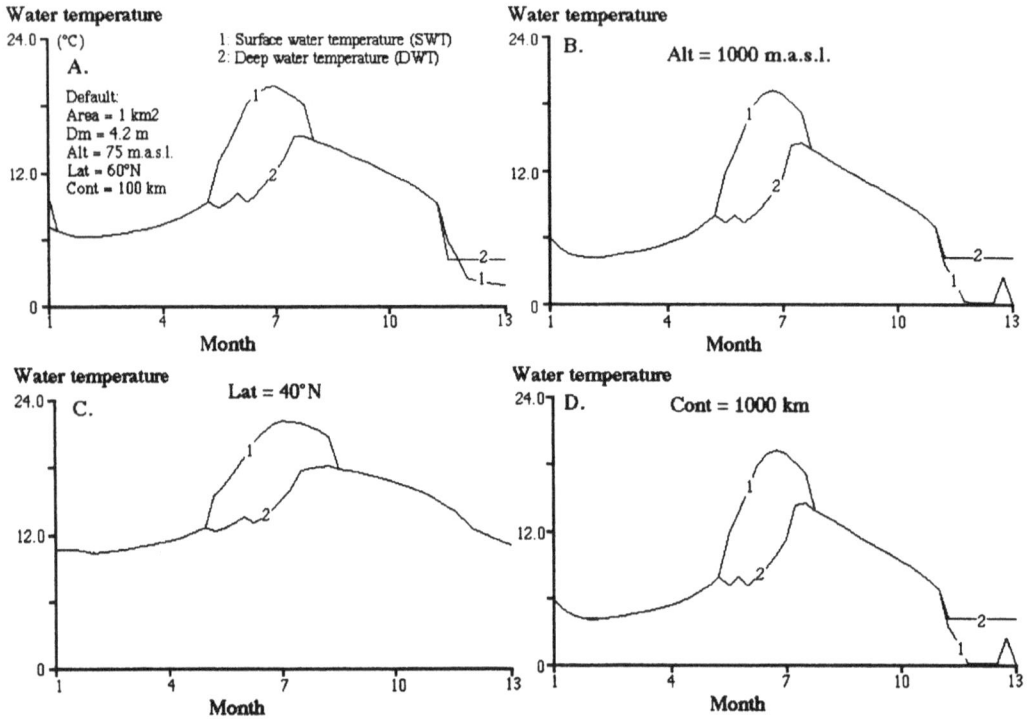

Fig. 9.13. Illustrations of the temperature submodel.
A. Predictions in the dimictic default lake.
B. Predictions if this lake is at altitude 1000 m.a.s.l.
C. Predictions if this lake is at latitude 40°N.
D: Predictions if this lake is 1000 km from the ocean.

A. Gives the surface and bottom temperatures for a lake at altitude 75 m.a.s.l., latitude 60°N, continentality 100 km, area 1 km^2 and mean depth 4.3 m. Note the difference between summer and winter values. The lake circulates during spring and fall when these two curves cross, otherwise the lake is stratified.

B. Simulations for a lake at altitude 1000 m (otherwise the same conditions as in fig. 9.13A). Note that the winter stratification is stronger in this case.

C. Gives the same for a lake at latitude 40°N. Note that the lake is not dimictic (it is warm monomictic, as it should be). It only stratifies during summer, and the temperatures are generally significantly higher, as compared to fig. 9.13A.

D. Simulations for a lake at 1000 km from the Sea. This lake is markedly dimictic.

To apply this model for other parts of the world, e.g., for China and/or Africa, one must re-calibrate the norm and the weight factors for latitude, altitude, and continentality using empirical data on water discharge from as many rivers as possible.

9.5. Characteristics of the lakes used in dynamic SPM modelling

This section will give brief descriptions of the lakes used in contexts of the dynamic SPM modelling (chapter 4). Some main features of these six lakes were given in table 4.9.

9.5.1. Naroch Lakes (Belarus)

The three Belarussian lakes Batorino, Miastro, and Naroch form a system of interconnected basins. The lakes are located in the northwest part of Belarus in a glacial landscape. The lakes have been intensely studied over many years (Winberg, 1970, 1977, 1980, 1985a, 1985b; Winberg et al., 1972, 1973, 1986). Lake Batorino, the first in the system (fig. 9.14), is a relatively shallow

Belarus & Naroch Lakes

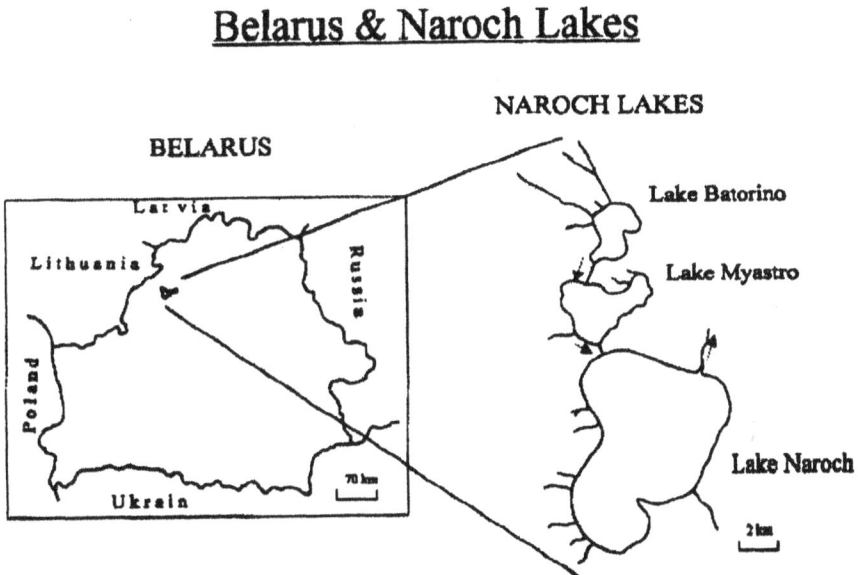

Fig. 9.14 Map of the three Belarussian lakes.

and eutrophic waterbody, connected with Lake Miastro via a narrow (5-6 m) channel, overgrown with higher plants. Also Lake Miastro is shallow (about 50% of the lake bed have depths 5 to 8 m). Lake Naroch is the largest lake in Belarus. About 14% of its area is shallower than 2 m. Lake Naroch has a clearer water, a higher Secchi depth, less SPM, a lower primary phytoplankton production than the other lakes, and significant esthetical and recreational potentials. Some factors influencing the water quality in Lake Naroch are:

1. The relatively small catchment area (the ratio of the catchment area to the lake is 3.5).

2. No major sources of pollution in the catchment.

3. The upstream Lakes Batorino and Miastro function as "purification plants" by entrapping substances.

Until the mid-80s, Lake Batorino was highly eutrophic, Lake Miastro eutrophic, and Lake Naroch mesotrophic. Following the oligotrophication during the 90s, the trophic state of the

Naroch lakes has been significantly altered. The following factors have influenced the oligotrophication of the Naroch Lakes (see Håkanson et al., 2000):

- A decrease of external nutrient loading from land-use measures in the watershed.

- In-lake processes triggered after introduction of the mollusc-filtrator *Dreissena polymorpha*.

- An economical crisis was followed by a decrease of the overall economical activity in the watershed.

- Natural multiannual cycles of the lake productivity associated with the respective climatic changes.

Recreation is the major use of the Naroch Lakes. Fishery (both industrial and leisure-time) is of minor importance. Narochanski National Park was established in 1999 as an important measure for the rational resources of this region. The agriculture and the stockbreeding are the most important factors for the pollution and nutrient loading in the Naroch region.

9.5.2. Lake Kinneret (Israel)

The limnology of Lake Kinneret is well documented (Serruya, 1978). The warm monomictic lake has a surface area of 168 km^2 and mean and maximum depths of 24 and 42 m, respectively. It is stratified between April and November and full homothermy occurs usually between December and February. The Jordan River is the major inflow (80% of the total inflow), while water pumped into the National Water Carrier System constitutes the main outflow. The lake altitude fluctuates between 209 and 213 m below mean sea level. Seiches with amplitudes up to 10 m are present during the entire stratified period, as a response to the daily westerly winds blowing in the afternoon (Serruya, 1978).

Since 1965, with the economic growth within the region, the main uses of Lake Kinneret waters have been domestic "drinking water" and agricultural irrigation. Progressively increasing water use together with unfavourable climatic conditions have lead to a lowering of the level.

A major concern is the relatively high salinity of the lake, which causes salinization of irrigated soils and underground aquifers. Prior to 1964, the lake waters were characterized by high concentrations of chloride (up to 400 mg Cl/l) caused by the inflows from several springs and seepages rich in sodium and chloride around the shores and on the bottom of the lake. The situation has been improved by the construction of a salt water channel that diverts some major saline springs at the northwest coast of the lake and removes about 60000 tons of chloride annually. Consequently, the chloride content fell from 370 mg/l in 1965-1966 to 204-221 mg/l in 1980-1995. Because the Cl concentrations are close to upper permissible limit for drinking water (250 mg/l) and much higher than those in the underground aquifers that causes their salinization, further reduction of lake water salinity is desirable.

There is a large difference between the chemical composition of the hypolimnion and the epilimnion in Lake Kinneret (Serruya, 1978; Berman et al., 1992, 1995). During the stratified period, the hypolimnion is enriched in H_2S, N, and P. The TP concentrations in the photic zone stay around 14 to 16 µg/l and monthly averages of algae biomass around 93-116 mg Chl/m^2 during summer and fall (Berman et al., 1992), when also sedimentation is high (Serruya, 1977) and the external nutrient supply limited. This illustrates the importance of an intense internal nutrient cycling (upwelling of nutrient enriched hypolimnetic water due to seiches activity).

The following features characterize the algal community of the lake (Pollingher, 1986):

- A winter-spring bloom of the dinoflagellate *Peridinium gatunense Nygaard*, comprise more than 90% of the algal biomass during the bloom and 59-90% on an annual basis.

- Chlorophytes, diatoms and nanoplanktonic cyanobacteria dominate during summer and fall.

- There are also sporadic winter blooms of the large diatom, *Aulacoseira granulata (Ehrbg.)* during periods of high turbulence.

9.5.3. Lake Erken (Sweden)

Lake Erken is well studied in terms of morphometry (Håkanson, 1978), sediments, resuspension, and suspended particulate matter (Weyhenmeyer, 1996) and general limnology (Pettersson, 1985). It is a moderately eutrophic lake (the characteristic TP concentration is 27 µg/l and of chlorophyll-a 3.7 µg/l) in Sweden (latitude 59.5°N). The lake is stratified during summer and winter (dimictic) and has a relatively small inflow of humic matter from the catchment. The mean Secchi depth is 4.5 m and the theoretical lake water retention time 6.5 years. The catchment area (140 km²) consists of 70% forest, 10% agricultural land and 20% lakes. The lake area is 24 km². The mean depth is 9 m and the maximum depth 21 m. The lake is located north of Stockholm close to the Swedish east coast. It is rather exposed to winds and the maximum fetch is 5-6 km in the east-west direction.

The seasonal phytoplankton development in Lake Erken has been studied by Istvanovics et al. (1993). Algal blooms often occur in July and August and cyanobacterium (like *Gloeotrichia echinulata*) can be dominant. *G. Echinulata* overwinters in the sediments and this means that the sediments in Lake Erken play an important role in determining the timing and magnitude of summer blooms as well as temporal patterns in photosynthesis (Pierson et al., 1992). Lake Erken is interesting in the way that many major and characteristic forms of algal successions can be seen, including diatoms and "blue greens." The pollution history of Lake Erken is closely related to classical theory and evolution of lake eutrophication related to increase in fertilizers used in agriculture during the 1950s and 60s, internal loading and different algal successions. Lake Erken has, however, never gone into a hypertrophic condition, like Lake Balaton.

9.5.4. Lake Balaton (Hungary)

Lake Balaton (596 km², mean depth 3.2 m) is a large shallow lake with historical water level fluctuations exceeding 10 m. The lake is 78 km long and 7.6 km wide and it can be divided into four subbasins (Balaton 1 to IV). The mean annual water temperature is 12.2°C and is normally above 20°C from the end of May to early September (Somlyódy and van Straten, 1986). Strong thermal stratification never develops so that the water remains well oxygenated throughout the year. The composition of the lake water is controlled by the inflow waters from the catchment area where limestone and dolomitic rocks predominate (average lake pH is 8.4). Calcite precipitation and frequently occurring sediment resuspension cause the continuously high SPM levels (Pettersson and Istvanovics, 1988).

Lake Balaton is primarily a recreational lake with important leisure-time and commercial fisheries. Algal production in Lake Balaton has changed drastically over the past decades. The history of this development has followed a classic pattern. Mass population expansion and economic and agricultural development led from the early fifties to increasing nutrient loads. Due to high external P loads from the Zala river, Basin I became hypertrophic in the 70s. The trophic level in the other basins also increased until the early 80s when the conditions in Basin IV changed from mesotrophic to eutrophic (Somlyódy and van Straten, 1986). The lake ecosystem responded with increased algal biomass and a species shift to nitrogen fixing cyanobacteria which after 1973 formed regularly blooms in Basin I. In Lake Balaton, more than 95% of the external P loading accumulated over the years in the sediments. Starting from the 1980s, remedial measures have reduced the external nutrient loading and have lead to improvements in the lake during the last two decades (Istvánovics and Somlyódy, 2001).

9.6. SPM emissions from fish farms

Fig. 9.15 illustrates the submodel to calculate the inflow of SPM to the aquatic system from a fish cage farm (rainbow trout). The inflow is calculated from annual fish production (AFP), feed coefficient (or the feed conversion ratio, FCR), and a seasonal (monthly) moderator, which accounts for the fact that there is generally a typical seasonal pattern in fish growth, fish feeding, and emissions of SPM such that high emissions generally occur in the fall, just before the harvest. The default values used for are given in the fig. 9.15. One can note that FCR is set to 1.5 (see Wallin et al., 1992). About 23% of the added feed (Johansson et al., 1998) is likely to be discharged from the farm as SPM. From Håkanson and Boulion (2002), one can also assume that 50% of these emissions go to the surface water compartment and 50% to the deep water compartment.

Fish cage farm sub-model

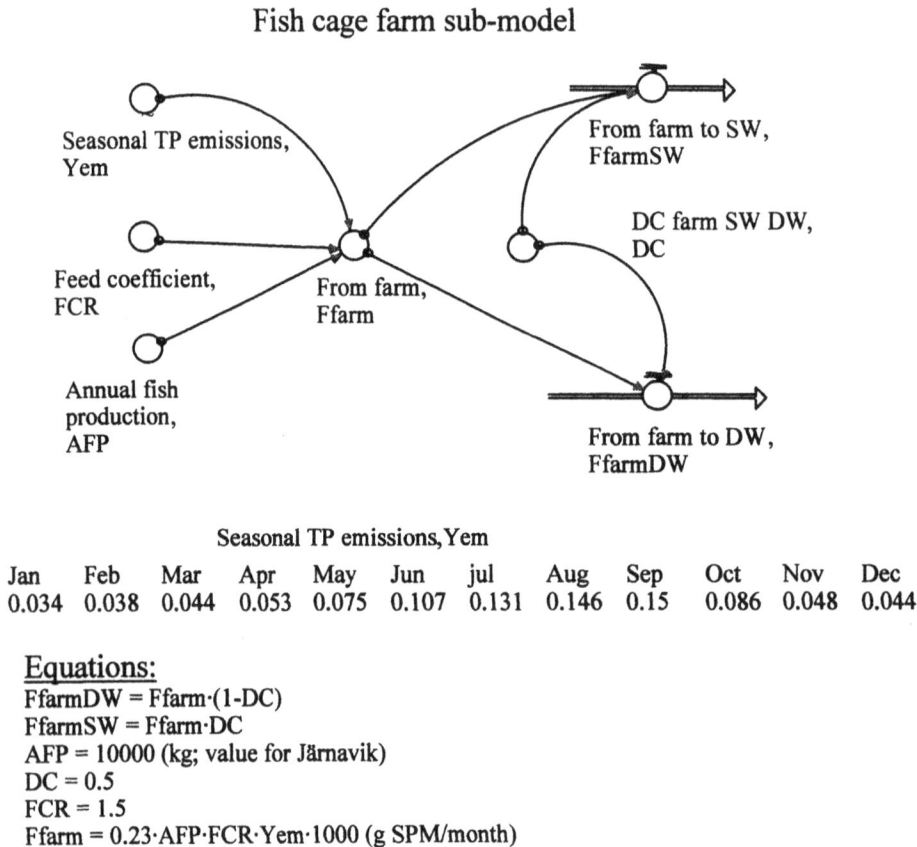

Seasonal TP emissions, Yem

Jan	Feb	Mar	Apr	May	Jun	jul	Aug	Sep	Oct	Nov	Dec
0.034	0.038	0.044	0.053	0.075	0.107	0.131	0.146	0.15	0.086	0.048	0.044

Equations:
FfarmDW = Ffarm·(1-DC)
FfarmSW = Ffarm·DC
AFP = 10000 (kg; value for Järnavik)
DC = 0.5
FCR = 1.5
Ffarm = 0.23·AFP·FCR·Yem·1000 (g SPM/month)

Fig. 9.15. The submodel for fish farm emissions.

The feed coefficient is the ratio between the amount of feed (kg wet weight) supplied to the farm relative to the amount of fish (kg ww) produced in the farm annually. This value can only be lower than 1 if the water content of the feed is significantly lower than the water content of the fish produced in the farm. A typical value for Swedish fish farms was 1.5 in 1990 and 1.2 in 2000.

The relationships betweens monthly load and annual load are obtained from Håkanson et al. (1988). The seasonal variability for the standing stock, the use of dry feed, fish respiration, and emissions from fish farms of feces and TP all follow the same general seasonal pattern.

9.7. SPM from land uplift

The amount of suspended particulate matter (SPM) always depends on two main causes: Allochthonous inflow and autochthonous production. In the northern part of the Baltic Sea, however, there is also another source, land uplift (see Voipio, 1981). Thousand-year-old sediments influence the Baltic ecosystem today. When the old bottom areas rise after being depressed by the glacial ice, they will eventually reach the critical depth above which the waves can exert a direct influence on and resuspend the sediments. The land uplift in the Baltic Sea (measured in relation to the sea surface) varies from about 9 mm/yr in the northern part of the Bothnian Bay to about 0 for the southern part of the Baltic Sea (fig. 9.1).

The transport of SPM from land uplift (F_{LU}) is estimated using the method discussed in chapter 4 (fig. 4.11) based on the hypsographic curve (as a function of Vd, the form factor) and the critical depth (D_{crit}). Instead of moving the critical depth downward when the effective fetch increases, D_{crit} is moved upward due to land uplift. When there is land uplift, the new supply of matter eroded from the glacial clays exposed to winds and waves does not emanate just from the newly raised areas but also from increased erosion of previously raised areas. In this modelling approach, it has been assumed that, on average, there would be no fine sediments on the erosion areas (E areas), which have been heavily influenced by wind-induced waves, and are dominated by coarse deposits. The depth separating E areas from T areas (D_{ET} in m) plus the depth related to land uplift (D_{LR} in m) is called D_{ELR} in fig. 9.16. It is estimated from an algorithm presented by Håkanson and Jansson (1983) as:

$$D_{ET} = 30.4 \cdot \sqrt{Area}/(34.2 + \sqrt{Area}) \qquad (9.16)$$

The depth separating T areas from A areas (D_{TA} plus D_{LR}) is estimated from a similar approach (fig. 9.16). The area between these two water depths ($Area_E - Area_T$ in m^2) may be calculated from the equation given in fig. 4.11.

It is also assumed that 50% of the new sediment volume (m^3) of glacial clays from land uplift will be accessible for resuspension - these clays have a relatively low water content (about 60% ww; Håkanson et al., 1984) and are not so easily resuspended, especially not in topographically sheltered areas or close to the wave base. This means that F_{LU} is given by:

$$F_{LU} = D_{LR} \cdot EF \cdot (Area_E - Area_T)(1-(W-15)/100) \cdot (d+0.2) \cdot 10^6 \qquad (9.17)$$

W is the water content of the sediments lifted above the critical depth. It is assumed that the water content of the compacted glacial clays is 15% lower than the recently deposited sediments and that the bulk density (d in g/cm^3) is 0.2 units higher than in the recently deposited sediments. The water content (W) and the bulk density (d) may be measured or estimated from the approach given in the following sections.

Land uplift submodel

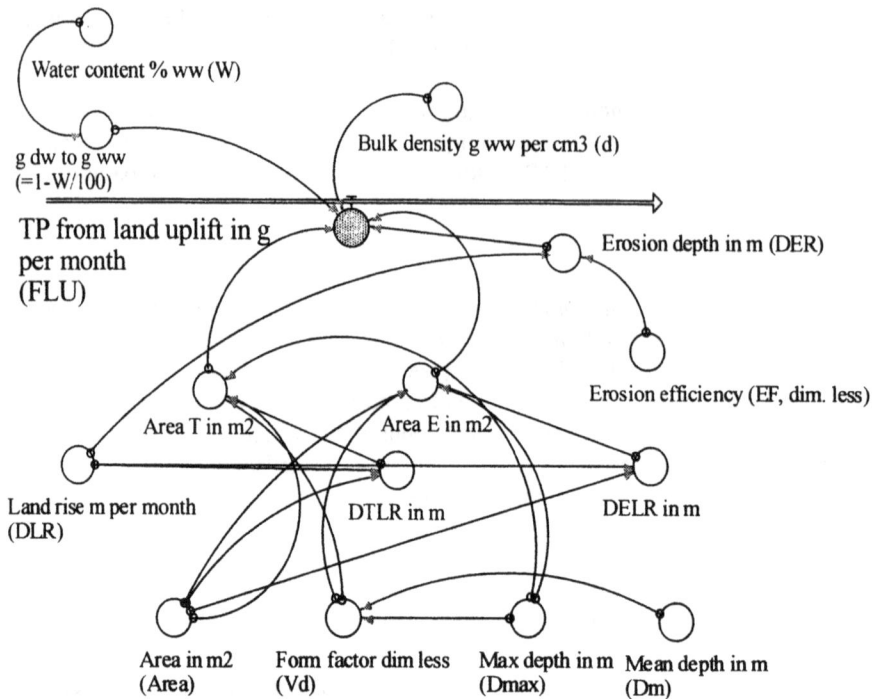

Equation:
FLU = DER·(AreaE-AreaT)·(1-W/100)·d·10^6
AreaE =Area·((Dmax-DELR)/(Dmax+DELR·EXP(3-Vd^1.5)))^(0.5/Vd)
AreaT =Area·((Dmax-DTLR)/(Dmax+DTLR·EXP(3-Vd^1.5)))^(0.5/Vd)
DELR = (30.4·(0.000001·Area)^0.5)/(34.2+(0.000001·Area)^0.5)+DLR
[DELR = water depth separating E-areas with coarse deposits from T-areas with mixed deposits; algorithm from Håkanson and Jansson, 1983]
DTLR = (45.7·(0.000001·Area)^0.5)/(21.4+(0.000001·Area)^0.5)+DLR
[DELR = water depth separating T-areas with mixed deposits from A-areas with fine deposits; algorithm from Håkanson and Jansson, 1983]
DER = DLR·EF
[DER = depth of sediment erosion from land uplift]
EF = 0.5
[50% of the deposits on from land uplift sediments are assumed to be resuspended]
Vd = 3·Dm/Dmax
LDLR = 0.004/12

Fig. 9.16. The land uplift submodel.

9.8. Estimation of sediment organic content

The data used to derive the empirical model for the sediment organic content (= loss on ignition) are given in table 9.8. The model has been presented by Håkanson and Boulion (2002) and is given in table 9.9. The methods used to derive this empirical model, including the stepwise multiple regression analysis, are given in Håkanson and Peters (1995). From table 9.9, it is clear that IG may be predicted from a model based on:

1. Lake pH. The higher the pH value, the lower the sediment organic content (IG) of surficial A sediments. pH influences the aggregation of suspended particles and many other lake processes. Low lake pH is common in humic lakes, which are also known to have very loose sediments with a high organic content (gyttja and dy).

2. The catchment area to lake area ratio (A_{DA}/Area). The higher the ratio, the lower sediment IG. This is logical and in good agreement with previous models for IG (see Håkanson and Peters, 1995).

3. The relative depth (D_{rel}), which is directly related to internal loading (resuspension) and the form of the lake.

4. Lake color concentration, which reflects allochthonous influences (humic matter, etc.).

This empirical model gave an r^2 value of 0.86 (for the 39 lakes given in table 9.9). The model applies if the model variables are within the ranges marked in bold in table 9.9. If the model is used outside these ranges, it should be done with caution, as a working hypothesis or a best estimate.

It is also evident that the sediment IG value will not change very much in relation to short-time alterations in lake pH or lake color values, so to obtain a realistic dynamic response to changes in the two water chemical variables in the model, one can applied a smoothing function (fig. 9.10), which will smooth weekly changes in lake pH and color values. The averaging time has been set to 1 year (= 12 months). IG is given by:

$$IG = SMTH((58.3-(9.69 \cdot pH)-(1.64 \cdot (ADA/Area)^{0.5})+(2.70 \cdot D_{rel})$$
$$+(17.6 \cdot \log(Col))), 12, 10) \qquad (9.18)$$

The initial IG value is set to 10%. Note that the model has only been tested for lakes in the pH-range from 4.8 to 7.6. Just like in the model for gross sedimentation, it seems logical that pH can influence the sedimentation of organic matter and the IG content of A sediments. However, it has not been proven that lake water pH would exert the same influence on IG for lakes with pH higher than 7.6. So, one applies a limiting if-then-else statement to calculate IG:

If pH> 7.6, then IG=SMTH(($58.3-(9.69 \cdot 7.6)-$
$(1.64 \cdot (ADA/Area)^{0.5})+(2.70 \cdot D_{rel})+(17.6 \cdot \log(Col))), 12, 10$) else use eq. 9.1.

Table 9.8. Organic content in surficial A sediments (IG, % dw) and factors assumed to influence IG. Data from 39 Swedish lakes (from Håkanson and Peters, 1995). A_{DA} = catchment area, D_m = mean depth, D_{max} = maximum depth, T = theoretical lake water retention time, D_{rel} = relative depth (= $D_{max} \cdot \sqrt{\pi}/(20 \cdot \sqrt{Area})$), TP = characteristic lake TP concentration.

Lake	IG % dw	A_{DA} km²	Area km²	D_m m	D_{max} m	T yr	D_{rel}	pH	TP µg/l	Color mg Pt/l
703	32.5	4.04	0.26	1.4	2.4	0.38	0.42	6.2	13	65
705	37.5	12.4	0.33	1.7	2.7	0.18	0.42	5.7	35	172
706	34.4	28.1	1.44	5.1	16.2	1.16	1.20	6.2	12	67
1804	34.7	44.2	0.74	8.7	25.2	0.47	2.60	5.7	16	68
1808	14.3	20.2	0.25	2.2	3.2	0.07	0.57	6.0	8	81
1814	52.2	1.63	0.22	1.3	2.9	0.5	0.55	4.8	32	258
1818	54.1	0.56	0.04	5.3	13.5	0.15	5.98	5.5	11	107
1819	48.7	0.39	0.06	6.2	18.4	1.86	6.66	6.0	10	82
1820	46.0	0.36	0.09	8.1	25	3.77	7.39	6.1	12	81
2110	36.9	13.3	0.44	4.7	19.2	0.48	2.57	6.6	12	121
2117	37.4	14.4	0.58	4.2	13	0.53	1.51	5.6	11	132
2119	37.8	1.00	0.15	2.7	6.8	1.2	1.56	5.9	9	79
2120	33.6	3.30	0.16	2.1	6.2	0.3	1.37	6.2	9	138
2121	29.4	1.90	0.12	2.4	6.8	0.47	1.74	6.4	12	126
2122	30.3	3.60	0.16	2.8	4.9	0.38	1.09	5.8	10	109
2201	28.2	20.1	0.25	8.3	16.3	0.33	2.89	5.2	10	130
2206	35.6	25.1	0.59	6.4	21.3	0.48	2.46	5.6	12	128
2212	40.3	17.6	0.28	4.9	19.5	0.24	3.27	6.0	9	81
2213	23.5	4.36	0.47	5.1	19.2	1.6	2.48	6.4	7	38
2214	20.1	38.5	0.07	2.9	9.2	0.02	3.08	5.1	10	123
2215	20.7	8.34	0.24	4.5	12.6	0.44	2.28	6.2	10	72
2216	16.3	6.04	0.16	4	11.2	0.37	2.48	6.5	11	35
2217	8.6	11.6	0.18	3.8	7.8	0.21	1.63	6.6	23	48
2218	14.6	16.8	0.25	2.1	4.3	0.1	0.76	6.5	14	73
7015	48.0	2.09	0.35	2.3	9	1.3	1.35	4.8	10	84
18021	41.6	8.09	0.96	4	17	1.33	1.54	5.2	14	54
18022	44.7	7.91	0.75	2.1	7.5	0.67	0.77	5.0	29	141
21107	26.2	2.90	0.39	2.4	5	0.96	0.71	6.1	13	65
21108	32.9	17.2	0.27	7.7	17	0.4	2.90	5.6	13	143
21109	41.7	1.46	0.11	3.4	8	0.79	2.14	5.8	22	108
21112	30.9	16.2	1.28	4.2	10	1.02	0.78	6.0	11	80
21114	31.6	3.60	0.59	5.1	12.5	2.4	1.44	6.2	9	90
21115	34.0	4.10	0.16	3.4	9	0.44	1.99	5.1	13	167
Vänern	9.7	46830	5650	27	106	8.8	0.14	7.1	8	40
Våtern	11.9	6359	1912	39	128	58	0.29	7.6	5	30
Målaren	11.0	22603	1140	13	61	2.8	0.18	7.5	38	50
Hjälmaren	12.3	4053	484	6.1	20	3.7	0.09	7.5	44	60
Bullaren	10	199	8.3	10.1	26.2	0.88	0.90	7.2	36	54
Våsman	16	84	38.6	10.5	53	1.08	0.85	6.8	15	30
Min.	8.6	0.4	0.0	1.3	2.4	0.0	0.1	4.8	5.0	30
Max.	54.1	46830	5650	39	128	58.0	7.4	7.6	44	258
Mean	30.0	2063.8	237.1	6.2	19.9	2.6	1.9	6.1	15.3	92.5

Table 9.9. An empirical model to predict lake characteristic values of the organic content (= loss on ignition, IG in % dw = y variable) of surficial A sediments based on data from 39 Swedish lakes (data from Håkanson and Peters, 1995; F > 4).

Step	r^2	x variable	Model
1	0.53	pH	$y=109-13.0 \cdot x_1$
2	0.71	$\sqrt{A_{DA}}/\text{Area}$	$y=128-14.9 \cdot x_1-1.52 \cdot x_2$
3	0.81	D_{rel}	$y=114-13.4 \cdot x_1-1.57 \cdot x_2+2.62 \cdot x_3$
4	0.861	og(Col)	$y=58.3-9.69 \cdot x_1-1.64 \cdot x_2+2.70 \cdot x_3+17.6 \cdot x_4$

9.9. Estimation of sediment water content

There exist logical relationships between all types of physical sediment parameters, like IG, water content (W), grain size, and bulk density (see Håkanson and Jansson, 1983). Fig. 9.17 gives the relationship between W and IG based on data from 122 sites in 59 lakes (see Håkanson and Boulion, 2002). The results in fig. 9.17 have been used to predict the requested W value for A sediments. This means that W is first calculated from IG by:

$$\text{If IG} \geq 6.5 \ (\% \text{ dw}), \text{ then } W = (207 \cdot \text{IG} - 1280)^{1/3} + 75 \qquad (9.19)$$
$$(r^2 = 0.76; \ n = 96)$$

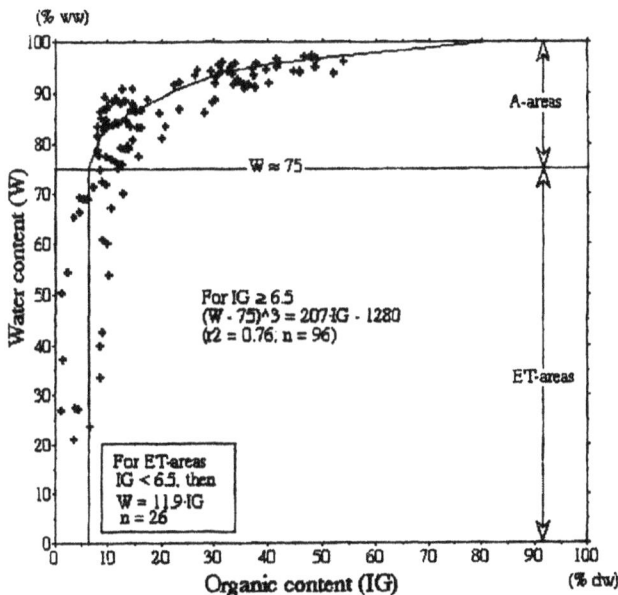

Fig. 9.17 The relationship between the organic content of surficial A sediments (IG) and the water content (W) based on data from 122 sites from 59 lakes covering a very wide range in lake sediment characteristics (from Hakanson and Boulion, 2002).

This regression is valid for surficial (0-10 cm) A sediments and is meant to give a mean value for the entire active A volume, which has an area of [(1-ET)·Area] and covers 10 cm of sediments. The water content in surface sediments is lower at smaller water depths in ET areas (see Håkanson and Jansson, 1983). At the critical water depth (D_{crit}) separating A sediments from T sediments, the water content is generally about 10% lower than the water content in the deepest part of the lake. Further, in a sediment core, the water content generally decreases vertically due to, e.g., compaction and mineralization.

9.10. Estimation of sediment bulk density

The bulk density of A sediments (d in g ww/cm^3) is calculated using a standard formula (from Håkanson and Jansson, 1983) based on water content (W) and IG (in % ww abbreviated as IG*). That is:

$$d = 260/(100 + 1.6 \cdot (W + IG \cdot ((100 - W)/100)))$$
(9.20)

10. Index

C

Z

www.ingramcontent.com/pod-product-compliance
Lightning Source LLC
Chambersburg PA
CBHW080924220326
41598CB00034B/5669